Probabilistic Extensions of Various Logical Systems

Zoran Ognjanović
Editor

Probabilistic Extensions of Various Logical Systems

Springer

Editor
Zoran Ognjanović
Mathematical Institute of the Serbian Academy
of Sciences and Arts (SANU)
Belgrade, Serbia

ISBN 978-3-030-52956-7 ISBN 978-3-030-52954-3 (eBook)
https://doi.org/10.1007/978-3-030-52954-3

This Springer imprint is published by the registered company Springer Nature Switzerland AG
The registered company address is: Gewerbestrasse 11, 6330 Cham, Switzerland

Preface

Let us assume that

$$P_{\approx 1} E_{\text{readers}}\ \text{standard\_definition\_of\_probability}$$

(i.e., it is very likely that readers know the standard definition of probability given by Kolmogorov). Some of them also accept that Kolmogorov's definition completely determines the notion of probability, which in turn means that, for them, probability is real-valued and based on the concept of classical logic (since it relies on (σ) additive algebras). In this book we are trying to challenge that view and to provide a broader understanding of probability: a number of formal systems are given which combine probabilistic and other logics, allow reasoning about probabilities with non-linearly ordered ranges, consider extensions of the usual probabilistic language, etc. So, we concentrate on logics with probabilistic operators (although one of the chapters presents a logic with probabilistic quantifiers à la Keisler) with the aim to give a systematic survey of our results published in research journal articles in the last ten years, and also a comparison with approaches of other authors.

The presented material continues our earlier book "Probabilistic logics (Probability based formalization of uncertain reasoning)" published by Springer in 2016. The ideas, results and techniques from "Probabilistic logics" are summarized here in the first chapter, so we recommend readers to start with it, while the rest of the text is largely independent and can be read in a more or less arbitrary order.

The authors would like to acknowledge the support obtained from the Ministry of Education, Science and Technological Development of the Republic of Serbia and from the Swiss National Science Foundation.

Authors

Contents

List of Contributors

Marija Boričić
Faculty of Organizational Sciences, University of Belgrade, Jove Ilića 154, Belgrade, Serbia, e-mail: marija.boricic@fon.bg.ac.rs

Dragan Doder
Utrecht University, Princetonplein 5, 3584 CC Utrecht, The Netherlands, e-mail: d.doder@uu.nl

Nebojša Ikodinović
Faculty of Mathematics, University of Belgrade, Studentski trg 16, 11000 Belgrade, Serbia, e-mail: ikodinovic@matf.bg.ac.rs

Angelina Ilić-Stepić
Mathematical Institute of the Serbian Academy of Sciences and Arts, Kneza Mihaila 36, 11000 Belgrade, Serbia, e-mail: angelina@mi.sanu.ac.rs

Ioannis Kokkinis
Department of Mathematics, Aristotle University of Thessaloniki, Thessaloniki, Greece, e-mail: ykokkinis@gmail.com

Zoran Marković
Mathematical Institute of the Serbian Academy of Sciences and Arts, Kneza Mihaila 36, 11000 Belgrade, Serbia, e-mail: zoranm@mi.sanu.ac.rs

Zoran Ognjanović
Mathematical Institute of the Serbian Academy of Sciences and Arts, Kneza Mihaila 36, 11000 Belgrade, Serbia, e-mail: zorano@mi.sanu.ac.rs

Aleksandar Perović
Faculty of Transport and Traffic Engineering, University of Belgrade, Vojvode Stepe 305, 11000 Belgrade, Serbia, e-mail: pera@sf.bg.ac.rs

Miodrag Rašković
Mathematical Institute of the Serbian Academy of Sciences and Arts, Kneza

Mihaila 36, 11000 Belgrade, Serbia, e-mail: miodragr@mi.sanu.ac.rs

Nenad Savić
Universität Bern, Neubrückstrasse 10, 3012 Bern, Switzerland, e-mail: nenad.savic@inf.unibe.ch

Thomas Studer
Universität Bern, Neubrückstrasse 10, 3012 Bern, Switzerland, e-mail: tstuder@inf.unibe.ch

General notations and conventions

$\mathbb{N}$	the set of all natural numbers 0, 1, 2, ...
$\mathbb{Z}$	the set of all integers 0, 1, −1, 2, −2, ...
$\mathbb{Q}$	the set of all rational numbers
$\mathbb{R}$	the set of all real numbers
$\mathbb{C}$	the set of all complex numbers
$^*\mathbb{N}$	the non-standard set of natural numbers
$^*\mathbb{R}$	the set of hyperreal numbers
$\mathbb{Q}(\varepsilon)$	Hardy field
$[0,1]_{\mathbb{Q}}$	the unit interval of rational numbers
$(0,1)_{\mathbb{Q}}$	the open unit interval of rational numbers
$[0,1]_{\mathbb{Q}(\varepsilon)}$	the unit interval of Hardy field
$\mathbb{P}(W)$	the power set of the set W
$\|W\|$	the cardinality of the set W
Subf(Φ)	the set of all subformulas of the formula Φ
$T \models \Phi$	The formula Φ is a semantical consequence of the set of formulas T
$T \vdash \Phi$	The formula Φ is a syntactical consequence of the set of formulas T
iff	if and only if
wrt.	with respect to

Chapter 1
Logics with Probability Operators

Zoran Ognjanović and Angelina Ilić-Stepić

Abstract This chapter presents the logic $LFOP_1$ which may be suitable to formalize reasoning about degrees of beliefs. The aim is that this chapter serves as an illustration for syntax, semantics and the main proof techniques used elsewhere in the book. The logic enriches first order calculus with probabilistic operators of the form $P_{\geq s}$ with the intended meaning "probability is at least s". We define a possible-world semantics with a finitely additive probability measure on sets of worlds. We provide an infinitary axiomatization which contains an infinitary rule with countable many premisses and one conclusion, related to the Archimedean property of real numbers. Other probability logics considered by the authors of this book are then presented, and an overview of related works of other authors is given.

1.1 Introduction

It is a reasonable question whether the probability that Homer is the author of the Iliad and the Odyssey is greater than $\frac{1}{2}$. But, what kind of probability is mentioned here? Obviously, there are no repeating experiments generating the relative frequency of such an event. Nor can one establish a well-defined set of equally likely cases and determine the quotient of the numbers of desired and all outcomes. The probability in question is epistemic, the kind of probability discussed by Gottfried Wilhelm Leibnitz[1] (1646 – 1716). Initially motivated by proofs of legal rights, he understood probability conclusions as conditional, i.e., as subjective and relative to

Zoran Ognjanović
Mathematical Institute of the Serbian Academy of Sciences and Arts, Kneza Mihaila 36, 11000 Belgrade, Serbia, e-mail: zorano@mi.sanu.ac.rs

Angelina Ilić-Stepić
Mathematical Institute of the Serbian Academy of Sciences and Arts, Kneza Mihaila 36, 11000 Belgrade, Serbia, e-mail: angelina@mi.sanu.ac.rs

[1] Versions of his last name: Lubeniecz, Leibniz.

Z. Ognjanović (ed.), *Probabilistic Extensions of Various Logical Systems*,
https://doi.org/10.1007/978-3-030-52954-3_1

the existing knowledge. Leibnitz promoted creation of a new kind of logic about degrees of probabilities which he saw as a powerful calculus which allows measuring of knowledge.

Leibnitz's influence is recognizable in works of Jacob Bernoulli (1654 – 1705), Johann Heinrich Lambert (1728 – 1777) and Bernard Bolzano (1781– 1848) who considered combinations of arguments to calculate probabilities. This development had a particular climax in [5], where George Boole (1815 – 1864) formulated so-called the general problem of the science: "Given the probabilities of any events, simple or compound, conditioned or unconditioned: required the probability of any other event equally arbitrary in expression and conception", and proposed a method to solve it. This method is practically the decision procedure for probability logic described in Section 1.2.5.

In the 20th century connections between mathematical logic and probability were considered in papers of important scholars like John Maynard Keynes (1883 – 1946), Hans Reichenbach (1891 – 1953), Rudolf Carnap (1891 – 1970) and their followers. However, it can be said that until the middle of the 1960s development of these disciplines were almost separated. The papers [25] and [99] changed it. They generalized the notion of models by considering probabilities defined on sentences instead of truth-value interpretations, and proved theorems about the existence of models of consistent sets of classical first order formulas and infinitary $L_{\omega_1 \omega}$-formulas. It is interesting that Scott and Krauss were pessimists about the possibility to create a deductive method for generating consequences from a given set of probability assertions [99, p. 243]. Theodore Hailperin (1915 – 2014) in a number of papers and books (e.g., [31–34]) extended Boole's approach and described semantics in which sentences have probability values instead of truth-values.

In this chapter and in the most part of the book we will present logics based on formal languages that involves probability operators $P_{\geq s}\alpha$ read as "the probability of α is at least s" which are suitable for reasoning about degrees of beliefs, i.e., about epistemic probabilities. Still, we have to emphasize that the major step in the development of the field of probability logics after Boole is related to logics more appropriate for reasoning about statistical information[2], see [38, 54, 55, 87–89]. In order to study models useful in applied mathematics Jerome Keisler introduced the first order logic $L_{\omega P}$ which has the probability quantifiers $(Px > r)$ and $(Px \geq r)$ instead of classical $(\exists x)$ and $(\forall x)$. The meaning of $(Px > r)\alpha(x)$ is that the probability of the set $\{x : \alpha(x)\}$ is greater than r. Keisler, his students and followers provided modal-theoretic analysis of a number of finitary and infinitary logics with probabilistic quantifiers and the corresponding strongly-complete axiomatizations. For example, among the considered axioms and rules are:

- monotonicity: $(Px \geq r)\alpha \rightarrow (Px \geq s)\alpha$, for $r \geq s$
- from $\{\beta \rightarrow (Px \geq r)(Py \geq s - \frac{1}{n})\alpha\}$, infer $\beta \rightarrow (Px \geq r)(Py \geq s)\alpha$.

[21, 91] give overviews of work in the field of logics with probability quantifiers. Such logics are also discussed in [1, 4, 35].

[2] For example: less than one half of patients recover after a certain critical illness.

We conclude this introductory section with two remarks. First, we admit that the selection of references in this chapter is subjective and that, probably, some important names and papers are omitted. Second, two excellent overviews of historical development of relationships between mathematical logic and probability theory are [30, 33]. An interested reader can also consult [82, Chapter 2].

In the rest of this chapter, Section 1.2 describes logics with probability operators proposed by the group of researchers who are the authors of this book, and Section 1.3 summarize related results of other authors.

1.2 The first order probability logic $LFOP_1$ and some related logics

$LFOP_1$ is a first order logic with probabilistic operators of the form $P_{\geq s}$ with the intended meaning "probability is at least s". It enables formalization of sentences like the above mentioned "The probability that Homer is the author of the Iliad and the Odyssey is greater than $\frac{1}{2}$" or "The probability that all patients recover after a certain critical illness is 1." We give syntax and semantics based on Kripke-like models with finitely additive probability measures defined on sets of worlds, and prove strong completeness of an axiom system which contains an infinitary rule related to the Archimedean property of real numbers. To our knowledge, this is the first published strongly complete axiomatization for first order probability logic with probabilistic operators and real valued probability functions. Our approach, presented here in detail for $LFOP_1$, will be used in the following chapters. This section also gives an overview of some logics more-or-less similar to $LFOP_1$.

1.2.1 Syntax

We will enrich classical first order language with unary operators of the form $P_{\geq s}$ with the intended meaning "the probability is at least s". More formally, let $[0,1]_{\mathbb{Q}}$ denote the set of all rational numbers from the real unit interval $[0,1]$. The first order language L of $LFOP_1$ contains:

- for each $k \in \mathbb{N}$, k-ary relation symbols $R_0^k, R_1^k, \ldots$,
- for each $k \in \mathbb{N}$, k-ary function symbols $f_0^k, f_1^k, \ldots$,
- the connectives $\wedge$, and $\neg$, and the universal quantifier $\forall$,
- a list of unary probability operators $P_{\geq s}\alpha$, for every $s \in [0,1]_{\mathbb{Q}}$,
- a list of individual variables $x, y, z, \ldots$, and
- auxiliary symbols comma and parentheses.

The function symbols of the arity 0 are called (individual) constant symbols. The set of terms of the language L is the smallest set:

- containing all variables and constant symbols, and

- closed under the formation rules: if f is a k-ary function symbol, and $t_1, \ldots, t_k$ are terms, then $f(t_1,\ldots,t_k)$ is a term.

Atomic formulas are expressions of the form $R_i^k(t_1,\ldots,t_k)$, where R_i^k is a k-ary relation symbol, and $t_1, \ldots, t_k$ are terms. The set $\mathrm{For}_{LFOP_1}(L)$ of all formulas of the language L (or simply For_{LFOP_1} if L is clear from the context) is the smallest set:

- containing all atomic formulas, and
- closed under the formation rules: if $\alpha, \beta \in \mathrm{For}_{LFOP_1}(L)$, then $\neg\alpha$, $P_{\geq s}\alpha$, $\alpha \wedge \beta$ and $(\forall x)\alpha \in \mathrm{For}_{LFOP_1}(L)$.

We will use letters (indexed if necessary):

- a, b, c, for constant symbols,
- f, g, h, for function symbols,
- P, Q, R, for predicate symbols,
- t, for terms, while
- α, β, γ range over formulas.

The following are standard abbreviations:

- $(\alpha \vee \beta)$ for $\neg(\neg\alpha \wedge \neg\beta)$,
- $(\alpha \rightarrow \beta)$ for $(\neg\alpha \vee \beta)$,
- $(\alpha \leftrightarrow \beta)$ for $((\alpha \rightarrow \beta) \wedge (\beta \rightarrow \alpha))$,
- $(\exists x)\alpha$ for $\neg(\forall x)\neg\alpha$,
- $P_{<s}\alpha$ for $\neg P_{\geq s}\alpha$,
- $P_{\leq s}\alpha$ for $P_{\geq 1-s}\neg\alpha$,
- $P_{>s}\alpha$ for $\neg P_{\leq s}\alpha$,
- $P_{=s}\alpha$ for $P_{\geq s}\alpha \wedge P_{\leq s}\alpha$, and
- $\perp$ for $\alpha \wedge \neg\alpha$.

In a formula of the form $(\forall x)\alpha$, α is called the scope of that quantifier. An occurrence of a variable x in a formula α is bound if it occurs in a part of α which is of the form $(\forall x)\beta$, otherwise, the occurrence is called free. If α is a formula and t is a term, then t is said to be free for x in α if no free occurrences of x lie in the scope of any quantifier $(\forall y)$, where y is a variable in t. $\alpha(x_1,\ldots,x_m)$ indicates that free variables of the formula α form a subset of $\{x_1,\ldots,x_m\}$, while $\alpha(t/x)$ denotes the result of substituting in α the term t for all free occurrences of x. We can also use the shorter form $\alpha(t)$ to denote the same substitution, if x is clear from the context. A formula α is a sentence if no variable is free in α. A set of sentences is called a theory.

Example 1.1. If $R(x)$ means that it rains in the region x, then

$$P_{\geq 0.5}(\exists x)R(x)$$

represents the statement that the probability is at least a half that there is a region x such that it rains there. Similarly, let $Q(x)$ mean that the ticket x wins the lottery. A formalization of the well known lottery paradox [60] (see also Section 8.4)

$$(\forall x)P_{\leq 0.001}Q(x) \wedge P_{\geq 0.999}(\exists x)Q(x)$$

says that, although for every thicket the winning chance is very low, there is a high chance that some ticket will win.

A more complicated formula

$$P_{\geq s}(\forall x)P(x) \to (Q_1(y) \wedge P_{\geq r}P_{<t}Q_2(f(b,z)))$$

which contains nested probability operators illustrates that it is possible to formalize reasoning about higher-order probabilities, e.g., "the probability is at least r that the probability of $Q_2(f(b,z))$ is less than t." ■

1.2.2 Semantics

The semantics for $\mathrm{For}_{LFOP_1}(L)$ is based on the possible-world approach similar to the widely accepted semantics for modal logics proposed in [59]. The main difference is that, instead of relations of visibility (accessibility) between worlds, $LFOP_1$-models are equipped with finitely additive probability measures defined on sets of worlds.

Definition 1.1. An *$LFOP_1$-model* is a structure $\mathbf{M} = \langle W, D, I, Prob\rangle$ where:

- W is a nonempty set of objects called worlds,
- D is a nonempty domain for every world $w \in W$,
- I associates an interpretation $I(w)$ with every world $w \in W$ such that:
  - if f is a k-ary function symbol, then for every $w \in W$, $I(w)(f)$ is the same function from D^k to D,
  - If P is a k-ary relation symbol, then $I(w)(P)$ is a relation over D^k,
- $Prob$ is a probability assignment which assigns to every $w \in W$ a probability space $Prob(w) = \langle W(w), H(w), \mu(w)\rangle$, such that:
  - $W(w)$ is a nonempty subset of W,
  - $H(w)$ is an algebra of subsets of $W(w)$, i.e.:
    - $W(w) \in H(w)$, and
    - if $A, B \in H(w)$, then $W(w) \setminus A \in H(w)$ and $A \cup B \in H(w)$,

    and
  - $\mu(w) : H(w) \to [0,1]$ is a finitely additive probability measure, i.e.,
    - $\mu(w)(W(w)) = 1$, and
    - $\mu(w)(A \cup B) = \mu(w)(A) + \mu(w)(B)$, whenever $A \cap B = \emptyset$. ■

To define values of terms and formulas we need also the notion of valuations.

Definition 1.2. Let $\mathbf{M} = \langle W, D, I, Prob\rangle$ be an $LFOP_1$-model. A *variable valuation* v is a mapping from the set of variables to D, i.e., for every variable x, $v(x) \in D$. If $d \in D$, then $v[d/x]$ is a valuation like v except that $v[d/x](x) = d$. ■

Definition 1.3. For an $LFOP_1$-model $\mathbf{M} = \langle W, D, I, Prob\rangle$, $w \in W$, and a valuation v, the *value of a term t in w* (denoted by $I(w)(t)_v$) is:

- if t is a variable x, then $I(w)(x)_v = v(x)$, and
- if $t = f(t_1, \ldots, t_m)$, then $I(w)(t)_v = I(w)(f)(I(w)(t_1)_v, \ldots, I(w)(t_m)_v)$. ■

Note that:

- worlds of $LFOP_1$-models are classical first order models,
- all worlds from an $LFOP_1$-model have the same domain, and
- terms are rigid, i.e., for every $LFOP_1$-model and every term, the value of the term is the same in all worlds of the model.

In other words, the class of $LFOP_1$-models contains constant domain models with rigid terms. These properties are essential in the presented approach, since they ensure that instances of classically valid formulas are valid in the discussed probabilistic framework which will be explained after the relevant notions are introduced.

Definition 1.4. The truth value of a formula α in a world $w \in W$ for a given $LFOP_1$-model $\mathbf{M} = \langle W, D, I, Prob\rangle$, and a valuation v (denoted by $I(w)(\alpha)_v$) is:

- if $\alpha = P(t_1, \ldots, t_k)$, then $I(w)(\alpha)_v = true$, if $\langle I(w)(t_1)_v, \ldots, I(w)(t_k)_v\rangle \in I(w)(P)$, otherwise $I(w)(\alpha)_v = false$,
- if $\alpha = \neg\beta$, then $I(w)(\alpha)_v = true$, if $I(w)(\beta)_v = false$, otherwise $I(w)(\alpha)_v = false$,
- if $\alpha = P_{\geq s}\beta$, then $I(w)(\alpha)_v = true$, if $\mu(w)(\{u \in W(w) : I(u)(\beta)_v = true\}) \geq s$, otherwise $I(w)(\alpha)_v = false$,
- if $\alpha = \beta \wedge \gamma$, then $I(w)(\alpha)_v = true$, if $I(w)(\beta)_v = true$, and $I(w)(\gamma)_v = true$, otherwise $I(w)(\alpha)_v = false$, and
- if $\alpha = (\forall x)\beta$, then $I(w)(\alpha)_v = true$, if for every $d \in D$, $I(w)(\beta)_{v_w[d/x]} = true$, otherwise $I(w)(\alpha)_v = false$. ■

One can observe that the previous definition does not guarantee that for a model $\mathbf{M}$, its world w, a valuation v and a formula β, the set of all worlds from $W(w)$ in which β is true ($[\beta]_{\mathbf{M},v,w} = \{u \in W(w) : I(u)(\beta)_v = true\}$) is measurable. Thus, it may happen that $\mu(w)$ is not defined for that set. To avoid this issue, we will consider the class of all $LFOP_1$-model which satisfy that for all worlds, formulas and valuations the set of worlds definable by β, is measurable, i.e., $[\beta]_{\mathbf{M},v,w} \in H(w)$. That class of all measurable constant domain $LFOP_1$-models with rigid terms will be denoted by $LFOP_{1,\text{Meas}}$.

Definition 1.5. A formula is *satisfied in a world w* from an $LFOP_{1,\text{Meas}}$-model $\mathbf{M} = \langle W, D, I, Prob\rangle$, denoted by

$$(\mathbf{M}, w) \models \alpha$$

(or simply $w \models \alpha$ if $\mathbf{M}$ is clear from the context) if for every valuation v, $I(w)(\alpha)_v = true$. If $d \in D$, $w \models \alpha(d)$ denotes that for every valuation v, $I(w)(\alpha(x))_{v_w[d/x]} = true$.

A formula is *valid in an $LFOP_{1,\text{Meas}}$-model* $\mathbf{M} = \langle W, D, I, Prob\rangle$, denoted by

$$\mathbf{M} \models \alpha$$

if it is satisfied in every world w from W.

A formula α is *valid*, denoted by

$$\models \alpha$$

if for every $LFOP_{1,\text{Meas}}$-model $\mathbf{M}$, $\mathbf{M} \models \alpha$.

A sentence α is *satisfiable* if there is a world w in an $LFOP_{1,\text{Meas}}$-model $\mathbf{M}$ such that $w \models \alpha$. A set T of sentences is satisfiable if there is a world w in an $LFOP_{1,\text{Meas}}$-model $\mathbf{M}$ such that for every $\alpha \in T$, $w \models \alpha$ (also denoted $w \models T$). ■

In the next examples we will illustrate some of the introduced notions.

Example 1.2. Let $\mathbf{M}$ be an $LFOP_{1,\text{Meas}}$-model $\langle W, D, I, Prob \rangle$ and $w \in W$. Suppose that $w \models P_{\geq s}R(x)$. By the definitions 1.4 and 1.5, the following holds for every valuation v:

- $I(w)(P_{\geq s}R(x))_v = \top$, iff
- $\mu(w)(\{u \in W(w) : I(u)(R(x))_v = true\}) \geq s$, iff
- $\mu(w)(\{u \in W(w) : v(x) \in I(u)(R)\}) \geq s$.

The last condition means that $v(x)$, the element of D assigned to x by v, should belong to the interpretation of R in u. In that way the formula $P_{\geq s}R(x)$ connects two different worlds. Since we consider constant domain models with rigid terms this will not cause any problems, but as Example 1.3 shows the situation changes without these constrains.

Also, note that a world w satisfies a formula $P_{\geq s}R(x)$ with a free variable x iff $w \models (\forall x)P_{\geq s}R(x)$. More generally, $w \models \alpha(x)$ iff $w \models (\forall x)\alpha(x)$. ■

Effects of abandoning the rigidness constraint for modal logic are discussed in [27], while the framework of probabilistic logic is considered in [35] where the next example regarding non-rigid terms is presented.

Example 1.3. Let $\mathbf{M}$ be the model (see Fig. 1.1):

- $W = W(w_1) = \{w_1, w_2\}$,
- $D = \{d_1, d_2\}$,
- $w_1 \models R(d_1)$, $w_1 \not\models R(d_2)$, $w_2 \not\models R(d_1)$, and $w_2 \models R(d_2)$, and
- $\mu(w_1)(\{w_1\}) = \mu(w_1)(\{w_2\}) = \frac{1}{2}$,

and the constant symbol c be interpreted such that the rigidness constraint fails:

- $I(w_1)(c) = d_2$, and
- $I(w_2)(c) = d_1$.

Then:

- $\mu(w_1)(\{w : w \models R(d_1)\}) = \mu(w_1)(\{w_1\}) = \frac{1}{2}$,
- $\mu(w_1)(\{w : w \models R(d_2)\}) = \mu(w_1)(\{w_2\}) = \frac{1}{2}$,

- $w_1 \not\models R(c)$,
- $w_2 \not\models R(c)$, and
- $w_1 \not\models P_{\geq \frac{1}{2}} R(c)$.

It follows that:

- $w_1 \models (\forall x) P_{\geq \frac{1}{2}} R(x)$, and
- $w_1 \not\models (\forall x) P_{\geq \frac{1}{2}} R(x) \to P_{\geq \frac{1}{2}} R(c)$.

So, an instance of the classical first order axiom $(\forall x)\alpha \to \alpha(c/x)$ is not valid. ∎

$$\boxed{w_1} \qquad\qquad \boxed{w_2} \qquad\qquad \mu(w_1)(\{w_1\}) = \mu(w_1)(\{w_2\}) = \tfrac{1}{2}$$

$$D = \{d_1, d_2\}$$

$I(w_1)(R) = \{d_1\}$	$I(w_2)(R) = \{d_2\}$	$[R(d_1)]_{w_1} = \{w_1\}$
$I(w_1)(c) = d_2$	$I(w_2)(c) = d_1$	
$w_1 \not\models R(c)$	$w_2 \not\models R(c)$	$[R(d_2)]_{w_1} = \{w_2\}$

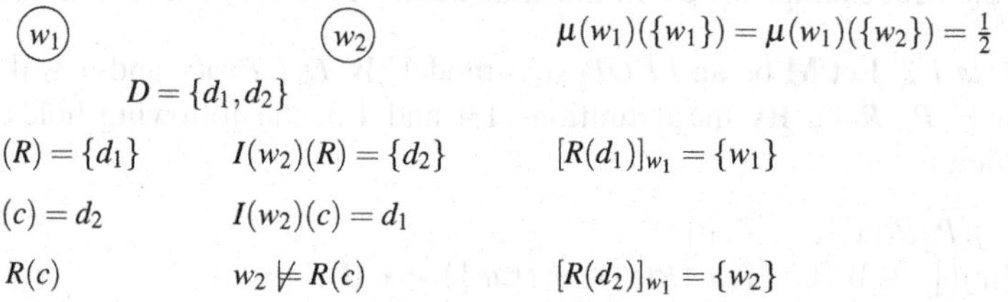

Fig. 1.1 Non-rigid terms.

On the other hand, although restrictive, the constant domain assumption enables us to avoid issues related to meanings of objects that do not exist in some worlds of a model. A possible alternative, i.e., models with world-relative domains, are also analyzed in [27].

Now, the main notions are defined which enables us to formulate two important results:

- [24, 37, 106]: The compactness theorem ("A set of formulas is satisfiable iff every finite subset of it is satisfiable") does not hold for probability logics that formalize probability functions that take values in $[0,1]$, the unit interval of reals.

- [1]: The set of valid formulas in the first order probabilistic logic is not recursively enumerable[3].

The former statement can be justified by the following example.

Example 1.4. Let us consider the set

$$T = \{\neg P_{=0} R(c)\} \cup \{P_{<1/n} R(c) : n \in \mathbb{N}\}.$$

For every finite $T' \subset T$, there is the largest $k \in \mathbb{N}$ such that $P_{<1/k} R(c) \in T'$. Then, in the following $LFOP_{1,\text{Meas}}$-model:

---

[3] Actually, in [1] a language more expressible than our For$_{LFOP_1}$ is considered, but it is shown in [86] how to extend our syntax to their level of expressiveness.

- $W = \{w_1, w_2\}, D = \{d\}$,
- $I(w_1)(c) = I(w_2)(c) = d$,
- $I(w_1)(R) = \emptyset, I(w_2)(R) = \{d\}$,
- $\mu(w_1)(\{w_1\}) = 1 - \frac{1}{k+1}$,
- $\mu(w_1)(\{w_2\}) = \frac{1}{k+1}$,

$w_1 \models T'$. So, every finite subset of T is $LFOP_{1,\text{Meas}}$-satisfiable. On the other hand, since for every $c > 0$, and an arbitrary model **M** and its world w, if $\mu(w)([R(c)]) = c$, there is a $k \in \mathbb{N}$ such that $\frac{1}{k} < c$, it follows that $w \not\models P_{<1/k}R(c)$, the set T is not $LFOP_{1,\text{Meas}}$-satisfiable. ■

The above mentioned statements influence axiomatizations of various probabilistic logics. More about that will be said in Section 1.3.

Definition 1.6. A formula $\alpha \in \text{For}_{LFOP_1}$ is a *(semantical) consequence* of a set T of For_{LFOP_1}-formulas, denoted

$$T \models \alpha$$

if for every model **M** in which all formulas from the set T are valid, $\mathbf{M} \models \alpha$. ■

This kind of relations of semantical consequence, where formulas from T can be seen as logical truths, is discussed for modal logics in [23].

1.2.3 Axiom system Ax_{LFOP_1}

Thanks to the above mentioned rigidness and constant domain constraints on models, it is possible to provide an axiom system for $LFOP_1$ which combines a classical first order axiomatization with axioms and rules for probabilistic reasoning. We will consider the system Ax_{LFOP_1} which involves the following axiom schemata:

1. all $\text{For}_{LFOP_1}(L)$-substitutional instances of the axioms of classical propositional logic
2. $(\forall x)(\alpha \to \beta) \to (\alpha \to (\forall x)\beta)$, where x is not free in α
3. $(\forall x)\alpha(x) \to \alpha(t/x)$, where $\alpha(t/x)$ is obtained by substituting all free occurrences of x in $\alpha(x)$ by the term t which is free for x in $\alpha(x)$
4. $P_{\geq 0}\alpha$
5. $P_{\leq r}\alpha \to P_{<s}\alpha,\ s > r$
6. $P_{<s}\alpha \to P_{\leq s}\alpha$
7. $(P_{\geq r}\alpha \wedge P_{\geq s}\beta \wedge P_{\geq 1}(\neg\alpha \vee \neg\beta)) \to P_{\geq \min(1,r+s)}(\alpha \vee \beta)$
8. $(P_{\leq r}\alpha \wedge P_{<s}\beta) \to P_{<r+s}(\alpha \vee \beta),\ r+s \leq 1$

and inference rules:

1. From α and $\alpha \to \beta$ infer β.
2. From α infer $(\forall x)\alpha$
3. From α infer $P_{\geq 1}\alpha$.

4. From $\beta \to P_{\geq s-\frac{1}{k}}\alpha$, for every integer $k \geq \frac{1}{s}$, infer $\beta \to P_{\geq s}\alpha$.

The axioms 1, 2 and 3 and the rules 1 and 2 represent a complete classical first-order axiom system, but any other equivalent system can be used as well. It means that classical first order logic is a sublogic of $LFOP_1$. The axioms 4 – 8 and the rules 2 and 3 form the probabilistic part of Ax_{LFOP_1}. Axiom 4 says that every formula is satisfied by a set of worlds of the measure at least 0, and, by substituting $\neg\alpha$ for α and according to the definition of $P_{\leq 1}$, it is equivalent to:

4'. $P_{\leq 1}\alpha$ $(= P_{\geq 1-s}\neg\alpha$, for $s = 1)$

which means that every formula is satisfied by a set of worlds of the measure at most 1. Relationships between probabilistic operators $P_{\leq \cdot}$ and $P_{<\cdot}$ are described by the axioms 5 and 6. Similarly as above, by corresponding substitutions, it is possible to obtain the equivalent forms of the axioms:

5'. $P_{\geq t}\alpha \to P_{>s}\alpha, t > s$
6'. $P_{>s}\alpha \to P_{\geq s}\alpha$

respectively. Finally, the axioms 7 and 8 correspond to the additivity of measures. If α and β are disjoint, Axiom 7 determines the lower bound of the probability of $\alpha \vee \beta$ which cannot be lesser than the sum of the probability of α and the probability of β. On the other hand, Axiom 8 gives the upper bound for the probability of $\alpha \vee \beta$. The rules 1 and 2 are the standard Modus Ponens, and generalization, and Rule 3 is the probabilistic counterpart of Necessitation rule of in modal logics.

Note that there is a countable set of assumptions and one consequence in Rule 4. It is an infinitary inference rule which corresponds to the Archimedean axiom for real numbers. Intuitively, the rule guarantees that if the probability is arbitrary close to s, then it is at least s. Rule 4 has an equivalent form:

4'. From $A \to P_{\leq s+\frac{1}{k}}\alpha$, for every integer $k \geq \frac{1}{1-s}$, infer $A \to P_{\leq s}\alpha$.

One can now consider the axiom and rule given in Section 1.1 and see that they are adapted to our language. The rule, written as an infinitary axiom, is presented in [38, 96]. Infinitary inference rules for modal logics are considered in [101, 103]. [2, 29] propose other infinitary inference rules for logics with probability operators.

Next, we define the notions of proofs and syntactical consequences. Since Ax_{LFOP_1} contains an infinitary rule, proofs may not be finite.

Definition 1.7. A formula α is *deducible from a set T of formulas* ($T \vdash \alpha$) if there is a sequence $\alpha_0, \alpha_1, \ldots, \alpha_{\lambda+1}$ (λ is a finite or countable ordinal[4]) of $\mathrm{For}_{LFOP_1}(L)$-formulas, such that:

- $\alpha_{\lambda+1} = \alpha$, and
- every α_i, $i \leq \lambda + 1$, is an axiom-instance, or $\alpha_i \in T$, or α_i is derived by an inference rule applied on some previous members of the sequence, with the proviso that *Rule 3 can be applied on the theorems only*.

[4] In other words, the length of a proof is an at most countable successor ordinal.

The sequence $\alpha_0, \alpha_1, \ldots, \alpha_{\lambda+1}$ is a *proof* for $\alpha_{\lambda+1}$ from T. If $T \vdash \alpha$, α is a *syntactical consequence* of T. A formula α is a *theorem* (denoted by $\vdash \alpha$) if it is deducible from the empty set. ■

Definition 1.8. A set T of $\mathrm{For}_{LFOP_1}(L)$-formulas is *consistent* if $T \nvdash \bot$, otherwise T is *inconsistent.*

A consistent set T of formulas is *maximal consistent* if for every $\alpha \in \mathrm{For}_{LFOP_1}(L)$, either $\alpha \in T$ or $\neg\alpha \in T$.

A set T of $\mathrm{For}_{LFOP_1}(L)$-formulas is *saturated* if it is maximal consistent and satisfies:

- if $\neg(\forall x)\alpha(x) \in T$, then for some term t, $\neg\alpha(t) \in T$. ■

Since (in)consistency of sets of formulas is a syntactical notion it is always related to particular axiom systems. The next example presents a proof that the set

$$T = \{\neg P_{=0}R(c)\} \cup \{P_{<1/n}R(c) : n \in \mathbb{N}\}$$

introduced in Example 1.4 is not consistent wrt. Ax_{LFOP_1}.

Example 1.5. The next sequence

1. $T \vdash \neg P_{=0}R(c)$, since $\neg P_{=0}R(c) \in T$
2. $T \vdash P_{<1/n}R(c)$, since $P_{<1/n}R(c) \in T$, for every $n \in \mathbb{N}$
3. $T \vdash P_{\leq 1/n}R(c)$, by Axiom 6 and Modus Ponens, for every $n \in \mathbb{N}$
4. $T \vdash P_{\leq 0}R(c)$, by Rule 4' from (3)
5. $T \vdash P_{=0}R(c)$, by Axiom 4 and definition of $P_{=0}$
6. $T \vdash \bot$ from (1) and (5)

shows that $\bot$ is deducible from a set T. The key step in the proof is an application of the infinitary rule 4.

A tree-like representation of the above proof is given in Fig. 1.2, such that nodes predecessors imply their successors. ■

1.2.4 Soundness and completeness

In this section we discuss soundness and completeness of the axiom system Ax_{LFOP_1} wrt. the class of $LFOP_{1,\mathrm{Meas}}$-models. It illustrate the procedure that is used in the proofs of the corresponding statements for other logics presented in the later chapters.

Soundness of Ax_{LFOP_1} is a consequence of the soundness of classical logic and properties of probabilistic measures.

Theorem 1.1 (Soundness theorem). *The axiom system Ax_{LFOP_1} is sound with respect to the $LFOP_{1,\mathrm{Meas}}$ class of models.*

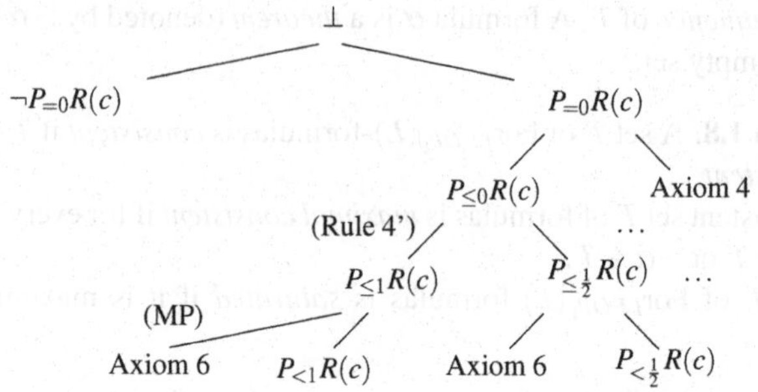

Fig. 1.2 Tree-like representation of the proof from Example 1.5.

Proof. As is mentioned above, worlds in $LFOP_{1,\mathrm{Meas}}$-models are classical first order models, and instances of classical axioms holds in every world. For example, let us consider Axiom 3. Let $\mathbf{M} = \langle W, D, I, Prob \rangle$ be an $LFOP_{1,\mathrm{Meas}}$-model, $\alpha(x)$ a formula, t a term free for x in $\alpha(x)$, and $w \in W$ such that:

- $w \models (\forall x)\alpha(x)$, i.e.,
- $I(w)((\forall x)\alpha(x))_v = true$ for every valuation v.

Let $I(w)(t)_v = d$, and $v' = v[d/x]$. Since $I(w)(\alpha(x))_{v'} = I(w)(\alpha(t/x))_v$, it follows that $I(w)(\alpha(t/x))_v = true$. We can see that, assuming the rigidness constraint, every instance of Axiom 3 is valid. Otherwise, Example 1.3 illustrates that without that constraint instances of Axiom 3 can be false. Validity of other axioms can be proved in a similar way.

Classical arguments show soundness for the rules 1 and 2, while explanation used for Necessitation in modal logics can be adapted for Rule 3. Finally, the properties of the set of real numbers guarantee that Rule 4 preserves validity. ∎

There are two usual forms of the completeness theorems:

- the weak (or simple) completeness: a formula is consistent iff it is satisfiable (i.e., a formula is valid iff it is provable), or
- the strong (or extended) completeness: a set of formulas is consistent iff it is satisfiable (a formula is a syntactical consequence of a set of formulas iff it is a semantical consequence of that set).

Since our approach suffices to prove the latter theorem, and it implies the former statement, we will focus on the strong completeness. In the proof we adapt and combine the Henkin style procedures for proving completeness for classical and modal logics:

- We first prove the Deduction theorem (Theorem 1.2).
- Then, we proceed with Lindenbaum's theorem, which guarantees that every consistent set of formulas can be extended to a saturated set (Theorem 1.3).

- Finally, the canonical model $\mathbf{M}_T$ is constructed using saturated sets (Lemma 1.4 and Lemma 1.5) such that for every world w, $w \models \alpha$ iff $\alpha \in w$ (Theorem 1.4).

Theorem 1.2 (Deduction theorem). *If T is a set of formulas, then*

$$T \cup \{\alpha\} \vdash \beta \quad \textit{iff} \quad T \vdash \alpha \to \beta.$$

Proof. The direction ($\Leftarrow$) can be proved exactly in the same way as in the classical case. For the ($\Rightarrow$)-direction we use the transfinite induction on the length of the inference. The cases:

- $\vdash \beta$,
- $\alpha = \beta$, or
- β is obtained by an application of Modus Ponens

are classical and follow as usual. If $T \cup \{\alpha\} \vdash P_{\geq 1}\gamma$ is obtained by an application of Rule 3, then γ and $\beta = P_{\geq 1}\gamma$ are theorems, and we have:

$T \vdash \beta$
$T \vdash \beta \to (\alpha \to \beta)$
$T \vdash \alpha \to \beta$, by Modus Ponens.

Finally, let $\beta = \gamma \to P_{\geq s}\delta$ be obtained from $T \cup \{\alpha\}$ by Rule 4. Then:

$T, \alpha \vdash \gamma \to P_{\geq s-\frac{1}{k}}\delta$, for every integer $k \geq \frac{1}{s}$
$T \vdash \alpha \to (\gamma \to P_{\geq s-\frac{1}{k}}\delta)$, for every integer $k \geq \frac{1}{s}$, by the induction hypothesis
$T \vdash (\alpha \wedge \gamma) \to P_{\geq s-\frac{1}{k}}\delta$, for every integer $k \geq \frac{1}{s}$
$T \vdash (\alpha \wedge \gamma) \to P_{\geq s}\delta$, by Rule 4
$T \vdash \alpha \to (\gamma \to P_{\geq s}\delta)$
$T \vdash \alpha \to \beta$. ■

The next lemmas 1.1, 1.2 and 1.3 will be used later on. In order to make this text self-contained, we give proofs for statements that are related to the probabilistic part of $LFOP_1$.

Lemma 1.1.

1. $\vdash P_{\geq s}\alpha \to P_{\geq s}\neg\neg\alpha$,
2. $\vdash P_{\geq 1}(\alpha \to \beta) \to (P_{\geq s}\alpha \to P_{\geq s}\beta)$,
3. if $\vdash \alpha \leftrightarrow \beta$, *then* $\vdash P_{\geq s}\alpha \leftrightarrow P_{\geq s}\beta$,
4. $\vdash P_{\geq s}\alpha \to P_{\geq r}\alpha$, *for* $s \geq r$,
5. $\vdash P_{< r}\alpha \to P_{< s}\alpha$, $s \geq r$.

Proof. (1) We have the following proof:

1. $\vdash P_{\geq 1}(\neg\alpha \vee \neg\bot)$, by Rule 3, from $\vdash \neg\alpha \vee \neg\bot$
2. $\vdash P_{\geq 1}((\neg\alpha \wedge \neg\bot) \vee \neg\neg\alpha)$, by Rule 3, from $\vdash (\neg\alpha \wedge \neg\bot) \vee \neg\neg\alpha$
3. $\vdash (P_{\geq s}\alpha \wedge P_{\geq 0}\bot \wedge P_{\geq 1}(\neg\alpha \vee \neg\bot)) \to P_{\geq s}(\alpha \vee \bot)$, an instance of Axiom 7
4. $\vdash P_{\geq s}\alpha \to P_{\geq s}(\alpha \vee \bot)$, from (1) and (3), since $P_{\geq 0}\bot$ is an instance of Axiom 4

5. $\vdash P_{\geq s}\alpha \to P_{\geq s}\neg(\neg\alpha \wedge \neg\bot)$, from (4) by the definition of $\vee$
6. $\vdash P_{\geq s}\alpha \to P_{\leq 1-s}(\neg\alpha \wedge \neg\bot)$, from (5) by the definition of $P_{\leq s}$
7. $(P_{\leq 1-s}(\neg\alpha \wedge \neg\bot) \wedge P_{<s}\neg\neg\alpha) \to P_{<1}((\neg\alpha \wedge \neg\bot) \vee \neg\neg\alpha)$, by Axiom 8
8. $(P_{\leq 1-s}(\neg\alpha \wedge \neg\bot) \wedge P_{<s}\neg\neg\alpha) \to \neg P_{\geq 1}((\neg\alpha \wedge \neg\bot) \vee \neg\neg\alpha)$, from (7) by the definition of $P_{<s}$
9. $\vdash (P_{\leq 1-s}(\neg\alpha \wedge \neg\bot) \wedge P_{<s}\neg\neg\alpha) \to \bot$, from (2) and (8)
10. $\vdash P_{\leq 1-s}(\neg\alpha \wedge \neg\bot) \to \neg P_{<s}\neg\neg\alpha$, by classical reasoning
11. $\vdash P_{\leq 1-s}(\neg\alpha \wedge \neg\bot) \to P_{\geq s}\neg\neg\alpha$, by the definition of $P_{<s}$
12. $\vdash P_{\geq s}\alpha \to P_{\geq s}\neg\neg\alpha$, from (6) and (11).

(2) The negation of the considered formula $P_{\geq 1}(\alpha \to \beta) \to (P_{\geq s}\alpha \to P_{\geq s}\beta)$ is equivalent to

$$P_{\geq 1}(\neg\alpha \vee \beta) \wedge P_{\geq s}\alpha \wedge P_{<s}\beta.$$

Using Lemma 1.1(1) that formula implies

$$P_{\geq 1}(\neg\alpha \vee \beta) \wedge P_{\geq s}\neg\neg\alpha \wedge P_{<s}\beta$$

which, using the definition of $P_{\leq 1-s}$ can be rewritten as

$$P_{\geq 1}(\neg\alpha \vee \beta) \wedge P_{\leq 1-s}\neg\alpha \wedge P_{<s}\beta.$$

By Axiom 8 we have

$$P_{\leq 1-s}\neg\alpha \wedge P_{<s}\beta \to P_{<1}(\neg\alpha \vee \beta)$$

and using the definition of $P_{<1}$ we obtain

$$\vdash \neg(P_{\geq 1}(\alpha \to \beta) \to (P_{\geq s}\alpha \to P_{\geq s}\beta)) \to P_{\geq 1}(\neg\alpha \vee \beta) \wedge \neg P_{\geq 1}(\neg\alpha \vee \beta).$$

It follows that

$$\vdash P_{\geq 1}(\alpha \to \beta) \to (P_{\geq s}\alpha \to P_{\geq s}\beta).$$

Note that this formula is a generalization of the modal axiom K: $\Box(\alpha \to \beta) \to (\Box\alpha \to \Box\beta)$.
(3) This is a consequence of Lemma 1.1(2).
(4) If $s = r$, the formula is of the form $\vdash \alpha \to \alpha$. If $s > r$, from Axiom 5' $P_{\geq s}\alpha \to P_{>r}\alpha$ (for $s > r$), and Axiom 6' $P_{>r}\alpha \to P_{\geq r}\alpha$, we obtain

$$\vdash P_{\geq s}\alpha \to P_{\geq r}\alpha.$$

Note that this formula expresses monotonicity of probabilities.
(5) Similarly as (4). ■

Lemma 1.2. *Let T be a consistent set of formulas.*

1. *For any formula α, either $T \cup \{\alpha\}$ is consistent or $T \cup \{\neg\alpha\}$ is consistent.*
2. *If $\neg(\alpha \to P_{\geq s}\beta) \in T$, then there is some $n > \frac{1}{s}$ such that $T \cup \{\alpha \to \neg P_{\geq s-\frac{1}{n}}\beta\}$ is consistent.*

Proof. (1) The proof is standard: if $T \cup \{\alpha\} \vdash \bot$, and $T \cup \{\neg\alpha\} \vdash \bot$, by the Deduction Theorem we have $T \vdash \bot$.
(2) Suppose the opposite. Then, we have:

$T, \{\alpha \rightarrow \neg P_{\geq s - \frac{1}{n}}\beta\} \vdash \bot$, for every $n > \frac{1}{s}$
$T \vdash \alpha \rightarrow P_{\geq s - \frac{1}{n}}\beta$, for every $n > \frac{1}{s}$, by the Deduction Theorem and classical reasoning
$T \vdash \alpha \rightarrow P_{\geq s}\beta$, by Rule 4,

which contradicts the fact that $\neg(\alpha \rightarrow P_{\geq s}\beta) \in T$. ∎

Lemma 1.3. *Let T be a saturated set. Then:*

1. *If $\alpha \in T$, then $\neg\alpha \notin T$.*
2. *$\alpha \wedge \beta \in T$ iff $\alpha \in T$ and $\beta \in T$.*
3. *If $T \vdash \alpha$, then $\alpha \in T$, i.e., T is deductively closed.*
4. *If $\alpha \in T$ and $\alpha \rightarrow \beta \in T$, then $\beta \in T$.*
5. *If $P_{\geq s}\alpha \in T$, and $s \geq r$, then $P_{\geq r}\alpha \in T$.*
6. *If r is a rational number and $r = \sup\{s : P_{\geq s}\alpha \in T\}$, then $P_{\geq r}\alpha \in T$.*

Proof. The statements (1) – (4) are classical, and (5) follows from Lemma 1.1.4.
(6) Let $r = \sup\{s : P_{\geq s}\alpha \in T\}$. By Rule 4 we obtain $T \vdash P_{\geq r}\alpha$. Since T is deductively closed, it follows that $P_{\geq r}\alpha \in T$. ∎

Theorem 1.3 (Lindenbaum's theorem). *Let T be a consistent set of $LFOP_1(L)$-formulas, and C be a countably infinite set of new constant symbols ($C \cap L = \emptyset$). Then T can be extended to a saturated set $\overline{\mathcal{T}}$ in the language $\overline{L} = L \cup C$.*

Proof. Let $\alpha_0, \alpha_1, \ldots$ be an enumeration of all formulas from $\mathrm{For}_{LFOP_1}(\overline{L})$. We define the following sequence of sets T_i, $i = 0, 1, 2, \ldots$:

1. $T_0 = T$
2. for every $i \geq 0$,
   a. if $T_i \cup \{\alpha_i\}$ is consistent, then $T_{i+1} = T_i \cup \{\alpha_i\}$, otherwise
      i. if α_i is of the form $\beta \rightarrow P_{\geq s}\gamma$, then $T_{i+1} = T_i \cup \{\neg\alpha_i, \beta \rightarrow \neg P_{\geq s - \frac{1}{n}}\gamma\}$, for some positive $n \in \mathbb{N}$, so that T_{i+1} is consistent, otherwise
      ii. $T_{i+1} = T_i \cup \{\neg\alpha_i\}$.
   b. If T_{i+1} is obtained by adding a formula of the form $\neg(\forall x)\beta(x)$ to T_i, then for some $c \in C$, $\neg\beta(c)$ is also added to T_{i+1}, so that T_{i+1} is consistent,
3. $\overline{\mathcal{T}} = \cup_{i=0}^{\infty} T_i$.

First note that the proof of consistency of the sets obtained by the steps (1), (2a) and (2(a)ii) is standard. Next, Lemma 1.2.2 guarantees consistency of the sets obtained by the step (2(a)i). Finally, we sketch the proof from [9] related to the step (2b). Suppose that for some $i > 0$, a formula of the form $\neg(\forall x)\beta(x)$ is added to T_i (in the steps (2a), or (2(a)ii)). If $\neg\beta(c) \in T_i$, for a constant symbol $c \in C$, then $T_i \cup \{\neg(\forall x)\beta(x), \neg\beta(c)\}$ is obviously consistent. Otherwise, there is a constant $c \in C$

which does not appear in $T_i \cup \{\neg(\forall x)\beta(x)\}$, since $T_i \cup \{\neg(\forall x)\beta(x)\}$ is obtained by adding only finitely many formulas to T, and constant symbols from C do not appear in T. Suppose that $\neg\beta(c)$ cannot be consistently added to $T_i \cup \{\neg(\forall x)\beta(x)\}$. Then:

$T_i \cup \{\neg(\forall x)\beta(x), \neg\beta(c)\} \vdash \bot$
$T_i, \{\neg(\forall x)\beta(x)\} \vdash \beta(c)$, by the Deduction theorem
$T_i, \{\neg(\forall x)\beta(x)\} \vdash (\forall x)\beta(x)$, since c does not appear in $T_i \cup \{\neg(\forall x)\beta(x)\}$
$T_i \vdash (\forall x)\beta(x)$,

which contradicts consistency of T_i. Thus, the step (2b) produces consistent sets.

It remains to prove that $\overline{\mathscr{T}}$ is saturated. Note that the above construction guarantees that:

- for any $\alpha \in \mathrm{For}_{LFOP_1}(\overline{L})$, either α or $\neg\alpha$ belongs to $\overline{\mathscr{T}}$,
- for any $\neg(\forall x)\beta(x) \in \overline{\mathscr{T}}$, there is a $c \in C$ such that $\neg\beta(c) \in \overline{\mathscr{T}}$.

We show that $\overline{\mathscr{T}}$ is deductively closed, i.e., if $\overline{\mathscr{T}} \vdash \alpha$, then $\alpha \in \overline{\mathscr{T}}$. Since $\overline{\mathscr{T}}$ is maximal, this implies its consistency. We will use the induction on the length of the inference of α from $\overline{\mathscr{T}}$. First, the following holds:

- if $\alpha = \alpha_j$ and $T_i \vdash \alpha$, it must be $\alpha \in \overline{\mathscr{T}}$ because $T_{\max(i,j)+1}$ is consistent.

Let the sequence $\gamma_1, \gamma_2, \ldots, \alpha$ be the proof of α from $\overline{\mathscr{T}}$. If the sequence is finite, there must be a set T_i such that $T_i \vdash \alpha$, and $\alpha \in \overline{\mathscr{T}}$.

So, let the sequence $\gamma_1, \gamma_2, \ldots, \alpha$ be countably infinite. We can show that for every i, if γ_i is obtained by an application of an inference rule, and all premisses belong to $\overline{\mathscr{T}}$, then also $\gamma_i \in \overline{\mathscr{T}}$. If the rule is a finitary one, then there is a set T_j which contains all premisses and $T_j \vdash \gamma_i$. By the above mentioned explanation, $\gamma_i \in \overline{\mathscr{T}}$. So, let $\gamma_i = \beta_1 \rightarrow P_{\geq s}\beta_2$ be obtained by Rule 4 from the set $\{\beta_1 \rightarrow P_{\geq s-\frac{1}{k}}\beta_2\}$ and every $\beta_1 \rightarrow P_{\geq s-\frac{1}{k}}\beta_2 \in \overline{\mathscr{T}}$. If $\beta_1 \rightarrow P_{\geq s}\beta_2 \notin \overline{\mathscr{T}}$, by the step (2(a)i) of the construction, there are some l and j such that $\neg(\beta_1 \rightarrow P_{\geq s}\beta_2)$, and $\beta_1 \rightarrow \neg P_{\geq s-\frac{1}{l}}\beta_2 \in T_j$. It means that for some $j' \geq j$:

- $\beta_1 \wedge \neg P_{\geq s}\beta_2 \in T_{j'}$,
- $\beta_1 \in T_{j'}$,
- $\neg P_{\geq s-\frac{1}{l}}\beta_2, P_{\geq s-\frac{1}{l}}\beta_2 \in T_{j'}$,

which contradicts consistency of $T_{j'}$. Hence, all γ_i and α belong to $\overline{\mathscr{T}}$. ■

Note that formulas of the form $\beta \rightarrow \neg P_{\geq s-\frac{1}{n}}\gamma$ which can be added to sets T_i's in the step (2(a)i) of the above procedure are witnesses that the corresponding formulas $\beta \rightarrow P_{\geq s}\gamma$ cannot be derived from the $\overline{\mathscr{T}}$. This is similar to the step (2b) and formulas of the forms $\neg\beta(c)$ and $\neg(\forall x)\beta(x)$.

We can now proceed by introducing the canonical model of the considered set T. Let the tuple $\mathbf{M}_T = \langle W, D, I, Prob\rangle$ be defined in the following way:

- W is the set of all saturated sets in the language $\overline{L} = L \cup C$,
- D is the set of all variable-free terms in $\overline{L}$,

- for every $w \in W$, $I(w)$ is an interpretation such that:
  - if f is a k-ary function symbol, $I(w)(f)$ is a function $I(f)$ from D^k to D such that for all variable-free terms $t_1, \ldots, t_k$ in $\overline{L}$, $I(f) : \langle t_1, \ldots, t_k \rangle \to f(t_1, \ldots, t_k)$, and
  - If P is a k-ary relation symbol, $I(w)(P) = \{\langle t_1, \ldots, t_k \rangle$ for all variable-free terms $t_1, \ldots, t_k$ in $\overline{L}$ such that $P(t_1, \ldots, t_k) \in w\}$.
- for every $w \in W$, $Prob(w) = \langle W(w), H(w), \mu(w) \rangle$ such that:
  - $W(w) = W$,
  - $H(w)$ is the class of sets $[\alpha] = \{w \in W : \alpha \in w\}$, and
  - for every set $[\alpha] \in H(w)$, $\mu(w)([\alpha]) = \sup_s \{P_{\geq s}\alpha \in w\}$.

The next lemma shows that the $Prob(w)$'s are probability spaces.

Lemma 1.4. *Let* $\mathbf{M}_T = \langle W, D, I, Prob \rangle$ *be defined as above and* $\alpha, \beta \in \mathrm{For}_{LFOP_1}(\overline{L})$. *Then, the following hold:*

1. *H is an algebra of subsets of W,*
2. *If $[\alpha] = [\beta]$, then $\mu([\alpha]) = \mu([\beta])$,*
3. *$\mu([\alpha]) \geq 0$.*
4. *$\mu(W) = 1$ and $\mu(\emptyset) = 0$.*
5. *$\mu([\alpha]) = 1 - \mu([\neg\alpha])$.*
6. *$\mu([\alpha] \cup [\beta]) = \mu([\alpha]) + \mu([\beta])$, for all disjoint $[\alpha]$ and $[\beta]$.*

Proof. (1) Let $\alpha, \alpha_1, \alpha_2, \ldots \alpha_n \in \mathrm{For}_{LFOP_1}(\overline{L})$. Then:

- $W = [\alpha \vee \neg\alpha]$, and $W \in H$,
- if $[\alpha] \in H$, then $[\neg\alpha]$ is its complement which belongs to H, and
- if $[\alpha_1], \ldots, [\alpha_n] \in H$, then $[\alpha_1] \cup \ldots \cup [\alpha_n] = [\alpha_1 \vee \ldots \vee \alpha_n]$ belongs to H.

It follows that H is an algebra of subsets of W.
(2) From $[\alpha] = [\beta]$ it follows that $\vdash \alpha \leftrightarrow \beta$. By Lemma 1.1.3 we have $\mu(w)([\alpha]) = \mu(w)([\beta])$.
(3) Since $P_{\geq 0}\alpha$ is an axiom, we have that $P_{\geq 0}\alpha \in w$, and $\mu(w)([\alpha]) \geq 0$.
(4) Since $W = [\alpha \vee \neg\alpha]$ and $P_{\geq 1}(\alpha \vee \neg\alpha) \in w$, $\mu(w)(W) = 1$. On the other hand, $\emptyset = [\alpha \wedge \neg\alpha]$, and $\mu(w)(\emptyset) \geq 0$. From $P_{\geq 1}(\alpha \vee \neg\alpha) \in w$, by the definition of $P_{<s}$, and the axioms 5, 6 and 5' we have $P_{<s}(\alpha \wedge \neg\alpha)$, $\neg P_{>s}(\alpha \wedge \neg\alpha)$, and $\neg P_{\geq t}(\alpha \wedge \neg\alpha) \in w$, for $s, t > 0$. Thus, $\sup_s\{P_{\geq s}(\alpha \wedge \neg\alpha) \in w\} = 0$, and $\mu(w)(\emptyset) = 0$.
(5) Let $r = \mu(w)([\alpha]) = \sup_s\{P_{\geq s}\alpha \in w\}$. If $r = 1$, by Lemma 1.3.6, $P_{\geq 1}\alpha \in w$. It means that w contains $\neg P_{>0}\neg\alpha (= P_{\leq 0}\neg\alpha = P_{\geq 1}\alpha)$. By Axiom 5' there is no $s > 0$ such that $P_{\geq s}\neg\alpha \in w$. Thus, $\mu(w)([\neg\alpha]) = 0$.

Let $r < 1$. Then, since $\neg P_{\geq r'}\alpha = P_{<r'}\alpha$, for every $r' > r$, $r' \in [0,1]_{\mathbb{Q}}$, we have $P_{<r'}\alpha \in w$. It follows by Axiom 6 that $P_{\leq r'}\alpha, P_{\geq 1-r'}\neg\alpha \in w$. If there is an $r'' < r$, $r'' \in [0,1]_{\mathbb{Q}}$, such that $P_{\geq 1-r''}\neg\alpha \in w$, then also $\neg P_{>r''}\alpha \in w$, which is a contradiction. Hence, $\sup_s\{P_{\geq s}(\neg\alpha) \in w\} = 1 - \sup_s\{P_{\geq s}\alpha \in w\}$, and $\mu(w)([\alpha]) = 1 - \mu(w)([\neg\alpha])$.

(6) Let $[\alpha] \cap [\beta] = \emptyset$, $\mu(w)([\alpha]) = r$ and $\mu(w)([\beta]) = s$. Since $[\beta] \subset [\neg\alpha]$, the above statements (2) and (5) guarantee that $r+s \leq r+(1-r) = 1$.

Let $r > 0$, and $s > 0$. By the well known properties of the supremum, for all $r', s' \in [0,1]_{\mathbb{Q}}$ such that $r' < r$ and $s' < s$, we have $P_{\geq r'}\alpha, P_{\geq s'}\beta \in w$. Then, Axiom 7 implies that $P_{\geq r'+s'}(\alpha \vee \beta) \in w$, which means that $r+s \leq t_0 = \sup_t\{P_{\geq t}(\alpha \vee \beta) \in w\}$. If $r+s = 1$, then the statement trivially holds. So, assume that $r+s < 1$. If $r+s < t_0$, then for every $t' \in [0,1]_{\mathbb{Q}} \cap (r+s, t_0)$ we have $P_{\geq t'}(\alpha \vee \beta) \in w$. We can choose $r'', s'' \in [0,1]_{\mathbb{Q}}$ such that $r'' > r$, $s'' > s$, and:

- $\neg P_{\geq r''}\alpha, P_{<r''}\alpha \in w$,
- $\neg P_{\geq s''}\beta, P_{<s''}\beta \in w$ and
- $r'' + s'' = t' \leq 1$.

We have

- $P_{\leq r''}\alpha \in w$, by Axiom 6,
- $P_{<r''+s''}(\alpha \vee \beta) \in w$, by Axiom 8,
- $\neg P_{\geq r''+s''}(\alpha \vee \beta) \in w$, and
- $\neg P_{\geq t'}(\alpha \vee \beta) \in w$,

a contradiction. Hence, $r+s = t_0$ and $\mu(w)([\alpha] \cup [\beta]) = \mu(w)([\alpha]) + \mu(w)([\beta])$.

Finally suppose that $r = 0$ or $s = 0$. Then we can reason as above, with the only exception that $r' = 0$ or $s' = 0$. ■

Note that $\mathbf{M}_T$ is *almost* an $LFOP_{1,\mathrm{Meas}}$-model. Here "almost" stands because the sets of the form $[\alpha]$ are defined using $\alpha \in w$, and not $(\mathbf{M}_T, w) \models \alpha$ as is required for $LFOP_{1,\mathrm{Meas}}$-models. This last step, i.e., $\mathbf{M}_T \in LFOP_{1,\mathrm{Meas}}$ follows from the next lemma.

Lemma 1.5. *$\mathbf{M}_T = \langle W, D, I, Prob \rangle$ is an $LFOP_{1,\mathrm{Meas}}$-model.*

Proof. As we noted above, we have to prove that for all $\alpha \in \mathrm{For}_{LFOP_1}(\overline{L})$ and $w \in W$, $w \models \alpha$ iff $\alpha \in w$, which implies that $[\alpha] = \{w : w \models \alpha\}$. We will use the induction on the complexity of formulas.

To begin the induction, let α be an atomic formula $P(t_1, \ldots, t_k)$. If α is a sentence, then by the definition of the interpretation I, $w \models P(t_1, \ldots, t_k)$ iff $P(t_1, \ldots, t_k) \in w$. Otherwise, if $x_1, \ldots, x_m$ are all free variables in α, then by Rule 2 and Axiom 3, $\alpha(x_1, \ldots, x_m) \in w$ iff $\alpha(t'_1, \ldots, t'_m) \in w$ for all variable-free terms t'_i. The formulas of the form $\alpha(t'_1, \ldots, t'_m)$ are atomic sentences, and for them the statement holds. It follows that $\alpha(x_1, \ldots, x_m) \in w$ iff for every valuation v, $I(w)(\alpha(x_1, \ldots, x_m))_v = true$, i.e., $w \models \alpha(x_1, \ldots, x_m)$.

If $\alpha = (\forall x)\beta$ and $\alpha \in w$, then, by Axiom 3, $\beta(t) \in w$ for every $t \in D$. By the induction hypothesis $w \models \beta(t)$ for every $t \in D$, and $w \models (\forall x)\beta$. On the other hand, let $\alpha \notin w$. Since w is saturated, there is some $t \in D$ such that $w \models \neg\beta(t)$. It follows that $(\mathbf{M}_T, w) \not\models (\forall x)\beta$.

The cases $\alpha = \neg\beta$ and $\alpha = \beta \wedge \gamma$ follow in the standard way. For example, $w \models \neg\beta$ iff $w \not\models \beta$ iff $\beta \notin w$ iff $\neg\beta \in w$.

Finally, let $\alpha = P_{\geq s}\beta$. If $P_{\geq s}\beta \in w$, then $\sup_r\{P_{\geq r}(\beta) \in w\} = \mu(w)([\alpha]) \geq s$, and $w \models P_{\geq s}\beta$. For the other direction, let $w \models P_{\geq s}\beta$. It means that $\sup_r\{P_{\geq r}\beta \in w\} \geq s$. If $\mu(w)([\beta]) = s$, then by Lemma 1.3.6, $P_{\geq s}\beta \in w$. Otherwise, let $\mu(w)([\beta]) > s$. By the well known property of supremum and monotonicity of μ, $P_{\geq s}\beta \in w$. ■

Now, from the previous lemmas and Theorem 1.3 we can prove:

Theorem 1.4 (Strong completeness theorem for $LFOP_{1,\mathrm{Meas}}$). *A set T of formulas is Ax_{LFOP_1}-consistent iff it is $LFOP_{1,\mathrm{Meas}}$-satisfiable.*

Proof. By Lindenbaum's theorem 1.3, T can be extended to a saturated set. The corresponding canonical model $\mathbf{M}_T$ is an $LFOP_{1,\mathrm{Meas}}$-model which contains at least one world $w \supset T$. By Lemma 1.5, $w \models T$ and T is $LFOP_{1,\mathrm{Meas}}$-satisfiable. ■

Theorem 1.4 can be equivalently formulated in the following way:

Theorem 1.5 (Strong completeness theorem for $LFOP_{1,\mathrm{Meas}}$). *Let $T \subset \mathrm{For}_{LFOP_1}$ and $\alpha \in \mathrm{For}_{LFOP_1}$. Then:*

$$T \models \alpha \text{ iff } T \vdash \alpha.$$

using the notions of semantical and syntactical consequences.

Strong completeness for σ-additive models is discussed in [39].

1.2.5 Versions of the logic $LFOP_1$

Several logics, presented in the literature ([70, 78–82, 90, 92–94]), can be obtained by restricting $LFOP_1$, e.g.:

- LPP_1 is a propositional counterpart of $LFOP_1$ (probability logic with iterations of the probability operators and real-valued probability functions),
- $LFOP_2$ and LPP_2 are first-order and propositional probability logics without iterations of the probability operators,
- $LFOP_1^S$ and LPP_1^S are the logics that formalize reasoning about probability functions whose ranges are restricted to a countable set S (e.g. $[0,1]_{\mathbb{Q}}$, or the unit interval of Hardy field $[0,1]_{\mathbb{Q}(\varepsilon)}$),
- the logics $LFOP_1^{\mathrm{Fr}(n)}$ and $LPP_1^{\mathrm{Fr}(n)}$ are similar to them, but the ranges of probability functions are finite, of the form $\{0, \frac{1}{n}, \dots, \frac{n-1}{n}, 1\}$.

In $LFOP_2$ and LPP_2 neither mixing of classical formulas and probability formulas, nor nesting of probability operators are allowed, so $\alpha \wedge P_{\geq s}\alpha$ and $P_{\geq s}P_{\geq r}\alpha$ are not well-defined formulas. As a consequence, models that correspond to those logics are simpler than $LFOP_{1,\mathrm{Meas}}$-models and $LPP_{1,\mathrm{Meas}}$-models, respectively. For example, measurable models for the propositional probability logic LPP_2 are of the following form:

Definition 1.9. A measurable *LPP_2-model* (an $LPP_{2,\mathrm{Meas}}$-model) is a structure $\mathbf{M} = \langle W, H, \mu, v\rangle$ where:

- W is a nonempty set of worlds,
- v associates with each world $w \in W$ a classical valuation,
- H is an algebra of subsets of the form $[\alpha] = \{w : w \models \alpha\}$, and
- μ is a finitely additive probability measure, $\mu : H \to [0,1]$. ■

It can be seen that worlds are classical propositional models, and that in each model there is only one probability measure defined over subsets of possible worlds, and not a particular probability measure associated to every world. Satisfiability is a relation between formulas and models.

Definition 1.10. *The satisfiability relation* $\models \subseteq LPP_{2,\mathrm{Meas}} \times \mathrm{For}_{LPP_2}$ fulfills the following conditions for every $LPP_{2,\mathrm{Meas}}$-model $\mathbf{M} = \langle W, H, \mu, v\rangle$:

- if α is a classical formula, $\mathbf{M} \models \alpha$ iff for every $w \in W$, $v(w,\alpha) = true$,
- $\mathbf{M} \models P_{\geq s}\alpha$ iff $\mu([\alpha]) \geq s$,
- if A and B are probabilistic formulas, $\mathbf{M} \models \neg A$ iff $\mathbf{M} \not\models A$, and $\mathbf{M} \models A \wedge B$ iff $\mathbf{M} \models A$ and $\mathbf{M} \models B$. ■

Despite these differences, Ax_{LFOP_1} is a strongly complete axiomatization for $LFOP_2$ (of course, formulas obey the above mentioned syntactical restrictions). On the other hand, strongly complete axiomatizations for propositional probability logics LPP_1 and LPP_2 can be obtained by removing the first order axioms 2 and 3 and Rule 2 from Ax_{LFOP_1}. Using linear programming, it can be proved that the satisfiability problem PSAT for these propositional probability logics are decidable.

For example, let us consider the logic LPP_2, and let A be a Boolean combination of formulas of the form $P_{\geq s}\alpha$. Then, A is equivalent to a formula in the disjunctive form

$$\mathrm{DNF}(A) = \bigvee_{i=1}^{m} \bigwedge_{j=1}^{k_i} X^{i,j}(p_1,\ldots,p_n)$$

where:

- $X^{i,j}$ is a probability operator from the set $\{P_{\geq s_{i,j}}, P_{<s_{i,j}}\}$, and
- $X^{i,j}(p_1,\ldots,p_n)$ denotes that the propositional formula which is in the scope of the probability operator $X^{i,j}$ is in the complete disjunctive normal form, i.e., the propositional formula is a disjunction of conjunctions (called atoms) of the form $a_t = \pm p_1 \wedge \ldots \wedge \pm p_n$, where $\pm p_i$ is a literal, i.e., either p_i or $\neg p_i$, and $\{p_1,\ldots,p_n\}$ are all primitive propositions of A.

Since DNF(A) is satisfiable iff at least one its disjunct is satisfiable, without loss of generality, we can assume that A is of the form $\bigwedge_{j=1}^{k} X^j(p_1,\ldots,p_n)$. Let y_t denote the measure of a_t and $a_t \in X(p_1,\ldots,p_n)$ mean that a_t appears in the propositional part of $X(p_1,\ldots,p_n)$. Then, since a_t's are mutually disjunctive, A is satisfiable iff the following system of linear equalities and inequalities is satisfiable:

$$\begin{aligned}
\Sigma_{i=1}^{2^n} y_i &= 1\\
y_i &\geq 0\,,\ \text{for } i = 1,\dots,2^n\\
\Sigma_{a_t \in X^1(p_1,\dots,p_n)} y_t &\begin{cases} \geq s_1 \text{ if } X^1 = P_{\geq s_1}\\ < s_1 \text{ if } X^1 = P_{< s_1}\end{cases}\\
&\dots\\
\Sigma_{a_t \in X^k(p_1,\dots,p_n)} y_t &\begin{cases} \geq s_k \text{ if } X^k = P_{\geq s_k}\\ < s_k \text{ if } X^k = P_{< s_k}\end{cases}
\end{aligned} \tag{1.1}$$

In this way, PSAT is reduced to the linear systems solving problem, which implies that PSAT for LPP_2-logic is decidable.

The strong completeness for probability logics with probability functions with the countable range S can be achieved by replacing the infinitary rule 4 from Ax_{LFOP_1} with the rule:

4. $\alpha \to P_{\neq s}\beta$, for every $s \in S$, infer $\alpha \to \bot$.

This rule is a generalization of the rule from [2] where $S = [0,1]_{\mathbb{Q}}$. On the other hand, if the range of probability functions is finite, i.e., of the form $\{0, \frac{1}{n}, \dots, \frac{n-1}{n}, 1\}$, it is enough to add the following axiom ([90, 106]):

9. $P_{>s}\alpha \to P_{\geq s+\frac{1}{n}}\alpha$, where $s \in \{0, \frac{1}{n}, \dots, \frac{n-1}{n}, 1\} \setminus \{1\}$

instead of the infinitary rule 4.

The proof technique for strong completeness introduced in Section 1.2.4 can be adapted for other probability logics. The following chapters present logics:

- that formalize reasoning about probability functions with ranges that are not linearly ordered ([41, 42, 44–50]),
- that combine probabilistic reasoning with other non-classical logics, e.g., intuitionistic ([6, 62, 63]), temporal ([11, 13, 15, 16, 61, 69, 71, 72, 76]), justification ([57, 58, 83]), epistemic logic ([104, 105]),
- that are suitable for reasoning about the uncertainty of events modeled by sets of probability measures using the boundaries called upper probability and lower probability ([17, 18, 97, 98]), or by conditional probabilities ([12, 40, 49, 65, 66, 73, 75]), qualitative probabilities ([10, 42, 43, 77]), probabilistic support ([14]), etc.

1.3 Related Work

Providing strongly complete axiom systems for probability logics was a question that attracted attention of researchers. [3] noted that in this framework (a propositional multi-agent logic with real-valued probabilities and operators – in our notation – $P^i_{\geq r}$, where i denotes the agent i), still there was no satisfactory syntactic definition of consistency which enables a proof of strong completeness[5]. This section gives an

[5] A possible approach to solve this problem for propositional and first order probability logics using infinitary axiomatizations is presented in Section 1.2.

overview of some logics related to the ones described in Section 1.2. The focus will be on the completeness and decidability issues. In the presentation we will use the above introduced syntax and semantics wherever possible.

[7] enriches the modal system S5 with the unary operator U interpreted as "probably". The semantics for the obtained logic S5U consists of an algebra on the power set $\mathbb{P}(I)$ of some nonempty set I with operations related to the classical, modal and probabilistic connectives, e.g.:

- for negation: $nx = I - x$,
- for the modal operator $\Box$: $lx = I$, if $x = I$, otherwise $lx = \emptyset$,
- for the probabilistic operator U: $ux = I$, if $x \in C_1$, otherwise $ux = \emptyset$,

where $C_1 \subset \mathbb{P}(I)$, such that:

- if $x \in C_1$ and $y \geq x$, then $y \in C_1$, and
- if $x \notin C_1$, then $I - x \in C_1$,

i.e., C_1 contains everything that is probable. A weakly complete axiomatization is provided. It contains the following axioms for the probabilistic operator:

- $\Box\alpha \rightarrow U\alpha$
- $U\alpha \rightarrow \Box U\alpha$
- $U\neg\alpha \rightarrow \neg U\alpha$
- $\Box(\alpha \rightarrow \beta) \rightarrow (U\alpha \rightarrow U\beta)$.

It is also proved that the logic is decidable.

[100] studies propositional modal calculus with the binary operator $\gtrsim$ read as "at least as probable as". The logic is characterized by a set of axioms which involves:

- $\Box(\alpha \leftrightarrow \alpha') \wedge \Box(\beta \leftrightarrow \beta') \rightarrow ((\alpha \gtrsim \beta) \rightarrow (\alpha' \gtrsim \beta'))$
- $\bot \gtrsim \bot$, etc.

It is proved that the axiom system is simply complete wrt. the class of models of the form $\langle W, R, v, B, M\rangle$, where $\langle W, R, v\rangle$ is a standard Kripke model for modal logics, $B(w)$ is a Boolean σ-algebra of subsets of $\{y : xRy\}$, and $M(w)$ is a probability measure on $B(w)$. Completeness is proved using filtration (which also implies decidability). Another axiom system for comparative probabilities is given in [8].

The papers [67, 68], motivated by development of medical expert systems with uncertain rules, consider a formal language in which it is possible to express probabilities of first order sentences, e.g., of a probabilistic generalization of modus ponens[6]: "if α holds with the probability s, and β follows from α with the probability t, then the probability of β belongs to $[s+t-1, 1]$". A technique for calculating bounds on probabilities of consequences given probabilities of assumptions is presented. It is interesting that this technique resembles the method proposed already in [5] and independently rediscovered several times (see [36]). In any case, from our point of view, [67, 68] are extremely important since the semantical approach presented there stimulated also a number of proof-theoretic oriented papers.

[6] In the language introduced in Section 1.2.1 the sentence can be written as $P_{\geq s}\alpha \wedge P_{\geq t}(\alpha \rightarrow \beta) \rightarrow (P_{\geq s+t-1}\beta \wedge P_{\leq 1}\beta)$.

To our knowledge, [26] is the first paper in which probabilistic operators with explicit numerical probabilities appear. Gaifman considers formulas of the form $PR(\alpha,\Delta)$ with the intended meaning "the probability of the propositional formula α is in the closed interval $\Delta = [\delta_1, \delta_2]$"[7]. α may contain PR, so higher order probabilities are allowed. Probability models (similar to measurable models $LPP_{1,\mathrm{Meas}}$ for the propositional logic with iterations of probability operators) are defined. It is proved that the modal logic S5 is a fragment of the presented logic, if $PR(., [1,1])$ is understood as $\Box$, the modal operator of necessity. The paper also discusses a temporal generalization of the probabilistic operators, $PR(\alpha,t,\Delta)$, interpreted as "the probability of α at t is in Δ", where time points can be partially ordered. In that way it is possible to describe changes of probabilities of events.

[22] considers modal operators[8] of the form M_r and L_r, $r \in [0,1]$. The corresponding probability Kripke-like models are $LPP^{\mathrm{Fr}(n)}_{1,\mathrm{Meas}}$ measurable models with probability functions with a finite range $F \subset [0,1]$, and with iterations of probability operators. The authors are able to provide a finitary axiomatization which is proved to be strongly complete. The important assumption is that the range of probability functions is finite, so that compactness holds. It is remarked that abandoning the finitary constrains on the range of probabilities produces a more complicated logic for which the problem of providing complete axiomatization is difficult. It is interesting that in the completeness proof it is showed that the infinitary inference rule 4 is deducible from the given axiomatization. This is used to define probabilities of sets definable by formulas via suprema, in the same way as is done in Section 1.2.4 for the canonical model. Note that this step can be performed in an easier way since the range of probability functions is finite [90, 106].

In [20] several logics suitable for reasoning about probabilities are studied. The paper begins by considering the class $LPP_{2,\mathrm{Meas}}$ of measurable propositional models with real-valued probability functions and without iteration of probability operators. A more expressive syntax than the one from Section 1.2 is introduced. The formal language allows linear combinations of probabilities in the form of basic probabilistic formulas

$$a_1 w(\alpha_1) + \ldots + a_n w(\alpha_n) \geq s,$$

where a_i's and s are rational numbers, α_i's are classical propositional formulas, and $w(\alpha_i)$'s denote probabilities[9]. Probability formulas are Boolean combinations of basic probabilistic formulas. The corresponding finitary axiom system AX_{MEAS} contains:

- the standard axioms and rules for propositional logic,
- axioms for reasoning about probability:
  - $w(\alpha) \geq 0$,
  - $w(true) = 1$,

[7] The sentence can be written as $P_{\geq \delta_1} \alpha \wedge P_{\leq \delta_2} \alpha$.

[8] They are, in our notation, $P_{>r}$, and $P_{\geq 1-r}$, respectively. Note that M_r and L_r are mutually definable.

[9] So, $w(\alpha) \geq s$ is our $P_{\geq s}\alpha$.

- $w(\varphi \wedge \psi) + w(\varphi \wedge \neg\psi) = w(\varphi)$, and
- if $\models \varphi \leftrightarrow \psi$, then $w(\varphi) = w(\psi)$, and

- axioms for reasoning about linear inequalities, e.g.:
  - $x \geq x$,
  - $(a_1x_1 + \ldots + a_nx_n \geq c) \wedge (a'_1x_1 + \ldots + a'_nx_n \geq c') \to ((a_1 + a'_1)x_1 + \ldots + (a_n + a'_n)x_n) \geq (c + c')$,
  - $(t \geq c) \to (t > d)$, for $c > d$, etc.

To prove the weak completeness, i.e., that every consistent formula A is satisfiable, it is first showed[10] that for every $w(\alpha)$, $\vdash w(\alpha) \leftrightarrow \sum_{a \in \mathrm{Atoms}(\alpha), \vdash a \to \alpha} w(a)$, where $\mathrm{Atoms}(\alpha)$ denotes the set of conjunctions of the form $\pm p_1 \wedge \ldots \wedge \pm p_n$, $\pm p_i$ is either p_i or $\neg p_i$, and $\{p_1, \ldots, p_n\}$ is the set of all primitive propositions that appear in α. Since every formula is equivalent to a finite disjunctions of finite conjunctions of (possibly negated) basic probabilistic formulas, A can be assumed to be a finite conjunction of (possibly negated) basic probabilistic formulas. That conjunction is satisfiable iff a linear system generated from A is satisfiable. Unsatisfiability of the system implies that $\vdash \neg A$, i.e., that A is inconsistent. Beside giving a sound and complete axiom system for reasoning about linear inequalities, another benefit from this approach is that the same ideas can be applied to prove the small model theorem (a formula is satisfiable iff it is satisfied in a model with the number of worlds limited by the size of the formula). That theorem allows the authors to prove decidability and NP-completeness of the satisfiability problem PSAT. The paper also discusses polynomial weight formulas that allow products of terms, e.g.,

$$w(p_1 \wedge p_2) \cdot (w(p_1) + w(p_2)) \geq w(p_1) \cdot w(p_2)$$

that are appropriate to represent conditional probabilities. For example, the above formula represents the sentence "the conditional probability of p_2 given p_1 plus the conditional probability of p_1 given p_2 is at least 1". A PSPACE decision procedure is given, but, similarly as above, it is noted that in obtaining the weak completeness additional expressiveness is needed. Thus, the authors introduce a first order language such that variables can appear in formulas:

$$(\forall x)(\exists y)[(3 + x)w(\varphi)w(\psi) + 2w(\varphi \vee \psi) \geq z].$$

The corresponding axiom system $AX_{FO-MEAS}$ contains the standard first order axiomatization and, axioms for real closed fields. In [84], [86] and [85], by adapting the approach from Section 1.2, the strong completeness was proved for the logics with polynomial weight formulas and $[0,1]$, and $[0,1]_{^*\mathbb{R}}$-valued probability functions. For the latter logic compactness was also showed. Finally, [20] also considers non measurable models. So, the probability measure μ is partial and, for some set A, only the inner measure $\mu_*(A) = \sup\{\mu(B) : B \subset A\}$ and the outer measure

[10] Here the axioms for reasoning about linear inequalities are crucial. The authors noted that they did not succeed to axiomatize the weaker language, i.e., the logic LPP_2.

$\mu^*(A) = \inf\{\mu(B) : B \supset A\}$ induced by μ can be defined. For this logic a weakly complete axiom system is provided, and decidability of PSAT is proved.

[35] considers two first order probability logics: in the first logic probabilities are defined on domains, similarly as in Keisler's approach (see [55]), while in the second one probabilities are defined on possible worlds. It is assumed that in each model for the second logic there is only one measure, so the models are similar to $LFOP_{2,\mathrm{Meas}}$-models. For both logics the corresponding axiomatization are provided, but completeness can be proved only if domains are bounded in size by some finite n. The paper [4] complains about that assumption and explains that, while domains may be finite, it is not obvious that there is a fixed upper bound on their size, and also that there are infinite domains interesting in applications. In any case, this restriction is very strong and actually reduces the considered systems to propositional logic. In [4] a finitary axiom system is given for a first order probabilistic logic with probabilities defined on domains. It is obtained by requiring that ranges of probability functions could be unit intervals of arbitrary ordered fields, instead of a particular field, e.g., the unit interval of reals $[0,1]$, or $[0,1]_{\mathbb{Q}}$. Thus, there are sentences on the real-valued probability functions that are not provable in the logic. [1] concerns expressiveness of logics discussed in [35]. The main results, under the assumption that probability functions are discrete, are:

- when the probability is on the domain, if the language contains one binary relational symbol, the validity problem is Π_1^2 complete,
- with equality in the language and with no other symbol, the validity problem for the logic with the probability on the domain, is at least Π_∞^1 hard,
- when the probability is on possible worlds, if the language contains at least one unary relational symbol, the validity problem is Π_1^2 complete, while
- with equality and only one constant symbol in the language, the validity problem is Π_∞^1 hard,
- in the case of $[0,1]_{\mathbb{Q}}$-valued probabilities, complexity decreases from Π_1^2 to Π_1^1,

which means that there is no finite (strongly or weakly) complete axiomatization for the considered probability logic. If arbitrary probabilities are considered, the corresponding upper bounds of complexity remained undetermined. As is mentioned above, if the sizes of domains in models are bounded by some fixed n, the validity problem becomes decidable. [1] defines a satisfiability-preserving translation between logics with probabilities defined on domains and logics with probabilities defined on possible worlds.

The paper [19] gives a finitary axiomatization of higher order probabilities in a propositional language that mixes the modal operator of knowledge K and linear combinations of terms representing probabilities and allows a set of agents. The considered models extend $LPP_{1,\mathrm{Meas}}$-models with the accessibility relations that correspond to modal notion of knowledge. Relationships between knowledge and probability are considered, for example:

1. the set of worlds accessible by the modal relation of accessibility from a world w has the probability 1 from w,

2. probabilities are objective in the sense that all agents in a world use the same probability, etc.

Characteristic axioms are proposed for each particular case, e.g., the former case is axiomatized by $K\alpha \rightarrow P_{=1}\alpha$. A procedure similar to the one from [20] is used to and prove decidability of PSAT.

Non-compactness in the framework of probability logics is discussed in [106, 107]. The papers address the logical issue caused by possible finitary axiomatizations of any non-compact probability logic: there are consistent but unsatisfiable sets of formulas (e.g. the set from Example 1.4), so the corresponding syntax and semantics do not match. Then, a finite axiom system which is sound and complete for the logic $LPP_{1,\mathrm{Meas}}^{\mathrm{Fr}(n)}$ is given. Decidability of the logic is proved, and also that, if $P_{\geq 1}\alpha$ is denoted by $\Box\alpha$, the modal system KD is embedded in $LPP_{1,\mathrm{Meas}}^{\mathrm{Fr}(n)}$.

Motivated by [67], the paper [24] proposes a deductive system which contains a set of inference rules. The considered formulas are of the form $P(\alpha|\beta) \in I$, where α and β are classical propositional formulas and I is a closed subinterval of $[0,1]$. One of the rules which calculates the conditional probability of β given δ is:

$$\frac{\begin{array}{c} P(\alpha|\delta) \in [x,y] \\ P(\alpha \vee \beta|\delta) \in [u,v] \\ P(\alpha \wedge \beta|\delta) \in [w,z] \end{array}}{P(\beta|\delta) \in [\max(w, u-y+w), \min(v, v-x+z)]}$$
$$\text{provided that } w \leq y,\ x \leq v,\ w \leq v$$

To avoid indefiniteness of the conditional probability $P(\alpha|\beta)$ when the probability of the conditioning formula β is 0, they stated that in that case $P(\alpha|\beta)$ holds in any model. A sound incomplete[11] iteration procedure which computes increasingly narrow probability intervals for propositional formulas in $LPP_{2,\mathrm{Meas}}$-models (in our notation) is described. The procedure can be stopped at any time, and generates proofs that give partial information about the probability of sentences, allowing one to make a tradeoff between precision and computational time. It is argued that such an approach is preferable to the method introduced in [67] since the former could be applicable in the case of first-order logic.

[2] gives a strongly complete axiomatization for the class of measurable first order probability models ($LPP_{1,\mathrm{Meas}}$, in our notation) with $[0,1]_{\mathbb{Q}}$-valued probabilities. The axiom system contains the infinitary inference rule:

- From $\Sigma \vdash \neg P_{=r}\varphi$, for every $r \in [0,1]_{\mathbb{Q}}$, infer $\Sigma \vdash \bot$

which implies that the probabilities take values in $[0,1]_{\mathbb{Q}}$. Similar rules are used in [79] and [92, 95] to ensure that probabilities are in decidable subsets of $[0,1]$ and in $[0,1]_{\mathbb{Q}(\varepsilon)}$, the unit interval of Hardy field, respectively. [29] provides strongly complete axiomatization for a propositional dynamic logic with qualitative probabilities using the infinitary inference rule which states that:

[11] The lack of finitary strongly complete axiomatization caused by non-compactness of the logic is explained.

if the set of rationals that are greater or equal to the probability of one proposition is contained in the set of rationals that are greater or equal to the probability of another proposition, then the probability of the first proposition is greater or equal to the probability of the second one.

Formally, the rule can be written:

- From $\theta \rightarrow (\Pi_j^t \wedge \bigwedge_{i=0}^{j-1}(\psi \leq_t \frac{i}{j} \rightarrow \varphi \leq_t \frac{i}{j}))$ for every j, infer $\theta \rightarrow (\psi \leq_t \varphi)$.

The paper [6] extends intuitionistic propositional logic with probabilistic operators of the form $P_{\geq r}$ and $P_{\leq r}$. These operators are not mutually expressible in the intuitionistic framework, so both types of operators are included in the formal language. The corresponding semantics consists of intuitionistic Kripke models in which each possible world w is equipped with two partial functions p^w and p_w representing upper and lower probabilities and satisfy:

- the range of all probability functions is finite,
- monotonicity: the upper probabilities cannot increase, and lower probability cannot decrease (which corresponds to the paradigm of increasing knowledge over time in intuitionistic logic): if wRu, and $X \subset W$ is measurable, then $p^w(X) \leq p^u(X)$, and $p_w(X) \geq p_u(X)$,
- subadditivity for upper probabilities: if $X, Y \subset W$ are measurable, then $p^w(X \cup Y) \leq \min\{p^w(X) + p^w(Y), 1\}$, and
- superadditivity for lower probabilities:: if $X, Y \subset W$ are measurable, then $p_w(X \cup Y) \geq \min\{p_w(X) + p_w(Y), 1\}$.

The presented axiom system, proved to be complete, besides the Heyting propositional calculus, contains:

- monotonicity axioms: $P_{\leq r}\alpha \rightarrow P_{\leq s}\alpha$, for $r \leq s$; $P_{\geq r}\alpha \rightarrow P_{\geq s}\alpha$, for $r \geq s$,
- $P_{\leq 1}\alpha$, $P_{\geq 0}\alpha$,
- finite semiadditivity: $(P_{\leq r}\alpha \wedge P_{\leq s}\beta) \rightarrow P_{\leq \min\{1,r+s\}}(\alpha \vee \beta)$, and $(P_{\geq r}\alpha \wedge P_{\geq s}\beta \wedge P_{\leq 0}(\alpha \wedge \beta)) \rightarrow P_{\geq \min\{1,r+s\}}(\alpha \vee \beta)$,
- Necessitation: from α infer $P_{\geq 1}\alpha$.

The paper [37] gives a finitary weakly complete axiom system for the multi-agent $LPP_{1,\mathrm{Meas}}$ discussed in [3]. The key inference rule guarantees compatible degrees of belief for any two sets of formulas that are equivalent in a particularly defined sense. Thus, the authors solve the problem from [20] about characterizing valid $LPP_{1,\mathrm{Meas}}$-formulas in a simpler language which does not contain linear combinations of probabilities, but they admit that "there is no hope of having strong completeness of the system".

[64] discusses propositional σ–additive $LPP_{1,\mathrm{Meas}}$-models, and assuming that:

- only finitary formulas can be in scopes of probability operators, and
- if λ denotes the cardinality of the disjoint union of the sets of probability operators and primitive propositions, and $\lambda \geq \aleph_0$, then the maximal length of formulas and inferences is equal to 2^λ,

gives the strong completeness theorem ($T \vdash \alpha$ iff $T \models \alpha$) for theories of cardinality $\leqslant 2^\lambda$.

The paper [28] considers For_{LPP_2}-formulas of the form

$$\bigwedge_{i=1}^{n} P_{=c_i}\alpha_i$$

and prove[12]:

- PSAT is NP-complete,
- if every α_i contains at most 2 primitive proposition, PSAT is NP-complete.

In [53] a heuristic procedure based on the column generation technique is proposed and problems with up to 50 primitive propositions and 80 propositional formulas α_i are solved. [51] further develops this heuristic approach to PSAT, enables intervals of probabilities and conditional probabilities, and again using the column generation technique solves instances of PSAT with up to 140 primitive propositions and 300 propositional formulas. Procedures for PSAT based on genetic algorithms and Variable neighborhood search, proposed in [52, 74], are able to solve PSAT instances with up to 200 primitive propositions and 1000 propositional formulas. The paper [102] considers a logic with approximate conditional probabilities, where conditional probabilities may take infinitesimal values from the unit interval of a recursive non-Archimedean Hardy field $[0,1]_{\mathbb{Q}(\varepsilon)}$. A bee colony optimization meta-heuristic is proposed for solving the corresponding PSAT. The considered logic is suitable for modeling default reasoning.

[56] discusses complexity of PSAT in non-iterated and iterated probability logics (see Chapter 8). Formal proofs are given for:

- PSAT for $LPP_{2,\mathrm{Meas}}$ is NP-complete, and
- PSAT for $LPP_{1,\mathrm{Meas}}$ is PSPACE-complete,

while for probabilistic extension over justification logics the following results are obtain:

- PSAT for non-iterated probabilistic logic over the justification logic J is Σ_2^p-complete, and
- PSAT for iterated probabilistic logic over the justification logic J is PSPACE-complete.

References

[1] M. Abadi and Y. Halpern. Decidability and expressiveness for first-order logics of probability. *Information and Computation*, 112:1–36, 1994.

[12] Note that [20] generalizes these results to the language containing linear combinations of probabilities of formulas.

[2] N. Alechina. Logic with probabilistic operators. In *Proceedings ACCOLADE '94*, pages 121–138, 1995.

[3] R. Aumann. Interactive epistemology II: Probability. *International Journal of Game Theory*, 28:301–314, 1999.

[4] F. Bacchus. *Lp*, A Logic for Representing and Reasoning with Statistical Knowledge. *Computational Intelligence*, 6 (4):209–231, 1990.

[5] G. Boole. *An investigation of the laws of thought on which are founded the mathematical theories of logic and probabilities*. London, 1854.

[6] B. Boričić and M. Rašković. A probabilistic validity measure in intuitionistic propositional logic. *Mathematica Balkanica*, 10:365–372, 1996.

[7] J.P. Burgess. Probability Logic. *The Journal of Symbolic Logic*, 34:264–274, 1969.

[8] J.P. Burgess. Axiomatizing the Logic of Comparative Probability. *Notre Dame Journal of Formal Logic*, 51(1):119–126, 2010.

[9] C.C. Chang and J.H. Keisler. *Model theory*. North-Holland, Amsterdam, 1977. Volume 73 of Studies in Logic and the Foundations of Mathematics.

[10] D. Doder. A logic with big-stepped probabilities that can model nonmonotonic reasoning of system P. *Publications de l'Institut Mathématique*, Ns. 90(104):13–22, 2011.

[11] D. Doder, J. Grant, and Z. Ognjanović. Probabilistic logics for objects located in space and time. *Journal of Logic and Computation*, 23(3):487–515, 2013.

[12] D. Doder, B. Marinković, P. Maksimović, and A. Perović. A Logic with Conditional Probability Operators. *Publications de l'Institut Mathématique*, Ns. 87(101):85–96, 2010.

[13] D. Doder and Z. Ognjanović. A probabilistic logic for reasoning about uncertain temporal information. In M. Meila and T. Heskes, editors, *Proceedings of the Thirty-First Conference on Uncertainty in Artificial Intelligence, UAI 2015, July 12-16, 2015, Amsterdam, The Netherlands*, pages 248–257. AUAI Press, 2015.

[14] D. Doder and Z. Ognjanović. Probabilistic logics with independence and confirmation. *Studia Logica*, 105(5):943–969, 2017.

[15] D. Doder, Z. Ognjanović, and Z. Marković. An Axiomatization of a First-order Branching Time Temporal Logic. *Journal of Universal Computer Science*, 16:1439–1451, 2010.

[16] D. Doder, Z. Ognjanović, Z. Marković, A. Perović, and M. Rašković. A probabilistic temporal logic that can model reasoning about evidence. In *Foundations of Information and Knowledge Systems, 6th International Symposium, FoIKS 2010, Sofia, Bulgaria, February 15-19, 2010. Proceedings*, pages 9–24. Springer, 2010.

[17] D. Doder, N. Savić, and Z. Ognjanović. A Decidable Multi-agent Logic with Iterations of Upper and Lower Probability Operators. In F. Ferrarotti and S. Woltran, editors, *Proceedings of the 10th International Symposium Foundations of Information and Knowledge Systems, FoIKS 2018, Budapest, Hungary, May 1418, 2018*, volume 10833 of *Lecture Notes in Computer Science*, pages 170–185. Springer, 2018.

[18] D. Doder, N. Savić, and Z. Ognjanović. Multi-agent Logics for Reasoning About Higher-Order Upper and Lower Probabilities. *Journal of Logic Language and information*, 29:77–107, 2020.

[19] R. Fagin and J.Y. Halpern. Reasoning about knowledge and probability. *Journal of the ACM*, 41 (2):340–367, 1994.

[20] R. Fagin, J.Y. Halpern, and N. Megiddo. A logic for reasoning about probabilities. *Information and Computation*, 87:78–128, 1990.

[21] S. Fajardo and J. Keisler. *Model Theory of Stochastic Processes. Lecture Notes in Logic*. Association for Symbolic Logic, 2002.

[22] M. Fattorosi-Barnaba and G. Amati. Modal operators with probabilistic interpretations I. *Studia Logica*, 46:383–393, 1989.

[23] M. Fitting. Basic modal logic. In D. M. Gabbay, C. Hogger, and J. Robinson, editors, *Handbook of logic in artificial intelligence and logic programming*, volume 1, pages 368–448. 1993.

[24] A. Frisch and P. Haddawy. Anytime deduction for probabilistic logic. *Artificial Intelligence*, 69:93–122, 1994.

[25] H. Gaifman. Concerning measures in first order calculi. *Israel Journal of Mathematics*, 2:1–18, 1964.

[26] H. Gaifman. A Theory of Higher Order Probabilities. In J. Y. Halpern, editor, *Proceedings of the Theoretical Aspects of Reasoning about Knowledge*, pages 275–292, San Mateo, California, 1986. Morgan Kaufmann.

[27] J.W. Garson. Quantification in modal logic. In D. Gabbay and F. Guenthner, editors, *Handbook of philosophical logic*, volume 2, pages 249–307. D. Reidel Publishing Comp, 1984.

[28] G. Georgakopoulos, D. Kavvadias, and C. Papadimitriou. Probabilistic satisfiability. *Journal of Complexity*, 4(1):1–11, 1988.

[29] D. Guelev. A propositional dynamic logic with qualitative probabilities. *Journal of Philosophical Logic*, 28:575–605, 1999.

[30] I. Hacking. *Emergence of probability. A philosophical study of early ideas about probability induction and statistical ineference.* Cambridge University Press, 2006.

[31] T. Hailperin. Best Possible Inequalities for the Probability of a Logical Function of Events. *The American Mathematical Monthly*, 72(4):343–359, 1965.

[32] T. Hailperin. Probability logic. *Notre Dame Journal of Formal Logic*, 25:198–212, 1984.

[33] T. Hailperin. *Sentential Probability Logic. Origins, development, current status and technical applications.* Lehigh University Press, 1996.

[34] T. Hailperin. *Logic with a probability semantics*. Lehigh University Press, 2011.

[35] J.Y. Halpern. An analysis of first-order logics of probability. *Artificial Intelligence*, 46:311–350, 1990.

[36] P. Hansen and B. Jaumard. Probabilistic satisfiability. In *Handbook of Defeasible Reasoning and Uncertainty Management Systems*, volume 5. Springer, 2000.

[37] A. Heifetz and P. Mongin. Probability Logic for Type Spaces. *Games and Economic Behavior*, 35:31–53, 2001.
[38] D. Hoover. Probability logic. *Annals of mathematical logic*, 14:287–313, 1978.
[39] N. Ikodinović, Z. Ognjanović, A. Perović, and M. Rašković. Completeness theorems for σ-additive probabilistic semantics. *Annals of Pure and Applied Logic*, 171(4):102755:1–102755:27, 2020.
[40] N. Ikodinović and Z. Ognjanović. A logic with coherent conditional probabilities. In L. Godo, editor, *Proceedings of the 8th European Conference Symbolic and Quantitative Approaches to Reasoning with Uncertainty, ECSQARU 2005; Barcelona; Spain; 6-8 July 2005*, volume 3571 of *Lecture Notes in Computer Science*, pages 726–736. Springer, 2005.
[41] N. Ikodinović, M. Rašković, Z. Marković, and Z. Ognjanović. Measure Logic. In *Proceedings of the 9th European Conference Symbolic and Quantitative Approaches to Reasoning with Uncertainty*, volume 4724 of *Lecture Notes in Computer Science*, pages 128–138. Springer, 2007.
[42] N. Ikodinović, M. Rašković, Z. Marković, and Z. Ognjanović. Logics with Generalized Measure Operators. *Journal of Multiple-Valued Logic and Soft Computing*, 20(5-6):527–555, 2013.
[43] A. Ilić-Stepić. A Logic for Reasoning About Qualitative Probability. *Publications de l'Institut Mathématique*, Ns. 87(101):97–108, 2010.
[44] A. Ilić-Stepić. *O formalizaciji p-adskih, kvalitativnih i uslovnih verovatnoća.* PhD thesis, University of Belgrade, Belgrade, 2013.
[45] A. Ilić-Stepić, Z. Ognjanović, N. Ikodinović, and A. Perović. A p-adic probability logic. *Mathematical Logic Quarterly*, 58(4-5):263–280, 2012.
[46] A. Ilić-Stepić and Z. Ognjanović. Complex valued probability logics. *Publications de l'Institut Mathématique*, Ns. 95(109):73–86, 2014.
[47] A. Ilić-Stepić and Z. Ognjanović. Logics for reasoning about processes of thinking with information coded by p-adic numbers. *Studia Logica*, 103:145–174, 2015.
[48] A. Ilić-Stepić and Z. Ognjanović. Logics to formalise p-adic valued probability and their applications. *International Journal of Parallel, Emergent and Distributed Systems*, 33:257–275, 2018.
[49] A. Ilić-Stepić, Z. Ognjanović, and N. Ikodinović. Conditional p-adic probability logic. *International Journal of Approximate Reasoning*, 55(9):1843–1865, 2014.
[50] A. Ilić-Stepić, Z. Ognjanović, N. Ikodinović, and A. Perović. p-Adic probability logics. *p-Adic Numbers, Ultrametric Analysis and Applications*, 8(3):177–203, 2016.
[51] B. Jaumard, P. Hansen, and M. P. de Aragao. Column generation methods for probabilistic logic. *ORSA Journal on Computing*, 3:135–147, 1991.
[52] D. Jovanović, N. Mladenović, and Z. Ognjanović. Variable neighborhood search for the probabilistic satisfiability problem. In K. F. Doerner, M. Gendreau, P. Greistorfer, W. Gutjahr, F. Hartl, and M. Reimann, editors, *Metaheuristics: Progress in Complex Systems Optimization, Springer Series: Op-*

erations Research/Computer Science Interfaces Series, volume 39, pages 173–188. Springer, Berlin, 2007.

[53] D. Kavvadias and C. Papadimitriou. Linear programming approach to reasoning about probabilities. *Annals of mathematics and artificial intelligence*, 1:189–205, 1990.

[54] H.J. Keisler. Hyperfinite model theory. In R.O. Gandy and J.M.E. Hyland, editors, *Logic Colloquium 76*, pages 5–110. North-Holland, 1977.

[55] H.J. Keisler. Probability quantifiers. In J. Barwise and S. Feferman, editors, *Model Theoretic Logic*, pages 509–556. Springer, Berlin, 1985.

[56] I. Kokkinis. The complexity of satisfiability in non-iterated and iterated probabilistic logics. *Annals of Mathematics and Artificial Intelligence*, 83(3-4):351–382, 2018.

[57] I. Kokkinis, P. Maksimović, Z. Ognjanović, and T. Studer. First steps towards probabilistic justification logic. *Logic Journal of the IGPL*, 23(4):662–687, 2015.

[58] I. Kokkinis, Z. Ognjanović, and T. Studer. Probabilistic Justification Logic. *Journal of Logic and Computation*, 30(1):257–280, 2020.

[59] S. Kripke. A completeness theorem in modal logic. *The Journal of Symbolic Logic*, 24:1–14, 1959.

[60] H.E. Kyburg Jr. *Probability and the Logic of Rational Belief.* Wesleyan University Press, 1961.

[61] B. Marinković, Z. Ognjanović, D. Doder, and A. Perović. A propositional linear time logic with time flow isomorphic to ω^2. *Journal of Applied Logic*, 12(2):208–229, 2014.

[62] Z. Marković, Z. Ognjanović, and M. Rašković. A probabilistic extension of intuitionistic logic. *Mathematical Logic Quarterly*, 49:415–424, 2003.

[63] Z. Marković, Z. Ognjanović, and M. Rašković. An intuitionistic logic with probabilistic operators. *Publications de l'Institut Mathématique*, Ns. 73 (87):31–38, 2003.

[64] M. Meier. An infinitary probability logic for type spaces. *Israel journal of mathematics*, 192:1–58, 2012.

[65] M. Milošević and Z. Ognjanović. A first-order conditional probability logic. *Logic Journal of the Interest Group in Pure and Applied Logics*, 20(1):235–253, 2012.

[66] M. Milošević and Z. Ognjanović. A First-Order Conditional Probability Logic With Iterations. *Publications de l'Institut Mathématique*, Ns. 93 (107):19–27, 2013.

[67] N. Nilsson. Probabilistic logic. *Artificial Intelligence*, 28:71–87, 1986.

[68] N. Nilsson. Probabilistic logic revisited. *Artificial Intelligence*, 59:39–42, 1993.

[69] Z. Ognjanović. A logic for temporal and probabilistic reasoning. In *Workshop on Probabilistic Logic and Randomised Computation, ESSLLI 1998, Saarbrücken, Germany*, 1998.

[70] Z. Ognjanović. *Some probability logics and their applications in computer sciences*. PhD thesis, University of Kragujevac, Kragujevac, 1999.

[71] Z. Ognjanović. Discrete Linear-time Probabilistic Logics: Completeness, Decidability and Complexity. *Journal of Logic and Computation*, 16(2):257–285, 2006.

[72] Z. Ognjanović, D. Doder, and Z. Marković. A Branching Time Logic with Two Types of Probability Operators. In S. Benferhat and J. Grant, editors, *Fifth International Conference on Scalable Uncertainty Management SUM-2011, October 10 - 12, 2011, Dayton, Ohio, USA*, volume 6929 of *Lecture Notes in Computer Science*, pages 219–232. Springer, 2011.

[73] Z. Ognjanović and N. Ikodinović. A logic with higher order conditional probabilities. *Publications de l'Institut Mathématique*, Ns. 82(96):141–154, 2007.

[74] Z. Ognjanović, J. Kratica, and M. Milovanović. A genetic algorithm for satisfiability problem in a probabilistic logic: A first report. 2143:805–816, 2001.

[75] Z. Ognjanović, Z. Marković, and M. Rašković. Completeness theorem for a Logic with imprecise and conditional probabilities. *Publications de l'Institut Mathématique*, Ns. 78(92):35–49, 2005.

[76] Z. Ognjanović, Z. Marković, M. Rašković, D. Doder, and A. Perović. A propositional probabilistic logic with discrete linear time for reasoning about evidence. *Annals of Mathematics and Artificial Intelligence*, 65(2-3):217–243, 2012.

[77] Z. Ognjanović, A. Perović, and M. Rašković. Logics with the Qualitative Probability Operator. *Logic Journal of IGPL*, 16(2):105–120, 2008.

[78] Z. Ognjanović and M. Rašković. A logic with higher order probabilities. *Publications de l'Institut Mathématique*, Ns. 60(74):1–4, 1996.

[79] Z. Ognjanović and M. Rašković. Some probability logics with new types of probability operators. *Journal of Logic and Computation*, 9(2):181–195, 1999.

[80] Z. Ognjanović and M. Rašković. Some first-order probability logics. *Theoretical Computer Science*, 247(1-2):191–212, 2000.

[81] Z. Ognjanović, M. Rašković, and Z. Marković. Probability logics. In Z. Ognjanović, editor, *Zbornik radova, subseries Logic in computer science*, volume 12(20), pages 35–111, Beograd, Serbia, 2009. Matematički institut SANU.

[82] Z. Ognjanović, M. Rašković, and Z. Marković. *Probability logics. Probability-Based Formalization of Uncertain Reasoning*. Springer, Cham, Switzerland, 2016.

[83] Z. Ognjanović, N. Savić, and T. Studer. Justification Logic with Approximate Conditional Probabilities. In A. Baltag, J. Seligman, and T. Yamada, editors, *Proceedings of the 6th International Workshop Logic, Rationality, and Interaction, LORI 2017, Sapporo, Japan, September 1114, 2017*, volume 10455 of *Lecture Notes in Computer Science*, pages 681–686. Springer, 2017.

[84] A. Perović. *Some applications of the formal method in set theory, model theory, probabilistic logics and fuzzy logics*. PhD thesis, University of Belgrade, Belgrade, 2008.

[85] A. Perović, Z. Ognjanović, M. Rašković, and Z. Marković. A Probabilistic Logic with Polynomial Weight Formulas. In L.S. Hartmann and G. Kern-Isberner, editors, *Proceedings of the 5th International Symposium Founda-*

tions of Information and Knowledge Systems FoIKS 2008, Pisa, Italy, February 11-15, 2008, volume 4932 of *Lecture Notes in Computer Science*, pages 239–252, Berlin, Heidelberg, 2008. Springer.

[86] A. Perović, Z. Ognjanović, M. Rašković, and Z. Marković. How to Restore Compactness into Probabilistic Logics? In S. Hölldobler, C. Lutz, and H. Wansing, editors, *Proceedings of the 11th European Conference, JELIA 2008, Dresden, Germany, September 28-October 1, 2008*, volume 5293 of *Lecture Notes in Computer Science*, pages 338–348, Berlin, Heidelberg, 2008. Springer.

[87] M. Rašković. Model theory for L_{AM} logic. *Publications de l'Institut Mathématique*, Ns. 37(51):17–22, 1985.

[88] M. Rašković. Completeness theorem for biprobability models. *The Journal of Symbolic Logic*, 51(3):586–590, 1986.

[89] M. Rašković. Completeness theorem for singular biprobability models. *Proceedings of the American Mathematical Society*, 102:389–392, 1988.

[90] M. Rašković. Classical logic with some probability operators. *Publications de l'Institut Mathématique*, Ns. 53(67):1–3, 1993.

[91] M. Rašković and R. Đorđević. *Probability Quantifiers and Operators*. VESTA, Beograd, 1996.

[92] M. Rašković, Z. Marković, and Z. Ognjanović. A logic with approximate conditional probabilities that can model default reasoning. *International Journal of Approximate Reasoning*, 49(1):52–66, 2008.

[93] M. Rašković and Z. Ognjanović. Some propositional probabilistic logics. In V. Kovačević-Vujičić and Ž. Mijajlović, editors, *Proceedings of the Kurepa's symposium 1996, Belgrade, Scientific review 19-20*, pages 83–90, Belgrade, 1996.

[94] M. Rašković and Z. Ognjanović. A first order probability logic, LP_Q. *Publications de l'Institut Mathématique*, Ns. 65(79):1–7, 1999.

[95] M. Rašković, Z. Ognjanović, and Z. Marković. A Logic with Conditional Probabilities. In *Proceedings of the European Workshop on Logics in Artificial Intelligence, JELIA'04*, volume 3229 of *Lecture Notes in Computer Science*, pages 226–238. Springer, 2004.

[96] M. Rašković and R. Živaljević. Barwise completeness theorems for some biprobability logics. *Zeitschrift für mathematische Logik und Grundlagen der Mathematik*, 31:133–135, 1986.

[97] N. Savić, D. Doder, and Z. Ognjanović. A Logic with Upper and Lower Probability Operators. In T. Augustin, S. Doria, E. Miranda, and E. Quaeghebeur, editors, *Proceedings of the 9th International Symposium on Imprecise Probability: Theories and Applications, ISIPTA '15, Pescara, Italy, 2015*, pages 267–276, 2015. SIPTA Society for Imprecise Probability: Theories and Applications. 2015.

[98] N. Savić, D. Doder, and Z. Ognjanović. Logics with lower and upper probability operators. *International Journal of Approximate Reasoning*, 88:148–168, 2017.

[99] D. Scott and P. Krauss. Assigning probabilities to logical formulas. In J. Hintikka and P. Suppes, editors, *Aspects of inductive logic*, pages 219–264. North-Holland, Amsterdam, 1966.

[100] K. Segerberg. Qualitative Probability in a Modal Setting. In E. Fenstad, editor, *Proceedings 2nd Scandinavian Logic Symposium*, volume 63 of *Studies in Logic and the Foundations of Mathematics*, pages 341–352, Amsterdam, 1971. North-Holland.

[101] K. Segerberg. A model existence theorem in infinitary propositional modal logic. *Jounal of philosophical logic*, 23:337–367, 1994.

[102] T. Stojanović, T. Davidović, and Z. Ognjanović. Bee Colony Optimization for the satisfiability problem in probabilistic logic. *Applied Soft Computing*, 1(31):339–347, 2015.

[103] G. Sundholm. A completeness proof for an infinitary tense-logic. *Theoria*, 43:47–51, 1977.

[104] S. Tomović, Z. Ognjanović, and D. Doder. Probabilistic Common Knowledge Among Infinite Number of Agents. In S. Destercke and T. Denoeux, editors, *Proceedings of the 13th European Conference Symbolic and Quantitative Approaches to Reasoning with Uncertainty ECSQARU 2015, Compiegne, France, July 15-17, 2015*, volume 9161 of *Lecture Notes in Computer Science*, pages 496–505, 2015.

[105] S. Tomović, Z. Ognjanović, and D. Doder. A First-Order Logic for Reasoning about Knowledge and Probability. *ACM Transactions on Computational Logic*, 21(2):16:1–16:30, 2020.

[106] W. van der Hoeck. Some consideration on the logics $P_F D$. In A. Voronkov, editor, *Proceeding of the 2nd Russian Conference on Logic Programming, 1991*, volume 592 of *Lecture notes in computer science*, pages 474–485, Berlin, 1992. Springer.

[107] W. van der Hoeck. Some consideration on the logics $P_F D$. *Journal of Applied Non-Classical Logics*, 7:287–307, 1997.

Chapter 2
Formalization of Probabilities with Non-linearly Ordered Ranges

Angelina Ilić-Stepić and Nebojša Ikodinović

Abstract This chapter is devoted to logical formalization of reasoning about probabilities with "non-standard" ranges, e.g., ranges that are not the unit interval of real numbers. The main part of the chapter considers probabilities whose ranges are fields of p-adic numbers $\mathbb{Q}_p$ for arbitrary prime number p. We describe the probability logic $L^D_{\mathbb{Q}_p}$ which allows statements in which probabilities are estimated by p-adic balls. We provide an example to illustrate how $L^D_{\mathbb{Q}_p}$ can be used to syntactically express statements about interference of waves in the double-slit experiment. Some variants of $L^D_{\mathbb{Q}_p}$ that formalize reasoning about p-adic conditional probabilities are also presented. Next, we consider the field of complex numbers as another unordered field and present two logics for reasoning about complex valued probability. Finally, we discuss formal systems for reasoning about probabilities whose ranges are partially ordered monoids.

2.1 Introduction

Motivation for the results that will be presented in this chapter comes from different sources. First we consider the notion of "immeasurable" and "incomparable probabilities". It can be said that Keynes opened a discussion on this issue in his book "A Treatise on Probability" [13]. Through several illustrative examples, Keynes made clear that it is not always correct to give numerical values to probabilities. He also considered situations when it is reasonable to say that one probability is greater than

Angelina Ilić-Stepić
Mathematical Institute of the Serbian Academy of Sciences and Arts, Kneza Mihaila 36, 11000 Belgrade, Serbia e-mail: angelina@mi.sanu.ac.rs

Nebojša Ikodinović
Faculty of Mathematics, University of Belgrade, Studentski trg 16, 11000 Belgrade, Serbia, e-mail: ikodinovic@matf.bg.ac.rs

Z. Ognjanović (ed.), *Probabilistic Extensions of Various Logical Systems*,
https://doi.org/10.1007/978-3-030-52954-3_2

the other, but not how much. For instance, a conclusion which is based on three experiments is more trustworthy than if it were based on two. However, it is not easy to determine which numerical value should be attributed to the difference. Keynes paid attention to considering the cases when the probabilities of events are not comparable and when they are comparable. His theory of comparison of probabilities relies on the following idea: beliefs that some probability is greater than another, arise out of an order in which it is possible to place them. Graphically speaking, certainty, impossibility, and a probability, which has an intermediate value, constitute an ordered series in which the probability lies between certainty and impossibility. Thus, when we say that one probability is greater than another, it means that the value of the first lies between certainty and the value of the second. It may happen that there exist more than one path between proof and disproof; between certainty and impossibility. It should be clear that probabilities that lie in different paths can not be compared. For instance consider two sentences: "When it is cloudy, it will rain" and "Caesar invaded Britain". It is reasonable to say that the probabilities of both sentences lie somewhere between $\frac{1}{2}$ and certainty, but on the different paths. Why different paths? The proofs for these statements are significantly different. The proof for the case of the rain would rely on voluminous statistical data that concerns: temperature, degree of cloudiness, air pressure, air humidity etc. On the other hand Caesar's invasion would be considered in relation to a wide range of historical facts related to the area and that period. It is difficult to unite and permeate these two fundamentally different sets. Keynes' theory of (non)comparative probabilities has some more reasonable properties: a path of degrees of probability may not be compact: it can happen that there is no probability between two probabilities on the path; the degree of probability can lie on more than one path.

After all it is reasonable to assume that probability is a map $Prob : \Omega \longrightarrow \Gamma$ where Ω is the set of events and Γ is a set with some additional structure. As the minimal requirements that probability should satisfy we consider:

- ranges of probabilities should be partially ordered, so that it is not always possible to compare probabilities, and
- probabilities should be (finitely) additive, which means that one binary operation is also defined on Γ.

These conditions help to determine Γ more precisely, and in this chapter we will study:

- In Section 2.2: partially ordered number fields of p-adic numbers. Namely, for every prime number p the field of p-adic numbers $\mathbb{Q}_p$ is a partially ordered field obtained by completing the field of rational numbers $\mathbb{Q}$ using p-adic norm $|\cdot|_p$.
- In Section 2.3: the field of complex numbers.
- In Section 2.4: partially ordered countable monoids.

2.2 Logics for reasoning about p-adic valued probabilities

In this section we discuss several logics that formalize reasoning with p-adic numbers [5, 7–10]. We present in detail the logic $L^D_{\mathbb{Q}_p}$. For the other logics we sketch only their specific properties. Let us emphasize that since there are infinite number of p-adic fields – $\mathbb{Q}_p$ for every prime number p, we actually present infinitely many logics.

Consideration of p-adic valued probability is not only encouraged by Keynes' ideas, and there are other motivations for using p-adic numbers for this purpose. The need for defining p-adic valued probably appeared in construction of different physical models. Successful application of p-adic numbers in physics started with construction of p-adic string amplitude. Investigations on the p-adic string theory induced investigations on p-adic quantum mechanics, with complex valued and also with p-adic valued wave functions. In quantum models, where a wave function takes values in $\mathbb{Q}_p$ or its quadratic extension [14–16], there is a need for a probabilistic interpretation of wave functions. For this purpose, a p-adic valued probability theory is presented and developed by Andrei Khrennikov [17–19]. He defined the p-adic probability space as a triple (W,H,μ), where W is a nonempty set, H is a subalgebra of the Boolean algebra $P(H)$ and $\mu : H \to \mathbb{Q}_p$ is an additive function such that $\mu(W) = 1$.

Beside applications in physics, p-adic valued probabilities appear in experiments where real valued probabilities do not exist, but it turns out that there exist p-adic valued probabilities. In the sequel we give an example from [16].

Let us consider an industrial process (denoted IND) which produces white and red balls. The balls are produced by series of $M = p^n$ $(n = 0, 1, 2, \dots)$ balls of the same type, where p is a fixed prime number. The process IND will be organized using two sequences of independent random variables, $\{\varepsilon_k\}$ and $\{\theta_k\}$, where:

- $\varepsilon_k = w$ or $\varepsilon_k = r$ ("w" stands for white and "r" for red), and
- $\theta_k = 1$ or $\theta_k = 2$.

Assume that $M_k = p^n$ balls of the same fixed color were produced in the step k. Then:

- if $\varepsilon_{k+1} = w(r)$, then IND will produce a series of white balls (red balls),
- the length of this new series is $M_{k+1} = M_k p^{\theta_{k+1}}$.

We suppose that the produced balls are mixed in a box. Now the question is:

- what will be the probabilities $P(w)$ and $P(r)$ to draw a white ball (red ball) from the box after a very long period of the IND-production (in a mathematical idealization we can consider an infinite period of IND-production).

Let $T_1, T_2, \dots, T_k, \dots$ be the moments when the corresponding previous series is produced, and next series of the IND-production starts. In order to calculate probabilities $P(w)$ and $P(r)$ we compute the relative frequencies $\nu_k(w)$ and $\nu_k(r)$ of white and red balls at the moments T_k $(k = 1, 2, \dots)$, and consider the limits of $\{\nu_k(w)\}$

and $\{v_k(r)\}$. The following results, the first one about the field of the p-adic numbers and the second one about the field of real numbers, are related to stabilization of relative frequencies:

Theorem 2.1. *Let $\{\varepsilon_k\}$ and $\{\theta_k\}$ be arbitrary sequences of random variables. Then the limits $P(w) = \lim^p_{k\to\infty} v_k(w)$, and $P(r) = \lim^p_{k\to\infty} v_k(r)$ exist in the field of the p-adic numbers. Moreover, if $\{\alpha_k\}$ is a sequence such that $\alpha_k = 1$ when $\varepsilon_k = w$ and $\alpha_k = 0$ when $\varepsilon_k = r$ then*

$$P(w) = \sum_{k=1}^{\infty} \alpha_k M_k / \sum_{k=1}^{\infty} M_k \qquad P(r) = \sum_{k=1}^{\infty} (1-\alpha_k) M_k / \sum_{k=1}^{\infty} M_k$$

■

Theorem 2.2. *Let $\{\theta_k\}$ be arbitrary sequences of random variables and let $\{\varepsilon_k\}$ be a sequences of random variables such that probabilities of $\varepsilon_k = 0$ and $\varepsilon_k = 1$ are not equal to 0. Then the relative frequencies oscillate in the field of real numbers.* ■

Logics for reasoning about p-adic valued probabilities can be also used to formalize processes of thinking where information are represented by p-adic numbers [7].

2.2.1 *p-adic numbers*

Here we provide some basic notions of p-adic numbers. For a detailed insight refer to [1, 16, 20].

Let p be a fixed prime number. We define the function $|\cdot|_p : \mathbb{Q} \to \{p^n | n \in \mathbb{Z}\} \cup \{0\}$ in the following way:

- If $n \in \mathbb{N}$, then n can be represented as a product of prime numbers

$$n = 2^{t_2} 3^{t_3} \ldots p^{t_p} \ldots s^{t_s}$$

  and $|n|_p = p^{-t_p}$, assuming $|0|_p = 0$.
- If $n \in \mathbb{Z}$, $n < 0$ then $|n|_p = |-n|_p$.
- Finally, if $\frac{n}{m} \in \mathbb{Q}$, $m \neq 0$, then $|\frac{n}{m}|_p = \frac{|n|_p}{|m|_p}$.

The function $|\cdot|_p$ satisfies the following properties (i.e., it is a norm on $\mathbb{Q}$):

1. $|x|_p = 0$ iff $x = 0$;
2. $|xy|_p = |x|_p |y|_p$;
3. $|x+y|_p \leq max\{|x|_p, |y|_p\}$ – strong triangle inequality.

The field of real numbers is constructed as a completion of the field of rational numbers $\mathbb{Q}$. The field of p-adic numbers, $\mathbb{Q}_p$, for a fixed prime p, is constructed in a similar way, using the norm $|\cdot|_p$. Therefore $\mathbb{Q}_p$ is complete with respect to the norm $|\cdot|_p$, i.e., every Cauchy sequence in $\mathbb{Q}_p$ has a limes in $\mathbb{Q}_p$.

Definition 2.1. The sequence of rational numbers $(a_n)_{n\in\mathbb{N}}$ tends to the limit a with respect to $|\cdot|_p$ (denoted $\lim^p_{n\to\infty} a_n = a$) if

$$(\forall\varepsilon > 0)(\exists n_0 \in \mathbb{N})(\forall n)[n \geq n_0 \to |a_n - a|_p \leq \varepsilon].$$

■

Let $a \in \mathbb{Q}_p$, $r \in \{p^n | n \in \mathbb{Z}\} \cup \{0\}$. $K^-(a,r) = \{x \in \mathbb{Q}_p : |x-a|_p < r\}$, and $K(a,r) = \{x \in \mathbb{Q}_p : |x-a|_p \leq r\}$ are open and closed balls in $\mathbb{Q}_p$ while $S(a,r) = \{x \in \mathbb{Q}_p : |x-a|_p = r\}$ is a sphere in $\mathbb{Q}_p$ with the center a and radius r. Any p-adic ball $K(0,r)$ is an additive subgroup of $\mathbb{Q}_p$, while the ball $K(0,1)$ is also a ring which is called ring of p-adic integers and is denoted by $\mathbf{Z}_p$. Balls and spheres in $\mathbb{Q}_p$ have some unusual properties:

- Two balls have nonempty intersection iff one of them is contained in the other.
- If $b \in K(a,r)$ then $K(a,r) = K(b,r)$. Thus any point of a p-adic ball can be chosen as its center.
- Balls $K(a,r_1)$ and $K(b,r_2)$ are disjoint iff $|a-b|_p > \max\{r_1,r_2\}$.

Every $p-$adic number a can be uniquely written in the following form

$$a = \sum_{j=n}^{\infty} a_j p^j$$

where each $a_j \in \{0,1,\ldots p-1\}$ and n is such that $|a|_p = p^{-n}$. The short p-adic representation of this form is $a_n a_{n+1} a_{n+2} \ldots$.

2.2.2 The logic $L^D_{\mathbb{Q}_p}$

An important condition for p-adic measures is the boundedness condition [18]: if H is a field of subsets of some set Ω then, for every $A \in H$

$$\sup\{|\mu(B)|_p : B \in H, B \subset A\} < \infty.$$

Obviously, if the range of a probability is an arbitrary large, but fixed p-adic ball, i.e., if $\mathrm{Im}(\mu) \subseteq K(0,p^n) = \{a \in \mathbf{Q}_p \ : \ |a|_p \leqslant p^n\}$ then the boundedness condition is satisfied.

The logic $L^D_{\mathbb{Q}_p}$ is primarily designed to allow expressing sentences of the form "the probability of an event (described by a formula) belongs to some specific p-adic ball", which can be written using a formula of the form $K_{r,\rho}\alpha$.

As Keynes noticed, there are situations in which we can compare two probabilities without knowing their numerical values. For example, in summer, everyone will say that it is more likely to be sunny than that snow will fall, although no one can find the exact probability of these events. Considering formalization of real valued probabilities there are papers about qualitative reasoning [4, 21] where we can make sentences as "the probability of α is greater than the probability of β" (see Chapter

5). Due to lack of total ordering we can not make these sentences in p-adic case[1]. However, even when we cannot compare probabilities of two events we can estimate how close these probabilities are. Thus, if the p-adic probabilities of two events are close enough, i.e., if their p-adic difference is close enough to 0, we can assume that these events are equally favorable.

In $L^D_{\mathbb{Q}_p}$ we can measure how close are probabilities of two events using operators of the form $D_\rho \alpha, \beta$ which enable us to make sentences of the form "the p-adic distance between the probabilities of α and β is less than or equal to ρ". For instance, consider the statement P_{11}: "the 11-adic distance between probabilities of α and β is less than or equal to $11^{10^{10}}$." In $\mathbb{Q}_{11}$ probabilities of α and β are very close to each other, while we do not have any information about their probabilities. In $L^D_{\mathbb{Q}_p}$ P_{11} is expressed as $D_{11^{10^{10}}} \alpha, \beta$.

2.2.2.1 Syntax and semantics

Let us fix a prime p and an arbitrary large $M \in \mathbb{N}$ and define the following sets:

1. $\mathbb{Q}_M = \{r \in \mathbb{Q} : |r|_p \leqslant p^M\}$;
2. $\mathbb{Z}_M = \{n \in \mathbb{Z} : n \leq M\}$, and
3. $R = \{p^{M-n} : n \in \mathbb{N}\} \cup \{0\} = \{p^n : n \in \mathbb{Z}_M\} \cup \{0\}$.

Let Var be a countable set of propositional letters and For_C the set of all propositional formulas over Var. We use α, β and γ, indexed if necessary, to denote propositional formulas. The set For_P of all probabilistic formulas is defined as the least set satisfying the following conditions:

- If $\alpha \in \text{For}_C$, $r \in \mathbb{Q}_M, \rho \in R$, then $K_{r,\rho}\alpha$ is a probabilistic formula,
- If $\alpha, \beta \in \text{For}_C$, $\rho \in R$, then $D_\rho \alpha, \beta$ is a probabilistic formula, and
- If φ, ϕ are probabilistic formulas, then $(\neg\varphi)$, $(\varphi \wedge \phi)$ are probabilistic formulas.

We denote probabilistic formulas by φ, ϕ and θ, indexed if necessary. The set For of all $L^D_{\mathbb{Q}_p}$-formulas is a union of For_C and For_P. $L^D_{\mathbb{Q}_p}$-formulas will be denoted by A, B, C ... (with or without an index). The other classical connectives $(\vee, \rightarrow, \leftrightarrow)$ can be defined as usual.

Definition 2.2. An $L^D_{\mathbb{Q}_p}$-model is a structure $\mathscr{M} = \langle W, H, \mu, v \rangle$ where:

- W is a nonempty set of elements called worlds;
- H is an algebra of subsets of W;
- $\mu : H \longrightarrow K(0, p^M)$ is a measure (additive function) such that $\mu(W) = 1$;
- $v : W \times Var \longrightarrow \{true, false\}$ is a valuation which associated with every world $w \in W$ a truth assignment $v(w, \cdot)$ on propositional letters; the valuation $v(w, \cdot)$ is extended to classical propositional formulas as usual. ■

[1] We can still consider logics that formalize p-adic probabilities, treating p-adic numbers with some partial order. In that formalism it is possible to make these sentences in some cases. However, the logic $L^D_{\mathbb{Q}_p}$ is not designed in such way.

Similarly as in Chapter 1, $[\alpha] = \{w \in W : v(w,\alpha) = true\}$.

Definition 2.3. Let $\mathcal{M} = \langle W, H, \mu, v\rangle$ be an $L^D_{\mathbb{Q}_p}$-model. The satisfiability relation is inductively defined as follows:

- If $\alpha \in \mathrm{For}_C$ then $\mathcal{M} \models \alpha$ iff $v(w,\alpha) = true$ for all $w \in W$.
- If $\alpha \in \mathrm{For}_C$ then $\mathcal{M} \models K_{r,\rho}\alpha$ iff $|\mu([\alpha]) - r|_p \leqslant \rho$.
- If $\alpha, \beta \in \mathrm{For}_C$ then $\mathcal{M} \models D_{r,\rho}\alpha, \beta$ iff $|\mu([\alpha]) - \mu([\beta])|_p \leqslant \rho$.
- If $\varphi \in \mathrm{For}_P$ then $\mathcal{M} \models \neg\varphi$ iff it is not $\mathcal{M} \models \varphi$.
- If $\varphi, \psi \in \mathrm{For}_P$ then $\mathcal{M} \models \varphi \wedge \psi$ iff $\mathcal{M} \models \varphi$ and $\mathcal{M} \models \psi$. ■

Also, we will consider only measurable models, i.e., $H = \{[\alpha] : \alpha \in \mathrm{For}_C\}$. Let us comment the cases $K_{r,\rho}\alpha$ and $D_\rho\alpha, \beta$-formulas. For arbitrary ρ, $\mathcal{M} \models K_{r,\rho}\alpha$ means that $\mu([\alpha])$ belongs to the p-adic ball with the center r and the radius ρ, while $\mathcal{M} \models D_\rho\alpha, \beta$ means that a p-adic distance between $\mu([\alpha])$ and $\mu([\beta])$ is less or equal to ρ. In the special case when $\rho = 0$ and $\mathcal{M} \models D_0\alpha, \beta$, $\mu([\alpha]) = \mu([\beta])$.

2.2.2.2 Axiomatization

The axiom system $Ax_{L^D_{\mathbb{Q}_p}}$ of the logic $L^D_{\mathbb{Q}_p}$ is given by the following list of axioms and inference rules.

Axioms

A1. Substitutional instances of tautologies
A2. $K_{r,\rho}\alpha \to K_{r,\rho'}\alpha$, whenever $\rho' \geqslant \rho$
A3. $K_{r_1,\rho_1}\alpha \wedge K_{r_2,\rho_2}\beta \wedge K_{0,0}(\alpha \wedge \beta) \to K_{r_1+r_2,\max(\rho_1,\rho_2)}(\alpha \vee \beta)$
A4. $K_{r_1,\rho_1}\alpha \to \neg K_{r_2,\rho_2}\alpha$, if $|r_1 - r_2|_p > \max(\rho_1, \rho_2)$
A5. $K_{r_1,\rho}\alpha \to K_{r_2,\rho}\alpha$, if $|r_1 - r_2|_p \leqslant \rho$
A6. $K_{r,\rho_1}\alpha \wedge D_{\rho_2}\alpha, \beta \to K_{r,\max\{\rho_1,\rho_2\}}\beta$
A7. $K_{r,\rho}\alpha \wedge K_{r,\rho}\beta \to D_\rho\alpha, \beta$

Inference rules

R1 From A and $A \to B$ infer B (A and B are either both propositional, or both probabilistic formulas).
R2 From α infer $K_{1,0}\alpha$.
R3 If $n \in \mathbb{N}$, then from $\varphi \to \neg K_{r,p^{M-n}}\alpha$ for every $r \in \mathbb{Q}_M$, infer $\varphi \to \perp$
R4 From $\alpha \to \perp$, infer $K_{0,0}\alpha$
R5 If $r \in \mathbb{Q}_M$, then from $\varphi \to K_{r,p^{M-n}}\alpha$ for every $n \in \mathbb{N}$, infer $\varphi \to K_{r,0}\alpha$
R6. From $\varphi \to D_{p^{M-n}}\alpha, \beta$ for every $n \in \mathbb{N}$, infer $\varphi \to D_0\alpha, \beta$
R7 From $\alpha \leftrightarrow \beta$ infer $(K_{r,\rho}\alpha \to K_{r,\rho}\beta)$

Most axioms and rules differ significantly from the axioms that formalize real valued probabilities because they reflect the characteristic properties of the field of p-adic numbers. For instance, Axiom A2 corresponds to the obvious property of p-adic balls: a ball of smaller radius is contained in a ball of a larger radius, provided the balls are not disjoint. Axiom A3 corresponds to the additivity of measures and it also reflects property of p-adic norm (strong triangle inequality). Axiom A4 ensures that the measure of a formula cannot belong to the two disjoint balls. Axiom A5 declares that any point of a p-adic ball can be its center. Axiom 6 corresponds to the following: Let B_1 be the ball such that the measure of the formula α belongs B_1 and let ρ_{max} be the maximum distance between measures of the formulas α and β, then we can determine a ball B_2 such that the measure of the formula β belongs to B_2. Axiom A7 says: If the measure of the formula α and the measure of the formula β belong to the same ball with the radius ρ then the distance between these measures is less or equal to ρ. Rule R3 provides that for every classical formula α and every radius ρ there must be some $r \in \mathbb{Q}_M$ such that the measure of α belongs to the ball $K(r,\rho)$. Rule R4 guaranties that a contradiction has the measure 0. Rule R5 says that if the measure of α is arbitrarily close to a rational number r, then the measure of α is equal to r. Similarly, Rule R6 means: if the distance between the measures of α and β is arbitrary close to 0 then it must be equal to 0. Finally, Rule R7 says that equivalent classical formulas have the same measures.

Without the axioms A6 and A7, and the rule R6 the logic $L^D_{\mathbb{Q}_p}$ becomes the logic $L_{\mathbb{Q}_p}$ presented in [5].

2.2.2.3 Soundness and completeness

Theorem 2.3. *(Soundness) The axiom system $Ax_{L^D_{\mathbb{Q}_p}}$ is sound with respect to the class of $L^D_{\mathbb{Q}_p}$-models.*

Proof. As an illustration, we will show that axioms A4 and A6 are valid and that R5 preserves validity.
Axiom A4: Suppose that $\mathcal{M} \models K_{r_1,\rho_1}\alpha$ and $|r_1 - r_2|_p > \max\{\rho_1,\rho_2\}$ for some $r_2 \in \mathbb{Q}_M, \rho_2 \in R$. Then $|\mu([\alpha]) - r_1|_p \leqslant \rho_1$ and if $\mathcal{M} \models K_{r_2,\rho_2}\alpha$ then $|\mu([\alpha]) - r_2|_p \leqslant \rho_2$, so $|r_1 - r_2|_p = |(r_1 - \mu([\alpha])) + (\mu([\alpha]) - r_2)|_p \leqslant \max\{|r_1 - \mu([\alpha])|_p, |r_2 - \mu([\alpha])|_p\} \leqslant \max\{\rho_1,\rho_2\}$, a contradiction. Therefore $\mathcal{M} \models \neg K_{r_2,\rho_2}\alpha$ and $\mathcal{M} \models K_{r_1,\rho_1}\alpha \to \neg K_{r_2,\rho_2}\alpha$.
Axiom A6: Let $\mathcal{M}$ be an $L^D_{\mathbb{Q}_p}$-model and assume that $\mathcal{M} \models K_{r,\rho_1}\alpha \wedge D_{\rho_2}\alpha,\beta$. Then, $|\mu([\alpha]) - r|_p \leq \rho_1$ and $|\mu([\alpha]) - \mu([\beta])|_p \leq \rho_2$. Thus, according to the strong triangle inequality which holds for p-adic numbers, $|\mu([\beta]) - r|_p = |(\mu([\beta]) - \mu([\alpha])) + (\mu([\alpha]) - r)|_p \leq \max\{|\mu([\beta]) - \mu([\alpha])|_p, |\mu([\alpha]) - r|_p\} = \max\{\rho_1,\rho_2\}$. Thus, $\mathcal{M} \models K_{r,\max\{\rho_1,\rho_2\}}\beta$.

Rule R5: Assume that $\varphi \to K_{r,p^{M-n}}\alpha$ holds in every model $\mathcal{M}$, for every $n \in \mathbb{N}$. Let $\mathcal{M}$ be an arbitrary $L^D_{\mathbb{Q}_p}$-model. If $\mathcal{M} \models \neg\varphi$, then $\mathcal{M} \models \varphi \to K_{r,0}\alpha$. On the contrary, if $\mathcal{M} \models \varphi$, then $\mathcal{M} \models K_{r,p^{M-n}}\alpha$ for every $n \in \mathbb{N}$. Thus $|\mu([\alpha]) - r|_p \leqslant p^{M-n}$ for every $n \in \mathbb{N}$. If for some n, $|\mu([\alpha]) - r|_p = p^{M-n}$, then $|\mu([\alpha]) - r|_p \leqslant p^{M-(n+1)}$ does not hold. Therefore there is no n such that $|\mu([\alpha]) - r|_p = p^{M-n}$. Hence, $|\mu([\alpha]) - r|_p = 0$. (since $|\cdot|_p : K(0,p^M) \to \{p^{M-n} | n \in \mathbb{N}\} \cup \{0\}$). ■

Theorem 2.4 (Deduction theorem). *Let T be a set of formulas and A and B both classical or both propositional formulas. Then, $T, A \vdash B$ implies $T \vdash A \to B$.*

Proof. The proof of the Deduction theorem is a straightforward induction on the length of the inference. For instance we will consider the cases when B is obtained by an application of R6.
R6: Assume that $B = (K_{r,\rho}\alpha \to K_{r,\rho}\beta)$ is obtained from $T \cup \{A\}$ using Rule R6, and $A \in For_P$. Then:

$T, A \vdash \alpha \leftrightarrow \beta$

Since $A \in For_P$ and $(\alpha \leftrightarrow \beta) \in For_C$, A does not affect the proof of $\alpha \leftrightarrow \beta$, therefore we have:

$T \vdash \alpha \leftrightarrow \beta$

$T \vdash (K_{r,\rho}\alpha \to K_{r,\rho}\beta)$ by Rule R6

Since $(K_{r,\rho}\alpha \to K_{r,\rho}\beta) \to (A \to (K_{r,\rho}\alpha \to K_{r,\rho}\beta))$ is an instance of tautology, we obtain:

$T \vdash A \to (K_{r,\rho}\alpha \to K_{r,\rho}\beta)$ using Rule R1. ■

Theorem 2.5 (Lindenbaum's theorem). *Every consistent set can be extended to a maximal consistent set.*

Proof. Suppose that T is a consistent set of formulas and $\overline{T}$ the set of all classical formulas that are consequences of T. Let's list all For_C-formulas in the sequence α_0, $\alpha_1, \ldots$ and all For_P-formulas in the series $\varphi_0, \varphi_1, \ldots$ Let $f : \mathbb{N} \to \mathbb{N}^2$ be any bijection (i.e., f is of the form $f(i) = (\pi_1(i), \pi_2(i))$). We define a sequence of theories T_i in the following way:

1. $T_0 = T \cup \overline{T} \cup \{K_{1,0}\alpha | \alpha \in \overline{T}\}$;
2. For every $i \geqslant 0$,
   a. If $T_{2i} \cup \{\varphi_i\}$ is consistent then $T_{2i+1} = T_{2i} \cup \{\varphi_i\}$;
   b. Otherwise, if $T_{2i} \cup \{\varphi_i\}$ is inconsistent then:
      (i) If $\varphi_i = (\psi \to K_{r,0}\alpha)$ then $T_{2i+1} = T_{2i} \cup \{\neg\varphi_i, \psi \to \neg K_{r,p^{M-n}}\alpha\}$ for some $n \in \mathbb{N}$ such that T_{2i+1} is consistent,
      (ii) If $\varphi_i = (\psi \to D_\rho \alpha, \beta)$ then $T_{2i+2} = T_{2i+1} \cup \{\neg\varphi_i, \psi \to \neg D_{p^{M-n}}\alpha\}$ for some $n \in \mathbb{N}$ such that T_{2i+1} is consistent,
      (ii) Otherwise $T_{2i+1} = T_{2i} \cup \{\neg\varphi_i\}$;
3. For every $i \geqslant 0$, $T_{2i+2} = T_{2i+1} \cup \{K_{r,p^{M-\pi_1(i)}}\alpha_{\pi_2(i)}\}$ for some $r \in \mathbb{Q}_M$ such that T_{2i+2} is consistent.

We show that for every i, T_i is consistent. The consistency of the sets obtained in steps 1 and 2 can be proven in a standard way.

Let us consider the step 3. Suppose that for every $r \in \mathbb{Q}_M$ the set $T_{2i+1} \cup \{K_{r,p^{M-\pi_1(i)}}\alpha_{\pi_2(i)}\}$ is inconsistent. Let $T_{2i+1} = T_0 \cup T^+_{2i+1}$, where T^+_{2i+1} is set of all formulas from For_P added to T_0 in the previous steps of the construction. Then:

$T_0, T^+_{2i+1}, K_{r,p^{M-\pi_1(i)}}\alpha_{\pi_2(i)} \vdash \bot$ for every $r \in \mathbb{Q}_M$

$T_0, T^+_{2i+1} \vdash \neg K_{r,p^{M-\pi_1(i)}}\alpha_{\pi_2(i)}$ for every $r \in \mathbb{Q}_M$, by the Deduction theorem

$T_0 \vdash (\bigwedge_{\varphi \in T^+_{2i+1}} \varphi) \to \neg K_{r,p^{M-\pi_1(i)}}\alpha_{\pi_2(i)}$ for every $r \in \mathbb{Q}_M$, by the Deduction theorem

$T_0 \vdash (\bigwedge_{\varphi \in T^+_{2i+1}} \varphi) \to \bot$ by Rule R3.

$T_{2i+1} \vdash \bot$, a contradiction.

Let $T^* = \bigcup_{i \in \mathbb{N}} T_i$. We have to show that T^* is maximal and consistent. The steps 1 and 2 of the above construction provide that T^* is maximal. Thus we will show that T^* is deductively closed set which does not contain all formulas, and, as a consequence, that T^* is consistent.

Note that T^* does not contain all formulas. Indeed, let $\alpha \in \mathrm{For}_C$. By the construction of T_0, α and $\neg\alpha$ cannot be simultaneously in T_0. Assume that $\varphi \in For_P$. Then for some i and j, $\varphi = \varphi_i$ and $\neg\varphi = \varphi_j$. Since $T_{\max(2i,2j)+1}$ is consistent, T^* cannot contain both φ and $\neg\varphi$.

We will show that T^* is deductively closed. Let $\alpha \in \mathrm{For}_C$ and $T^* \vdash \alpha$. Then by the construction of T_0, $\alpha \in T^*$ and $K_{1,0}\alpha \in T^*$. Let $\varphi \in For_P$. If $\varphi = \varphi_j$ and $T_i \vdash \varphi_j$, it must be $\varphi \in T^*$ because $T_{\max(i,2j)+1}$ is consistent. Suppose that the sequence $\varphi_1, \varphi_2, \ldots \varphi$ forms the proof of φ from T^*. If the sequence is finite, there must be a set T_i such that $T_i \vdash \varphi$. Thus, as above, $\varphi \in T^*$. Therefore we assume that the sequence is countably infinite. Next, we show that for every i, if φ_i is obtained by an application of an inference rule, and all premises belong to T^*, then there must be $\varphi_i \in T^*$. If the rule is a finitary one, then there must be a set T_j which contains all premises and $T_j \vdash \varphi_i$. Thus, in the same way as above, we conclude that $\varphi_i \in T^*$. Therefore we need to consider the infinitary rules.

Consider Rule R5. Assume that $\varphi_i = (\psi \to K_{r,0}\alpha)$ is obtained from the set of premises $\{\varphi_n^i = (\psi \to K_{r,p^{M-n}}\alpha) | n \in \mathbb{N}\}$. By the induction hypothesis $\varphi_n^i \in T^*$ for every n. Suppose that $\varphi_i \notin T^*$. According to the step 2b(i) of the above construction there are n and j such that $\psi \to \neg K_{r,p^{M-n}}\alpha \in T_j$. Since $\psi \to K_{r,p^{M-n}}\alpha$ is a premise there is some j' such that $\psi \to K_{r,p^{M-n}}\alpha \in T_{j'}$. If $l = \max(j, j')$ then $\psi \to \neg K_{r,p^{M-n}}\alpha, \psi \to K_{r,p^{M-n}}\alpha \in T_l$. Thus $T_l \vdash \psi \to K_{r,p^{M-n}}\alpha$ and $T_l \vdash \psi \to \neg K_{r,p^{M-n}}\alpha$, so $T_l \vdash \psi \to \bot$ and hence $T_l \vdash \psi \to K_{r,0}\alpha$. Thus $\psi \to K_{r,0}\alpha \in T^*$, a contradiction. Finally, the other infinitary rules can be considered in a similar way. ∎

In order to construct the canonical model consider a maximal consistent set T^* obtained from a consistent set T using the above construction. The step (3) of this construction allows us to define the probability of any propositional formula in the model. Indeed, by this step, T^* has the next property: For arbitrary formula $\alpha_n \in \mathrm{For}_C$ and every $m \in \mathbb{N}$ there is (at least one) $r \in \mathbb{Q}_M$ such that $K_{r,p^{M-m}}\alpha_n \in T^*$. T^* is deductively closed, so using Axiom A5, we can obtain countably many rational numbers $r \in \mathbb{Q}_M$ such that $K_{r,p^{M-m}}\alpha_n \in T^*$. For each formula $\alpha_n \in \mathrm{For}_C$ we make a sequence of rational numbers r_m^n and have the following steps:

- For every $m \in \mathbb{N}$ we arbitrarily chose any number r such that $K_{r,p^{M-m}}\alpha_n \in T^*$.
- This r will be the m-th number of the sequence, i.e., $r_m^n = r$.
- We obtain the sequence $r_0^n, r_1^n, r_2^n \ldots$ where $K_{r_j^n,p^{M-j}}\alpha_n \in T^*$.
- When the index n of a formula α_n is irrelevant, we will omit it, i.e., for a formula α we will consider a sequence of the form $r(\alpha) = r_0, r_1, \ldots,$ where $K_{r_j,p^{M-j}}\alpha \in T^*$.

Lemma 2.1. *If $r(\alpha)$ is a sequence defined as above, then, $r(\alpha)$ is a Cauchy sequence with respect to the p-adic norm, and* $\lim_{n\to\infty}^p r_n \in K(0, p^M)$.

Proof. Take some arbitrary ε and choose n_0 such that $p^{M-n_0} \leqslant \varepsilon$. If $n,m \geqslant n_0$ then there are members $r_n, r_m \in \mathbb{Q}_M$ of the sequence such that $K_{r_n,p^{M-n}}\alpha, K_{r_m,p^{M-m}}\alpha \in T^*$. We prove that $|r_n - r_m|_p \leqslant \max\{p^{M-n}, p^{M-m}\}$. Suppose that $|r_n - r_m|_p > \max\{p^{M-n}, p^{M-m}\}$. Then:

$T^* \vdash K_{r_n,p^{M-n}}\alpha$
$T^* \vdash K_{r_m,p^{M-m}}\alpha$
$T^* \vdash K_{r_n,p^{M-n}}\alpha \to \neg K_{r_m,p^{M-m}}\alpha$ by Axiom A4
$T^* \vdash \neg K_{r_m,p^{M-m}}\alpha$ by Rule R1, which contradicts the consistency of T^*. Thus,

$$|r_n - r_m|_p \leqslant \max\{p^{M-n}, p^{M-m}\} \leqslant p^{M-n_0} \leqslant \varepsilon.$$

Hence, there exists $\lim^p_{n\to\infty} r_n$ in $\mathbf{Q}_p$. Let $a = \lim^p_{n\to\infty} r_n$. Then there is $n_0 \in \mathbb{N}$ such that for every $n \geqslant n_0$, $|a - r_n|_p \leqslant p^M$.

Thus $|a|_p = |a - r_n + r_n|_p \leqslant \max\{|a - r_n|_p, |r_n|_p\} \leqslant p^M$. ■

Lemma 2.2. *Let $\alpha \in T^*$, $\alpha \in \mathrm{For}_C$. Suppose that $(r_n)_{n\in\mathbb{N}}$, and $(r'_n)_{n\in\mathbb{N}}$ are two different sequences obtained by the above construction (i.e., for at least one m, $r_m \neq r'_m$). Then $\lim^p_{n\to\infty} r_n = \lim^p_{n\to\infty} r'_n$.* ■

The proof is performed using similar techniques as in the previous lemma.

Now we are ready to define a canonical model. Let $\mathscr{M}_{T^*} = \langle W, H, \mu, v\rangle$, where:

- $W = \{w | w \models \overline{T}\}$ contains all classical propositional interpretations that satisfy the set $\overline{T}$ of all classical consequences of the set T,
- $H = \{[\alpha] \mid \alpha \in \mathrm{For}_C\}$
- $\mu : H \to \mathbf{Q}_p$: Let $r(\alpha) = (r_n)_{n\in\mathbb{N}}$. Then

$$\mu([\alpha]) = \begin{cases} r & \text{if } K_{r,0}\alpha \in T^* \\ \lim^p_{n\to\infty} r_n & \text{otherwise} \end{cases}$$

- for every world w and every propositional letter $p \in Var$, $v(w,p) = true$ iff $w \models p$.

The following Theorem says that $\mathscr{M}_{T^*}$ is indeed an $L^D_{\mathbb{Q}_p}$-model.

Theorem 2.6. *Let $\mathscr{M}_{T^*} = \langle W, H, \mu, v\rangle$ be defined as above. Then for every $\alpha, \beta \in \mathrm{For}_C$ the following holds:*

1. *if $[\alpha] = [\beta]$ then $\mu([\alpha]) = \mu([\beta])$;*
2. *if $[\alpha] \cap [\beta] = \emptyset$ then $\mu([\alpha \vee \beta]) = \mu([\alpha]) + \mu([\beta])$;*
3. *$\mu(W) = 1$ and therefore $\mu(\emptyset) = 0$;*
4. *$\mu([\neg\alpha]) = 1 - \mu([\alpha])$.*

Proof. (1) If $[\alpha] = [\beta]$ then $\{w | v(w,\alpha) = true\} = \{w | v(w,\beta) = true\}$. Thus, for every world w, $v(w, \alpha \leftrightarrow \beta) = true$ and hence $\alpha \leftrightarrow \beta \in \overline{T}$, i.e., $\alpha \leftrightarrow \beta \in T^*$. Therefore $T^* \vdash \alpha \leftrightarrow \beta$. Let $\mu([\alpha]) = r$. We distinguish two cases: $K_{r,0}\alpha \in T^*$, and $K_{r,0}\alpha \notin T^*$ First consider situation when $K_{r,0}\alpha \in T^*$. In this case:

$T^* \vdash K_{r,0}\alpha$
$T^* \vdash \alpha \leftrightarrow \beta$

$T^* \vdash K_{r,0}\alpha \to K_{r,0}\beta$ by Rule R6
$T^* \vdash K_{r,0}\beta$ by Rule R1
Thus $K_{r,0}\beta \in T^*$ and hence $\mu([\beta]) = r$.

On the other hand, if $K_{r,0}\alpha \notin T^*$ then $\lim^p_{n\to\infty} r_n = r$, where $(r_n)_{n\in\mathbb{N}} = r(\alpha)$. Therefore, for every element of this sequence the following holds:
$T^* \vdash \alpha \leftrightarrow \beta$
$T^* \vdash K_{r_n,p^{M-n}}\alpha$
$T^* \vdash K_{r_n,p^{M-n}}\alpha \to K_{r_n,p^{M-n}}\beta$
$T^* \vdash K_{r_n,p^{M-n}}\beta$.
Thus, for every n, $K_{r_n,p^{M-n}}\beta \in T^*$ and by Lemma 2.1, $\mu([\beta]) = \lim^p_{n\to\infty} r_n = r$.

(2) Assume that $[\alpha]\cap[\beta] = \emptyset$. Then there is no w such that $v(w,\alpha) = true$ and $v(w,\beta) = true$, i.e., there is no w such that $v(w,\alpha\wedge\beta) = true$. Therefore for every w, $v(w,\neg(\alpha\wedge\beta)) = true$ and hence by the completeness of classical propositional logic, $\neg(\alpha\wedge\beta) \in T^*$. Thus $T^* \vdash (\alpha\wedge\beta) \to \bot$ and so by Rule R4. $T^* \vdash K_{0,0}(\alpha\wedge\beta)$.

Let $\mu([\alpha]) = r_1$ and $\mu([\beta]) = r_2$. Here we need to consider three cases.

- First suppose that $K_{r_1,0}\alpha \in T^*$ and $K_{r_2,0}\beta \in T^*$. Then:
  $T^* \vdash K_{r_1,0}\alpha \wedge K_{r_2,0}\beta \wedge K_{0,0}(\alpha\wedge\beta)$
  $T^* \vdash (K_{r_1,0}\alpha \wedge K_{r_2,0}\beta \wedge K_{0,0}(\alpha\wedge\beta)) \to K_{r_1+r_2,0}(\alpha\vee\beta)$, by Axiom A3
  $T^* \vdash K_{r_1+r_2,0}(\alpha\vee\beta)$ by Rule R1.
  Therefore $\mu([\alpha\vee\beta]) = r_1 + r_2$.
- Further assume that $K_{r_1,0}\alpha \notin T^*$ and $K_{r_2,0}\beta \notin T^*$. Then $r_1 = \lim^p_{n\to\infty} r^1_n$, $r_2 = \lim^p_{n\to\infty} r^2_n$ where $r(\alpha) = (r^1_n)_{n\in\mathbb{N}}$ and $r(\beta) = (r^2_n)_{n\in\mathbb{N}}$. Moreover, for every $n\in\mathbb{N}$:
  $T^* \vdash K_{r^1_n,p^{M-n}}\alpha \wedge K_{r^2_n,p^{M-n}}\beta \wedge K_{0,0}(\alpha\wedge\beta)$
  $T^* \vdash (K_{r^1_n,p^{M-n}}\alpha \wedge K_{r^2_n,p^{M-n}}\beta \wedge K_{0,0}(\alpha\wedge\beta)) \to K_{r^1_n+r^2_n,p^{M-n}}(\alpha\vee\beta)$
  $T^* \vdash K_{r^1_n+r^2_n,p^{M-n}}(\alpha\vee\beta)$.
  Therefore for every $n\in\mathbb{N}$, we have $K_{r^1_n+r^2_n,p^{M-n}}(\alpha\vee\beta) \in T^*$ and hence
  $\mu([\alpha\vee\beta]) = \lim^p_{n\to\infty}(r^1_n + r^2_n) = \lim^p_{n\to\infty} r^1_n + \lim^p_{n\to\infty} r^2_n = r_1 + r_2$.
- Finally, let $K_{r_1,0}\alpha \in T^*$ and $K_{r_2,0}\beta \notin T^*$(or reverse). Note that the constant sequence $r^1_n = r_1$ may be selected for $r(\alpha)$ since $K_{r_1,p^{M-n}} \in T^*$ for every $n\in\mathbb{N}$ (using Axiom A2 and the fact that $K_{r_1,0} \in T^*$). Now, if $r_2 = \lim^p_{n\to\infty} r^2_n$, $(r^2_n)_{n\in\mathbb{N}} = r(\beta)$, thus case is reduced to the previous one.

(3) If $p \in Var$ then $p\vee\neg p \in \overline{T}$ and hence $K_{1,0}(p\vee\neg p) \in T^*$. Thus $\mu([p\vee\neg p]) = 1$. Moreover, for every w, $v(w,p\vee\neg p) = true$ so $[p\vee\neg p] = W$. In the same way, for every $p \in Var$, $T^* \vdash (p\wedge\neg p) \to \bot$. Thus, by Rule R4, $T^* \vdash K_{0,0}(p\wedge\neg p)$, so $K_{0,0}(p\wedge\neg p) \in T^*$ and $\mu([p\wedge\neg p]) = 0$. On the other hand there is no w such that $v(w,p\wedge\neg p) = true$, so $[p\wedge\neg p] = \emptyset$.

(4) Since there is no w such that $v(w,\alpha) = true$ and $v(w,\neg\alpha) = true$ we have $[\alpha]\cap[\neg\alpha] = \emptyset$ and $\mu([\alpha\vee\neg\alpha]) = 1$. So by the step (3), 1= $\mu([\alpha]) + \mu([\neg\alpha])$. ■

Theorem 2.7 (Strong completeness). *A set of formulas T is consistent iff it has an $L^D_{\mathbb{Q}_p}$-model.*

Proof. ($\Leftarrow$) This direction follows from the soundness of the above axiom system (Theorem 2.3).

($\Rightarrow$) In order to prove this direction we construct $\mathscr{M}_{T^*} = (W, H, \mu, v)$ as above, and show, by induction on complexity of formulas, that for every formula A, $\mathscr{M}_{T^*} \models A$ iff $A \in T^*$. Since the cases of propositional formulas and the Boolean operators are considered in the usual way, we will focus on probabilistic formulas of the forms $K_{r,\rho}\alpha$ and $D_\rho \alpha, \beta$.

Let $r \in \mathbb{Q}_M$, $\rho \in R$ and $\alpha \in \mathrm{For}_C$ and suppose that $K_{r,\rho}\alpha \in T^*$. If $\rho > 0$ then $\rho = p^{M-t}$ for some $t \in \mathbb{N}$. We choose $r(\alpha) = (r_n)_{n\in\mathbb{N}}$ such that $r_t = r$. Thus $\mu([\alpha]) = \lim^p_{n\to\infty} r_n$. Therefore $(\forall\varepsilon)(\exists n_0)(\forall n)(n \geqslant n_0 \to |r_n - \mu([\alpha])|_p \leqslant \varepsilon)$. Take $\varepsilon = p^{M-t}$ and assume first that $t \geqslant n_0$. Then $|r_t - \mu([\alpha])|_p \leqslant p^{M-t}$ and thus $\mathscr{M}_{T^*} \models K_{r_t,p^{M-t}}\alpha$, i.e., $\mathscr{M}_{T^*} \models K_{r,p^{M-t}}\alpha$.

Now assume $t < n_0$ and choose some $k \geqslant n_0$. Then $K_{r_k,p^{M-k}}\alpha \in T^*$ and $|r_k - \mu([\alpha])|_p \leqslant p^{M-t}$. Therefore:

$T^* \vdash K_{r_t,p^{M-t}}\alpha$
$T^* \vdash K_{r_k,p^{M-k}}\alpha$
$T^* \vdash K_{r_k,p^{M-k}}\alpha \to K_{r_k,p^{M-t}}\alpha$ by Axiom A2 since $p^{M-t} > p^{M-k}$
$T^* \vdash K_{r_k,p^{M-t}}\alpha$
$T^* \vdash K_{r_t,p^{M-t}}\alpha \wedge K_{r_k,p^{M-t}}\alpha$.

If we assume that $|r_t - r_k| > p^{M-t}$ then by the Axiom A4 it follows that $T^* \vdash K_{r_t,p^{M-t}}\alpha \to \neg K_{r_k,p^{M-t}}\alpha$ and so $T^* \vdash \neg K_{r_k,p^{M-t}}\alpha$. This contradicts the consistency of T^*. Hence, we must conclude that $|r_t - r_k| \leqslant p^{M-t}$. It follows that

$$|r_t - \mu([\alpha])|_p = |(r_t - r_k) + (r_k - \mu([\alpha]))|_p \leqslant \max\{|r_t - r_k|_p, |r_k - \mu([\alpha])|_p\} \leqslant p^{M-t}.$$

Therefore $\mathscr{M}_{T^*} \models K_{r,p^{M-t}}\alpha$.

Now assume that $\rho = 0$, i.e., $K_{r,0}\alpha \in T^*$. Then, according to the definition of μ, $\mu([\alpha]) = 0$, hence $|\mu([\alpha]) - r|_p = 0$, we have $\mathscr{M}_{T^*} \models K_{r,0}\alpha$.

For the other direction, assume that $\mathscr{M}_{T^*} \models K_{r,\rho}\alpha$ and $r(\alpha) = (r_m)_{m\in\mathbb{N}}$. Then, $\mu([\alpha]) = \lim_{m\to\infty} r_m$ and for every $m \in \mathbb{N}$, $K_{r_m,p^{M-m}}\alpha \in T^*$. Let $\rho = p^{M-t}$ for some $t \in \mathbb{N}$. Thus $|\mu([\alpha]) - r|_p \leqslant p^{M-t}$. For $\varepsilon = p^{M-t}$ we have

$(\exists m_0)(\forall m)(m \geqslant m_0 \to |r_m - \mu([\alpha])|_p \leqslant \varepsilon)$.

Let $m \geqslant \max\{t, m_0\}$. Then $p^{M-m} \leqslant p^{M-t}$ and $|r_m - \mu([\alpha])|_p \leqslant p^{M-t}$. Also $|\mu([\alpha]) - r|_p \leqslant p^{M-t}$ and hence

$$|r_m - r|_p = |(r_m - \mu([\alpha])) + (\mu([\alpha]) - r)|_p \leqslant \max\{|r_m - \mu([\alpha])|_p, |\mu([\alpha]) - r|_p\} \leqslant p^{M-t}.$$

Therefore:

$T^* \vdash K_{r_m,p^{M-m}}\alpha$
$T^* \vdash K_{r_m,p^{M-m}}\alpha \to K_{r_m,p^{M-t}}\alpha$, by Axiom A2, since $p^{M-t} \geqslant p^{M-m}$
$T^* \vdash K_{r_m,p^{M-t}}\alpha$ by Rule R1
$T^* \vdash K_{r_m,p^{M-t}}\alpha \to K_{r,p^{M-t}}\alpha$, by Axiom A5, since $|r_m - r|_p \leqslant p^{M-t}$
$T^* \vdash K_{r,p^{M-t}}\alpha$ by Rule R1
$K_{r,p^{M-t}}\alpha \in T^*$, since T^* is deductively closed.

Finally, take $\rho = 0$. Then $\mathscr{M}_{T^*} \models K_{r,0}\alpha$, that is $|\mu([\alpha]) - r|_p = 0$. If $n \in \mathbb{N}$ then $|\mu([\alpha]) - r|_p \leqslant p^{M-n}$ and so $\mathscr{M}_{T^*} \models K_{r,p^{M-n}}\alpha$. Thus, according to the above considerations for every $n \in \mathbb{N}$, $K_{r,p^{M-n}}\alpha \in T^*$. Therefore, by Rule R5, $T^* \vdash K_{r,0}\alpha$, i.e., $K_{r,0}\alpha \in T^*$.

Assume that $D_\rho \alpha, \beta \in T^*$ and $\mu([\alpha]) = r$ where $r = \lim_{n\to\infty}^p r_n$, $r(\alpha) = (r_n)_{n\in N}$. Let $\rho > 0$. Choose $n \in \mathbb{N}$ such that $p^{M-n} < \rho$. According to the previous step, from $T^* \vdash K_{r_n,p^{M-n}}\alpha$ we obtain $\mathscr{M}_{T^*} \models K_{r_n,p^{M-n}}\alpha$, so $|\mu([\alpha]) - r_n|_p \leq p^{M-n} < \rho$. Therefore:

$T^* \vdash D_\rho \alpha, \beta$

$T^* \vdash K_{r_n,p^{M-n}}\alpha$

$T^* \vdash D_\rho \alpha, \beta \wedge K_{r_n,p^{M-n}}\alpha \to K_{r_n,\rho}\beta$, by Axiom 6,

$T^* \vdash K_{r_n,\rho}\beta$, by Rule 1,

$\mathscr{M}_{T^*} \models K_{r_n,\rho}\beta$, i.e., $|\mu([\beta]) - r_n|_p \leq \rho$.

We obtain: $|\mu([\alpha]) - \mu([\beta])|_p = |(\mu([\alpha]) - r_n) + (r_n - \mu([\beta]))|_p \leq \max\{|\mu([\alpha]) - r_n|_p, |r_n - \mu([\beta])|_p\} \leq \rho$, and $\mathscr{M}_{T^*} \models D_\rho \alpha, \beta$.

If $\rho = 0$, i.e., if $D_0\alpha, \beta \in T^*$, then, by Axiom 6, for every $n \in \mathbb{N}$ we have $T^* \vdash K_{r_n,p^{M-n}}\alpha \wedge D_0\alpha, \beta \to K_{r_n,p^{M-n}}\beta$, i.e., for every $n \in \mathbb{N}$, $K_{r_n,p^{M-n}}\beta \in T^*$. Thus $\mu([\beta]) = \lim_{n\to\infty}^p r_n = r = \mu([\alpha])$. Hence $|\mu([\alpha]) - \mu([\beta])|_p = 0$, we have $M_{T^*} \models D_0\alpha, \beta$.

Now suppose that $\mathscr{M}_{T^*} \models D_\rho \alpha, \beta$. Then $|\mu([\alpha]) - \mu([\beta])|_p \leq \rho$. Let $\mu([\alpha]) = r$ and $\mu([\beta]) = q$, where $r = \lim_{n\to\infty}^p r_n$, $q = \lim_{n\to\infty}^p q_n$, $r(\alpha) = (r_n)_{n\in\mathbb{N}}$ and $r(\beta) = (q_n)_{n\in\mathbb{N}}$. Let $\rho > 0$. We choose n_0' such that for $n \geq n_0'$, $|r - r_n|_p \leq \rho$ and $|q - q_n|_p \leq \rho$. From $|r - q|_p \leq \rho$ and $|r - r_n|_p \leq \rho$ we obtain $|q - r_n|_p \leq \rho$. Similarly, from $|q - r_n|_p \leq \rho$ and $|q - q_n|_p \leq \rho$ we obtain $|r_n - q_n|_p \leq \rho$. Now, select n_0'' such that $p^{M-n_0''} \leq \rho$ and let $n_0 \geq \max\{n_0', n_0''\}$ and $n \geq n_0$. Then, $p^{M-n} \leq \rho$ and $|r_n - q_n|_p \leq \rho$. Hence:

$T^* \vdash K_{r_n,p^{M-n}}\alpha$

$T^* \vdash K_{q_n,p^{M-n}}\beta$

$T^* \vdash K_{r_n,\rho}\alpha$ using Axiom 2,

$T^* \vdash K_{q_n,\rho}\beta$ using Axiom 2,

$T^* \vdash K_{r_n,\rho}\beta$, using Axiom 5, since $|r_n - q_n|_p \leq \rho$,

$T^* \vdash K_{r_n,\rho}\alpha \wedge K_{r_n,\rho}\beta \to D_\rho \alpha, \beta$, by Axiom 7.

$T^* \vdash D_\rho \alpha, \beta$, by Rule 1.

Finally, assume that $\rho = 0$, that is $\mathscr{M}_{T^*} \models D_0\alpha, \beta$. Then, for every $n \in \mathbb{N}$, $\mathscr{M}_{T^*} \models D_{p^{M-n}}\alpha, \beta$. According to the previous considerations, for every $n \in \mathbb{N}$, $T^* \vdash D_{p^{M-n}}\alpha, \beta$. Thus, by Rule 6, we obtain $T^* \vdash D_0\alpha, \beta$. ■

2.2.2.4 Decidability

Since there is a well-known procedure for deciding satisfiability of classical propositional formulas, the problem of decidability of the satisfiability problem for $L^D_{\mathbb{Q}_p}$-formulas can be reduced to considering only For_P-formulas.

Let $\varphi \in For_P$. Since φ is equivalent to a formula of the form

$$DNF(\varphi) = \bigvee_{i=1,m} \Big(\big(\bigwedge_{j=1,k_i} \pm K_{r_{i,j},p^{n_{i,j}}}\alpha_{i,j}\big) \wedge \big(\bigwedge_{l=1,s_i} \pm D_{p^{n_{i,l}}}\beta_{i,l}, \gamma_{i,l}\big)\Big).$$

φ is satisfiable iff at least one its disjunct D_i of the form

$$(\bigwedge_{j=1,k_i} \pm K_{r_{i,j},p^{n_{i,j}}}\alpha_{i,j}) \wedge (\bigwedge_{l=1,s_i} \pm D_{p^{n_{i,l}}}\beta_{i,l},\gamma_{i,l})$$

is satisfiable.

Assume that $p_1,\dots,p_n$ are all propositional letters appearing in D_i. If $\models (\alpha \leftrightarrow \beta)$ then, by Rule R7, for every model $\mathcal{M}$ and all $r \in \mathbb{Q}_M$, $\rho \in R$: $\mathcal{M} \models K_{r,\rho}\alpha$ iff $\mathcal{M} \models K_{r,\rho}\beta$. Therefore we can still simplify the disjunct that we consider, i.e., D_i is satisfiable iff the formula

$$(\bigwedge_{j=1,k_i} \pm K_{r_{i,j},p^{n_{i,j}}} FDNF(\alpha_{i,j})) \wedge (\bigwedge_{l=1,s_i} \pm D_{p^{n_{i,l}}} FDNF(\beta_{i,l}), FDNF(\gamma_{i,l}))$$

is satisfiable[2]. Note that for different atoms[3] a_i and a_j, $[a_i] \cap [a_j] = \emptyset$. Thus for every model $\mathcal{M}$ we have that $\mu[a_i \vee a_j] = \mu[a_i] + \mu[a_j]$. Also, D_i is satisfiable iff the following system Σ^D is satisfiable:

$$\sum_{i=1}^{2^n} y_t = 1$$

$$\Upsilon_1 = \begin{cases} |\sum_{a_t \in \alpha_{i,1}} y_t - r_1|_p \leqslant p^{n_1} \text{ if } \pm K_{r_1,p^{n_1}}\alpha_{i,1} = K_{r_1,p^{n_1}}\alpha_{i,1} \\ |\sum_{a_t \in \alpha_{i,1}} y_t - r_1|_p > p^{n_1} \text{ if } \pm K_{r_1,p^{n_1}}\alpha_{i,1} = \neg K_{r_1,p^{n_1}}\alpha_{i,1} \end{cases}$$

$$\dots$$

$$\Upsilon_{k_i} = \begin{cases} |\sum_{a_t \in \alpha_{i,k_i}} y_t - r_{k_i}|_p \leqslant p^{n_{k_i}} \text{ if } \pm K_{r_{k_i},p^{n_{k_i}}}\alpha_{i,k_i} = K_{r_{k_i},p^{n_{k_i}}}\alpha_{i,k_i} \\ |\sum_{a_t \in \alpha_{i,k_i}} y_t - r_{k_i}|_p > p^{n_{k_i}} \text{ if } \pm K_{r_{k_i},p^{n_{k_i}}}\alpha_{i,k_i} = \neg K_{r_{k_i},p^{n_{k_i}}}\alpha_{i,k_i} \end{cases}$$

$$\dots$$

$$\Gamma_1 = \begin{cases} |\sum_{a_t \in \beta_{i,1}} y_t - \sum_{a_t \in \gamma_{i,1}} y_t|_p \leqslant p^{n_1} \text{ if } \pm D_{p^{n_1}}\beta_{i,1},\gamma_{i,1} = D_{p^{n_1}}\beta_{i,1},\gamma_{i,1} \\ |\sum_{a_t \in \beta_{i,1}} y_t - \sum_{a_t \in \gamma_{i,1}} y_t|_p > p^{n_1} \text{ if } \pm D_{p^{n_1}}\beta_{i,1},\gamma_{i,1} = \neg D_{p^{n_1}}\beta_{i,1},\gamma_{i,1} \end{cases}$$

$$\dots$$

$$\Gamma_{s_i} = \begin{cases} |\sum_{a_t \in \beta_{i,s_i}} y_t - \sum_{a_t \in \gamma_{i,s_i}} y_t|_p \leqslant p^{n_{s_i}} \text{ if } \pm D_{p^{n_{s_i}}}\beta_{i,s_i},\gamma_{i,s_i} = D_{p^{n_{s_i}}}\beta_{i,s_i},\gamma_{i,s_i} \\ |\sum_{a_t \in \beta_{i,s_i}} y_t - \sum_{a_t \in \gamma_{i,s_i}} y_t|_p > p^{n_{s_i}} \text{ if } \pm D_{p^{n_{s_i}}}\beta_{i,s_i},\gamma_{i,s_i} = \neg D_{p^{n_{s_i}}}\beta_{i,s_i},\gamma_{i,s_i} \end{cases}$$

---

[2] $FDNF(\alpha)$ is the full disjunctive normal form of α.

[3] Recall that an atom of a formula φ is a formula of the form $\pm p_1 \wedge \dots \wedge \pm p_n$, where $\pm p_i$ is either p_i or $\neg p_i$ and $p_1,\dots,p_n$ are all propositional letters appearing in φ.

where $a_t \in \alpha_{i,j}$ denote that the atom a_t appears in $FDNF(\alpha_{i,j})$ and $y_t = \mu([a_t])$.

We can show that inequalities of the form $|\sum_{a_t \in \alpha_{i,j}} y_t - \sum_{a_t \in \beta_{i,j}} y_t|_p \leqslant p^{n_j}$ are reducible to inequalities $|\sum_{a_t \in \alpha_{i,j} \triangle \beta_{i,j}} \pm y_t - 0|_p \leqslant p^{n_j}$. In order to check satisfiability of the system Σ^D we can consider only (in)equalities: $\sum_{i=1}^{2^n} y_t = 1, \Upsilon_1, \ldots, \Upsilon_k$.

We denote this system by Σ_i. In order to further simplify this system, we need to consider some details related to p-adic norm, p-adic numbers and their p-adic representation:

- Let $n \in \mathbb{Z}_M$, and the p-adic representation of r be $r_{-M}p^{-M} + \ldots + r_{-n-1}p^{-n-1} + r_{-n}p^{-n} + r_{-n+1}p^{-n+1} + \ldots$, i.e., $r = r_{-M}r_{-M+1} \ldots r_{-n-1}r_{-n}r_{-n+1} \ldots$.
- The inequality $|\sum_{a_t \in \alpha} y_t - r|_p \leqslant p^n$ means that $\sum_{a_t \in \alpha} y_t$ and r have a common initial piece. Precisely, if $\sum_{a_t \in \alpha} y_t = (\sum_{a_t \in \alpha} y_t)_{-M}(\sum_{a_t \in \alpha} y_t)_{-M+1} \cdots (\sum_{a_t \in \alpha} y_t)_{-n} \cdots$ and $r = r_{-M}r_{-M+1} \ldots r_{-n-1}r_{-n}r_{-n+1}$, then $|\sum_{a_t \in \alpha} y_t - r|_p \leqslant p^n$ is equivalent to:

$$(\sum_{a_t \in \alpha} y_t)_{-M} = r_{-M} \text{ and}$$
$$(\sum_{a_t \in \alpha} y_t)_{-M+1} = r_{-M+1} \text{ and}$$
$$\ldots$$
$$(\sum_{a_t \in \alpha} y_t)_{-n-1} = r_{-n-1}$$

- Regarding inequalities with the sign $>$, we have the following situation: $|\sum_{a_t \in \alpha} y_t - r|_p > p^n$ is satisfied iff at least one of the following equations is satisfied

$$(\Lambda^1) : |\sum_{a_t \in \alpha} y_t - r|_p = p^{n+1},$$
$$(\Lambda^2) : |\sum_{a_t \in \alpha} y_t - r|_p = p^{n+2},$$
$$\ldots$$
$$(\Lambda^{M-n}) : |\sum_{a_t \in \alpha} y_t - r|_p = p^M.$$

- The following lemma is useful for checking the satisfiability of the above equations.

Lemma 2.3. *Let $n \in \mathbb{Z}_M$, $r = r_{-M}r_{-M+1} \ldots r_{-n-1}r_{-n}r_{-n+1} \ldots$ be the short p-adic representation of r. If $r' = r_{-M}r_{-M+1} \ldots r_{-n-1}r_{-n}$ is the finite part of r and $\alpha \in$ For$_C$, then*

$$|\sum_{a_t \in \alpha} y_t - r|_p = p^n \text{ iff } |\sum_{a_t \in \alpha} y_t - r'|_p = p^n$$

Proof. Suppose that $|\sum_{a_t \in \alpha} y_t - r|_p = p^n$. Thus
$\sum_{a_t \in \alpha} y_t = r + c_{-n}p^{-n} + c_{-n+1}p^{-n+1} + \ldots$, where $c_{-n} \neq 0$. We can rewrite r as $r = r' + r_{-n+1}p^{-n+1} + r_{-n+2}p^{-n+2} + \ldots$ and hence
$\sum_{a_t \in \alpha} y_t - r' = c_{-n}p^{-n} + (c_{-n+1} + r_{-n+1})p^{-n+1} + (c_{-n+2} + r_{-n+2})p^{-n+2} + \ldots$,
so $|\sum_{a_t \in \alpha} y_t - r'|_p = p^n$.

Now assume that $|\sum_{a_t\in\alpha} y_t - r'|_p = p^n$. Therefore $\sum_{a_t\in\alpha} y_t = r' + c_{-n}p^{-n} + c_{-n+1}p^{-n+1} + \ldots$, where $c_{-n} \neq 0$. Thus
$\sum_{a_t\in\alpha} y_t = r + c_{-n}p^{-n} + (c_{-n+1} - r_{-n+1})p^{-n+1} + (c_{-n+2} - r_{-n+2})p^{-n+2} + \ldots$
and hence $|\sum_{a_t\in\alpha} y_t - r|_p = p^n$. ∎

- Using the previous lemma we can conclude:

$$|\textstyle\sum_{a_t\in\alpha} y_t - r|_p = p^n \text{ iff } |\sum_{a_t\in\alpha} y_t - r'|_p = p^n$$

iff $\sum_{a_t\in\alpha} y_t - r' = c_i p^{-n} + c_{-n+1}p^{-n+1} + \ldots$ for some $c_i \in \{1, 2 \ldots, p-1\}$ iff

$$(\textstyle\sum_{a_t\in\alpha} y_t)_{-M} = r_{-M}, \ldots, (\sum_{a_t\in\alpha} y_t)_{-n-1} = r_{-n-1}, (\sum_{a_t\in\alpha} y_t)_{-n} = r_{-n} + c_i.$$

Now we are ready to consider the system Σ_i. Assume that there are a inequalities of the form $|\cdot|_p > p^{n^i}$, $1 \leqslant i \leqslant a$ and b inequalities of the form $|\cdot|_p \leqslant p^{n^j}$, $a+1 \leqslant j \leqslant a+b$, where $a+b = k_i$. If we replace y^{2^n} with $1 - \sum_{i=1}^{2^n} y_t$ and put it in the inequalities $\Upsilon_1, \ldots, \Upsilon_{k_i}$ we obtain the system:

$$
\begin{aligned}
(\Lambda_1) &: |y_1^1 + y_2^1 + \ldots + y_{n_1}^1 - r^1|_p > p^{n^1}\\
&\ldots\\
(\Lambda_a) &: |y_1^a + y_2^a + \ldots + y_{n_a}^a - r^a|_p > p^{n^a}\\
(\Lambda_{a+1}) &: |y_1^{a+1} + y_2^{a+1} + \ldots + y_{n_{a+1}}^{a+1} - r^{a+1}|_p \leqslant p^{n^{a+1}}\\
&\ldots\\
(\Lambda_{a+b}) &: |y_1^{a+b} + y_2^{a+b} + \ldots + y_{n_{a+b}}^{a+b} - r^{a+b}|_p \leqslant p^{n^{a+b}}
\end{aligned}
$$

By the above considerations, this system is satisfiable iff at least one of the following $(M-n^1)(M-n^2)\ldots(M-n^a)$ systems is satisfiable:

$$
\begin{aligned}
(\Lambda'_1) &: |y_1^1 + y_2^1 + \ldots + y_{n_1}^1 - r^1|_p = p^{m^1}\\
&\ldots\\
(\Lambda'_a) &: |y_1^a + y_2^a + \ldots + y_{n_a}^a - r^a|_p = p^{m^a}\\
(\Lambda_{a+1}) &: |y_1^{a+1} + y_2^{a+1} + \ldots + y_{n_{a+1}}^{a+1} - r^{a+1}|_p \leqslant p^{n^{a+1}}\\
&\ldots\\
(\Lambda_{a+b}) &: |y_1^{a+b} + y_2^{a+b} + \ldots + y_{n_{a+b}}^{a+b} - r^{a+b}|_p \leqslant p^{n^{a+b}}
\end{aligned}
$$

where $n^i < m^i \leqslant M$. Each of these systems is satisfiable iff at least one among $(p-1)^a$ systems of the following form is satisfiable:

$$
\Theta_1 = \begin{cases}
(y_1^1 + y_2^1 + \ldots + y_{n_1}^1)_{-M} = r^1_{-M}\\
\ldots\\
(y_1^1 + y_2^1 + \ldots + y_{n_1}^1)_{-m^1-1} = r^1_{-m^1-1}\\
(y_1^1 + y_2^1 + \ldots + y_{n_1}^1)_{-m^1} = r^1_{-m^1} + c_1
\end{cases}
$$

$$\dots$$

$$\Theta_a = \begin{cases} (y_1^a + y_2^a + \ldots + y_{n_a}^a)_{-M} = r_{-M}^a \\ \ldots \\ (y_1^a + y_2^a + \ldots + y_{n_a}^a)_{-m_a - 1} = r_{-m_a - 1}^a \\ (y_1^a + y_2^a + \ldots + y_{n_a}^a)_{-m_a} = r_{-m_a}^a + c_a \end{cases}$$

$$\Theta_{a+1} = \begin{cases} (y_1^{a+1} + y_2^{a+1} + \ldots + y_{n_{a+1}}^{a+1})_{-M} = r_{-M}^{a+1} \\ \ldots \\ (y_1^{a+1} + y_2^{a+1} + \ldots + y_{n_{a+1}}^{a+1})_{-n^{a+1}-1} = r_{-n^{a+1}-1}^{a+1} \end{cases}$$

$$\dots$$

$$\Theta_{a+b} = \begin{cases} (y_1^{a+b} + y_2^{a+b} + \ldots + y_{n_{a+b}}^{a+b})_{-M} = r_{-M}^{a+b} \\ \ldots \\ (y_1^{a+b} + y_2^{a+b} + \ldots + y_{n_{a+b}}^{a+b})_{-n^{a+b}-1} = r_{-n^{a+b}-1}^{a+b} \end{cases}$$

where $c_i \in \{1, 2 \ldots p-1\}$.

We can now provide an explicit procedure for checking satisfiability for systems of the previous form.

Let $G = \max\{-m^1, \ldots, -m^a, -n^{a+1}, \ldots, -n^{a+b}\}$. Since for every i, j, k, $(y_j^i)_k \in \{0, 1, \ldots, p-1\}$ there are p^{G+M+1} possibilities for each representation of the form $y_j^i = (y_j^i)_{-M}(y_j^i)_{-M+1} \cdots (y_j^i)_G$. Moreover, the system $\Theta_1, \ldots, \Theta_a, \Theta_{a+1}, \ldots, \Theta_{a+b}$ has at most $2^n - 1$ variables y_j^i so we can choose representations of the form $y_j^i = (y_j^i)_{-M}(y_j^i)_{-M+1} \cdots (y_j^i)_G$ for all variables appearing in the system in at most $p^{(G+M+1)(2^n-1)}$ ways. Let $R_1, R_2 \ldots R_{p^{(G+M+1)(2^n-1)}}$ denote these representations. Precisely:

- R_1 denotes $000\ldots0, 000\ldots0, \ldots 000\ldots0$, i.e., $(y_j^i)_k = 0$ for all i, j, k,
- R_2 denotes $100\ldots0, 000\ldots0, \ldots 000\ldots0$, i.e., $(y_1^1)_{-M} = 1$, while all other $(y_j^i)_k$ are equal to 0, etc.

Thus:

1. We assign the representation R_1 to the variables and check whether the system is satisfiable.
2. If R_1 does not satisfy the system, we can try with R_2 and so on.
3. After a finite number of steps, we will find a representation which satisfies the system, or we can conclude that no representation satisfies the system $\Theta_1, \ldots, \Theta_a, \Theta_{a+1}, \ldots, \Theta_{a+b}$.

Finally, we can make the following conclusion about testing satisfiability of an arbitrary probabilistic formula φ. The number M (which appears in the definition of the set $\mathbb{Q}_M$), all n^i's from the inequalities J_i's, and the number n of atoms of φ are fixed integers. Thus, the number $p^{(G+M+1)(2^n-1)}$ is a fixed integer. In order to check the satisfiability of φ we need to test at most $p^{(G+M+1)(2^n-1)}$ representations. Testing each representation can be done by finitely many additions of finite numbers. Therefore, the whole procedure finishes in a finite number of steps, which implies decidability of our logic:

Theorem 2.8 (Decidability). *The satisfiability problem for $L^D_{\mathbb{Q}_p}$ is decidable.*

2.2.3 Modeling the double-slit experiment

We present in this section how the known double-slit experiment and the corresponding interference can be described using the logic $L^D_{\mathbb{Q}_p}$ and p-adic numbers.

We provide a brief description of the double-slit experiment. The setup of the experiment consists of a barrier with two slits with a small separation d and an observation screen at a distance D from the barrier. When the barrier is illuminated, it results in the appearance of certain dark and certain bright regions on the observation screen. This is because the light propagates as a single wave and passes through two slits splitting into two waves that interfere with each other. If waves interfere positively the corresponding regions are bright, otherwise they are dark. If we perform the experiment with a singe photon (or electron) at a time, the same interference pattern appears. This shows the wave nature of a particle. But, if one slit is closed there is no such interference. The pattern of bright and dark regions can indicate the probability of certain region on the screen to get hit by a photon. In quantum mechanical formalism probability is given as $|c|^2$ where c is a complex number[4] and $0 \leq |c|^2 \leq 1$ (rather than just a real number between 0 and 1). In contrast to real valued probability that can only increase when added, complex numbers can cancel each other, which results in a lower probability of their sum. Thus complex numbers play a major rule in expression of the principle of interference, but later in this section we illustrate how p-adic numbers can replace them. However, to begin with, we will use complex numbers in the description.

We use a simplified version of the double slit experiment (see Fig. 2.1) from [22] in order to illustrate interference in a clear way.

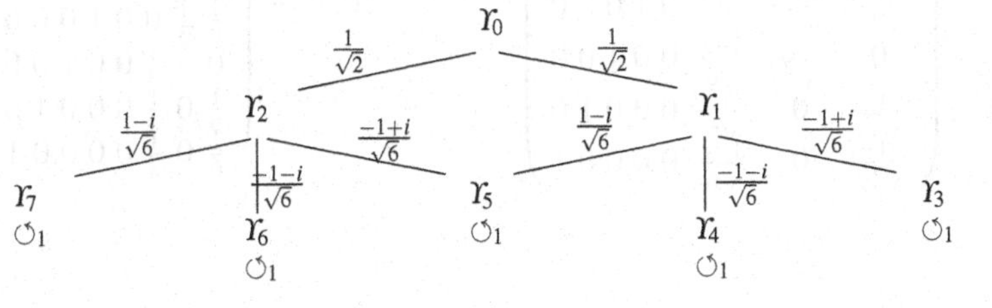

Fig. 2.1 Double slit experiment

4 Wave function of quantum particle is complex valued.

Photon moves from Υ_0 and has equal chances to pass through slits Υ_1 and Υ_2 on the barrier. Thus the arrows from Υ_0 to Υ_1 and Υ_2 are associated with the weight $\frac{1}{\sqrt{2}}$ which corresponds to the probability $\frac{1}{2} = |\frac{1}{\sqrt{2}}|^2$. Similarly, there are equal chance for pass from Υ_1 to Υ_3, Υ_4 and Υ_5 (from Υ_2 to Υ_5, Υ_6 and Υ_7), i.e., $|\frac{\pm 1 \pm i}{\sqrt{6}}|^2 = \frac{1}{3}$. $\Upsilon_3, \ldots \Upsilon_7$ are some specific targets at the observation screen[5]. The circular arrows labeled with 1 show that the photon stays at the specific point on the screen after it hits on it. It is assumed that it takes one time click to pass from Υ_0 to barrier and one time click to pass from the barrier to the screen.

The numerical values of weights are given by the transition matrix[6] P, while $|P|^2$ shows the corresponding probabilities, $|P[i,j]|^2 =$ probability of going from Υ_{j-1} to Υ_{i-1} in one time click:

$$P = \begin{pmatrix} 0 & 0 & 0 & 0&0&0&0&0 \\ \frac{1}{\sqrt{2}} & 0 & 0 & 0&0&0&0&0 \\ \frac{1}{\sqrt{2}} & 0 & 0 & 0&0&0&0&0 \\ 0 & \frac{-1+i}{\sqrt{6}} & 0 & 1&0&0&0&0 \\ 0 & \frac{-1-i}{\sqrt{6}} & 0 & 0&1&0&0&0 \\ 0 & \frac{1-i}{\sqrt{6}} & \frac{1+i}{\sqrt{6}} & 0&0&1&0&0 \\ 0 & 0 & \frac{-1-i}{\sqrt{6}} & 0&0&0&1&0 \\ 0 & 0 & \frac{1-i}{\sqrt{6}} & 0&0&0&0&1 \end{pmatrix} \qquad |P|^2 = \begin{pmatrix} 0&0&0&0&0&0&0&0 \\ \frac{1}{2}&0&0&0&0&0&0&0 \\ \frac{1}{2}&0&0&0&0&0&0&0 \\ 0&\frac{1}{3}&0&1&0&0&0&0 \\ 0&\frac{1}{3}&0&0&1&0&0&0 \\ 0&\frac{1}{3}&\frac{1}{3}&0&0&1&0&0 \\ 0&0&\frac{1}{3}&0&0&0&1&0 \\ 0&0&\frac{1}{3}&0&0&0&0&1 \end{pmatrix}$$

and P^2 and $|P^2|^2$ are the corresponding transition matrices after two time clicks.

$$P^2 = \begin{pmatrix} 0 & 0 & 0 & 0&0&0&0&0 \\ 0 & 0 & 0 & 0&0&0&0&0 \\ 0 & 0 & 0 & 0&0&0&0&0 \\ \frac{-1+i}{\sqrt{12}} & \frac{-1+i}{\sqrt{6}} & 0 & 1&0&0&0&0 \\ \frac{-1-i}{\sqrt{12}} & \frac{-1-i}{\sqrt{6}} & 0 & 0&1&0&0&0 \\ 0 & \frac{-1+i}{\sqrt{6}} & \frac{1-i}{\sqrt{6}} & 0&0&1&0&0 \\ \frac{-1-i}{\sqrt{12}} & 0 & \frac{-1-i}{\sqrt{6}} & 0&0&0&1&0 \\ \frac{-1+i}{\sqrt{12}} & 0 & \frac{1-i}{\sqrt{6}} & 0&0&0&0&1 \end{pmatrix} \qquad |P^2|^2 = \begin{pmatrix} 0&0&0&0&0&0&0&0 \\ 0&0&0&0&0&0&0&0 \\ 0&0&0&0&0&0&0&0 \\ \frac{1}{6}&\frac{1}{3}&0&1&0&0&0&0 \\ \frac{1}{6}&\frac{1}{3}&0&0&1&0&0&0 \\ 0&\frac{1}{3}&\frac{1}{3}&0&0&1&0&0 \\ \frac{1}{6}&0&\frac{1}{3}&0&0&0&1&0 \\ \frac{1}{6}&0&\frac{1}{3}&0&0&0&0&1 \end{pmatrix}$$

---

[5] In the quantum mechanical formalism a list of mutually exclusive states (orthonormal basis) $\varphi_1, \varphi_2, \ldots, \varphi_n$ is associated with some specific measurement. Each state φ can be expressed as $\varphi = c_1\varphi_1 + c_2\varphi_2 + \ldots + c_n\varphi_n$ where $c_i \in \mathbf{C}$. Then, the probability of passing from φ to φ_i is equal to $|c_i|^2$. Here, for instance, we can measure whether a photon passed through Υ_1 or through Υ_2 and $\Upsilon_0 = \frac{1}{\sqrt{2}}\Upsilon_1 + \frac{1}{\sqrt{2}}\Upsilon_2$.

[6] Actual complex numbers depend on d and D. Square modules of the chosen complex numbers correspond to the simple symmetric case, which clearly demonstrates the interference.

Note that $P^2[5,0]=0$. This is the consequence of canceling of complex numbers, i.e., of the following addition and multiplication:

$$\frac{1}{\sqrt{2}}(\frac{-1+i}{\sqrt{6}})+\frac{1}{\sqrt{2}}(\frac{1-i}{\sqrt{6}})=\frac{-1+i+1-i}{\sqrt{12}}=0.$$

Thus, the probability of passing from Υ_0 to Υ_5 is zero. This corresponds to a dark region in the experiment.

Now we will show how the p-adic numbers can be used to explain the results of the experiment. Precisely, we use quadratic extensions of the fields of p-adic numbers. These quadratic extensions have the form $\mathbf{Q_p}(\sqrt{\tau})$ with elements $z=x+\sqrt{\tau}y$ where $x,y\in\mathbb{Q}_p$ while the square of their norm is $\| z \|^2=x^2-\tau y^2$. The number τ depends on p.

If $p\neq 2$,τ_p is not a square element of $\mathbb{Q}_p$ ($x^2=\tau_p$ has no solutions in $\mathbb{Q}_p$) and $|\tau_p|_p=1$, then $\mathbf{Q_p}(\sqrt{\tau_p})$, $\mathbf{Q_p}(\sqrt{\mathbf{p}})$, $\mathbf{Q_p}(\sqrt{\tau_p\cdot\mathbf{p}})$ are three different[7] extensions of $\mathbb{Q}_p$. If $p=2$ there are seven non-isomorphic extensions which may be characterized by $\tau_2=\pm1,\pm2,\pm3,\pm6$.

Existence of the square root in p-adic fields is characterized by the following theorem.

Theorem 2.9. *Let p be an odd prime number, $a\in\mathbb{Z}$ and $p^2\nmid a$. Then, the equation $x^2-a=0$ has:*

- *no solution in $\mathbb{Q}_p$, if $p\mid a$,*
- *two solution in $\mathbb{Q}_p$, if $y^2\equiv_p a$ has solution in $\mathbb{Z}$ and $p\nmid a$, and*
- *no solution in $\mathbb{Q}_p$, if $y^2\equiv_p a$ has no solution in $\mathbb{Z}$ and $p\nmid a$.* ■

For instance, $\sqrt{1-p}\in\mathbb{Q}_p$ for every prime p and $\sqrt{2}\notin\mathbf{Q_5}$. Thus we can choose $\mathbf{Q_5}(\sqrt{\mathbf{2}})$, $\mathbf{Q_5}(\sqrt{\mathbf{5}})$ and $\mathbf{Q_5}(\sqrt{\mathbf{10}})$ for quadratic extensions of $\mathbf{Q_5}$.

Let $P_p=(a_{ij}^p=x_{ij}^p+\tau_p y_{ij}^p)_{i,j=\overline{1,8}}$ be the corresponding matrix in $\mathbb{Q}_p(\sqrt{\tau_p})$

Table 2.1 gives transition matrices[8] for various p-adic fields $\mathbb{Q}_p$. In order to calculate τ_p we have classified the p-adic fields, depending on whether $\sqrt{2}$, $\sqrt{3}$, $\sqrt{6}$ belong or do not belong to the field $\mathbb{Q}_p$.

Note that for all $a_{i,j}$ form the table:

$$\| a_{21} \|^2=\| a_{31} \|^2=\frac{1}{2}$$

while

$$\| a_{42} \|^2=\| a_{52} \|^2=\| a_{63} \|^2=\| a_{73} \|^2=\| a_{62} \|^2=\| a_{83} \|^2=\frac{1}{3}$$

Here are some first p-adic fields meeting the requirements of the table.

- $\sqrt{2},\sqrt{3}\in\mathbb{Q}_{23}$

[7] Actually there are three non-isomorphic extensions of $\mathbb{Q}_p$.

[8] We did not extend the example by searching extensions for all fields. It is easy to verify that for every such p, $P_p^2[5,0]=0$. Also we point out that these a_{ij} from the table are not the only solutions.

$\sqrt{2},\sqrt{3}\in\mathbb{Q}_p$	$\tau_p=p$	$a_{21}=a_{31}=\frac{1}{\sqrt{2}}$ $a_{52}=a_{73}=$ $\frac{-1}{\sqrt{3(1-p)}}-\frac{\sqrt{p}}{\sqrt{3(1-p)}}$	$a_{42}=a_{63}=$ $\frac{-1}{\sqrt{3(1-p)}}+\frac{\sqrt{p}}{\sqrt{3(1-p)}}$ $a_{62}=a_{83}=$ $\frac{1}{\sqrt{3(1-p)}}-\frac{\sqrt{p}}{\sqrt{3(1-p)}}$
$\sqrt{2}\notin\mathbb{Q}_p,\sqrt{3}\in\mathbf{Q_p}$	$\tau_p=2$	$a_{21}=a_{31}=1+\frac{\sqrt{2}}{2}$ $a_{52}=a_{73}=-1-\frac{\sqrt{2}\sqrt{3}}{3}$	$a_{42}=a_{63}=-1+\frac{\sqrt{2}\sqrt{3}}{3}$ $a_{62}=a_{83}=1-\frac{\sqrt{2}\sqrt{3}}{3}$
$\sqrt{2}\in\mathbb{Q}_p,\sqrt{3}\notin\mathbf{Q_p}$	$\tau_p=3$	$a_{21}=a_{31}=\frac{1}{\sqrt{2}}$ $a_{52}=a_{73}=-1-\frac{\sqrt{2}\sqrt{3}}{3}$	$a_{42}=a_{63}=-1+\frac{\sqrt{2}\sqrt{3}}{3}$ $a_{62}=a_{83}=1-\frac{\sqrt{2}\sqrt{3}}{3}$
$\sqrt{2},\sqrt{3}\notin\mathbb{Q}_p,\sqrt{6}\in\mathbb{Q}_p$	$\tau_p=2$	$a_{21}=a_{31}=1+\frac{\sqrt{2}}{2}$ $a_{52}=a_{73}=\frac{-2}{\sqrt{6}}-\frac{\sqrt{2}}{\sqrt{6}}$	$a_{42}=a_{63}=\frac{-2}{\sqrt{6}}+\frac{\sqrt{2}}{\sqrt{6}}$ $a_{62}=a_{83}=\frac{2}{\sqrt{6}}-\frac{\sqrt{2}}{\sqrt{6}}$

Table 2.1 Transition matrices

- $\sqrt{2}\in\mathbb{Q}_7,\mathbb{Q}_{17}$, $\sqrt{3}\notin\mathbb{Q}_7,\mathbb{Q}_{17}$
- $\sqrt{2}\notin\mathbb{Q}_{11},\mathbb{Q}_{13}$, $\sqrt{3}\in\mathbb{Q}_{11},\mathbb{Q}_{13}$
- $\sqrt{2},\sqrt{3}\notin\mathbb{Q}_5,\mathbb{Q}_{19}$, $\sqrt{6}\in\mathbb{Q}_5,\mathbb{Q}_{19}$.

Let N and M be arbitrary large but fixed natural numbers. We build a theory T over the finite set of propositional letters $Var_f=\{p_{ij}\mid i\in\{0,1,\dots N\}, j\in\{N,N+1,\dots M\}\}$.

For the purpose of expressing interference over $\mathbb{Q}_p(\sqrt{\tau})$ we look at the p-adic numbers that appear in the formulas with "more detailed glasses". Let $\overline{Q_p^1}=\{(\pm\sqrt{r})\mid r\in Q_p^1\}$. We allow formulas of the form: $K_{a^2-\tau b^2,\rho}\alpha$ where $a,b\in\overline{Q_p^1}\cup Q_p^1$. Although formulas $K_{(-\sqrt{a})^2,\rho}\alpha$, $K_{(\sqrt{a})^2,\rho}\alpha$ and $K_{a,\rho}\alpha$ have the same meaning, we syntactically distinguish them.

The following two axioms schemas reflect the principle of interference in situations like double slit experiment, where it can be more slits and more targets on the wall.

$$A_{\tau_p,p,t}: K_{1,0}(p_{01}\vee p_{02}\vee\ldots\vee p_{0t})\wedge\bigwedge_{i=1}^{t}K_{c_i^2-\tau_p d_i^2,0}p_{0i}\bigwedge_{i=1}^{t}K_{a_i^2-\tau_p b_i^2,0}p_{ij}\Rightarrow$$

$$\Rightarrow K_{(\sum_{i=1}^t(a_ic_i+\tau_p b_id_i))^2-\tau_p(\sum_{i=1}^t(b_ic_i+\tau_p a_id_i))^2,0)^2}p_{0j}$$

when $t\le N$, $j\in\{N,N+1,\dots M\}$ and

$$(\sum_{i=1}^{t}(a_ic_i+\tau_p b_id_i))^2-\tau_p(\sum_{i=1}^{t}(b_ic_i+\tau_p a_id_i))^2\in Q_p^1$$

$A_{\tau_p,p,t}$ describes how we can use the transition probabilities in one time click to calculate transition probabilities after two time click, while the probabilities of the first type are given as hypotheses. Below there are hypothesis that concern the case $\tau_p=2$, $p=5$, $t=2$:

- $H_1 = p_{01} \vee p_{02}$;
- $H_2 = K_{(\frac{2}{\sqrt{6}})^2 - 2\cdot(\frac{-1}{\sqrt{6}})^2, 0} p_{15} \wedge K_{(\frac{-2}{\sqrt{6}})^2 - 2\cdot(\frac{1}{\sqrt{6}})^2, 0} p_{25}$;
- $H_3 = K_{(1)^2 - 2\cdot(\frac{1}{\sqrt{2}})^2, 0} p_{01} \wedge K_{(1)^2 - 2(\frac{1}{\sqrt{2}})^2, 0} p_{02}$

  H_1 says that a photon must go through one of the slits. The rest of the hypothesis carries information about transitions $\Upsilon_0 \longrightarrow \Upsilon_{1,2}$ and $\Upsilon_{1,2} \longrightarrow \Upsilon_5$.

In the sequel we give a formal proof which explains interference using p-adic numbers. Particularly, the result is the probability of passing from Υ_0 to Υ_5.

- $A_{2,5,2}, H_1 \vdash K_{1,0}(p_{01} \vee p_{02})$ by Rule $R2$ of the logic $L^D_{\mathbb{Q}_p}$,
- $A_{2,5,2}, H_1, H_2, H_3 \vdash K_{(1\cdot\frac{2}{\sqrt{6}} + 2\cdot\frac{-1}{\sqrt{6}}\frac{1}{\sqrt{2}} + 1\cdot\frac{-2}{\sqrt{6}} + 2\cdot\frac{1}{\sqrt{6}}\frac{1}{\sqrt{2}})^2 - 2(\frac{-1}{\sqrt{6}} + 2\cdot\frac{2}{\sqrt{6}}\cdot\frac{1}{\sqrt{2}} + \frac{1}{\sqrt{6}} + 2\cdot\frac{-2}{\sqrt{6}}\cdot\frac{1}{\sqrt{2}})^2, 0} p_{05}$
- $A_{2,5,2}, H_1, H_2, H_3 \vdash K_{0,0} p_{05}$

2.2.4 Conditional p-adic probability logics

Conditional p-adic probability logics [9] are created by extending classical propositional logic with binary probability operators of the form $CK_{r,\rho}$, with the intended meaning of $CK_{r,\rho}\alpha, \beta$ that the conditional probability of α given β is in the p-adic ball $K(r, \rho)$. When we worked with the logic $L^D_{\mathbb{Q}_p}$ in Section 2.2.2 we used an arbitrary large but fixed p-adic ball $K(0, p^M)$ as the range of probabilities to ensure the essential boundedness condition. However, when we consider conditional probabilities we have to multiply p-adic numbers, the balls of the form $K(0, p^M)$ are not closed under multiplication, and such rangers of probabilities are no longer useful. To overcome this problem we can:

- consider the unit ball $K(0, 1)$ which is closed under multiplication, or
- build formulas over a finite set of propositional letters.

Using the former option we construct the logic[9] $CPL_{\mathbb{Z}_p}$, while the latter option leads us to the logic $CPL^{fin}_{\mathbb{Q}_p}$ in which case we can use $\mathbb{Q}_p$ as the range of probabilities. Having the above mentioned constraints, one can follow the proofs presented in the previous section and obtain the corresponding strong completeness and decidability theorems for the conditional p-adic probability logics.

2.3 Complex valued probability logics

As we mentioned above in quantum mechanics wave functions are complex valued. More precisely, a wave function can be represented as $\varphi = c_1\varphi_1 + \ldots + c_1\varphi_k$, where $c_i \in \mathbb{C}$ are complex numbers and $\varphi_1, \ldots, \varphi_k$ are eigenvectors of some observable[10]

[9] $\mathbb{Z}_p$ denotes the ring of p-adic integers.

[10] Observables are represented by linear operators.

O with the eigenvalues $a_1, \ldots, a_k$. Then the probability of obtaining the value a_i as the result of measuring O in the state φ, is $\|c_i\|^2$. One of the main reasons for using complex numbers in this formalism is the fact that it may happen that: $\| z_1 + z_2 \|$ is strictly less than $\| z_1 \|$ and $\| z_2 \|$. This property has the key role in expressing negative interference of waves, and although quantum probabilities are not complex valued, complex numbers are probability amplitudes used in calculations. However, there are also other problems involving complex valued probabilities. For example, they appear in Markov Stochastic Processes. In [2] the authors consider the transition probability matrix P of a discrete Markov chain. They calculate P^r where $n < r < n+1$, for some $n \in \mathbb{Z}$ and show that P^r is a complex matrix. The papers [11, 12] consider complex probability set $\mathscr{C} = \mathscr{R} + \mathscr{M}$ where $\mathscr{R}$ is the real set with its corresponding real probability and $\mathscr{M}$ is the imaginary set with its corresponding imaginary probability. The classical Kolmogorov's approach is extended so that the probability of an event $m \in \mathscr{M}$ associated to an event $r \in \mathscr{R}$ is $P_m = i \cdot (1 - P_r)$, while the probability of an event $c \in \mathscr{C}$ is $P_c^2 = (P_r + \frac{P_m}{i})^2$. This novel complex valued probability is applied to different concepts such as: degradation and the remaining useful lifetime of a vehicle suspension system, the law of large numbers, unburied petrochemical pipelines analytic prognostic, Chebyshev's inequality and Ludwig Boltzmann's entropy. It is shown that some phenomena that seem to be random and stochastic in $\mathscr{R}$ become deterministic and certain in $\mathscr{C}$. In [23] a complex valued probability density function is used to analyze Boltzmann equation.

Again, to formalize reasoning about probabilities, due the lack of ordering in this framework, we cannot use the standard probability operators of the form $P_{\geq r}\alpha$. In [6] the following probability formulas are considered:

- $B_{z,\rho}\alpha$ with the meaning "The probability of α is in the ball with the center z and the radius ρ", and
- $S_{z,\rho}\alpha$ with the meaning "The probability of α is in the square with the center z and the side $2 \cdot \rho$".

The logic $LCOMP_B$ uses the operators $B_{z,\rho}$, while the operators $S_{z,\rho}$ are used in the logic $LCOMP_S$. In Section 2.3.1 we present the first logic. We use the following notation:

- $|\cdot|$ denotes the standard real absolute value, while $\|\cdot\|$ is the complex norm.
- $\mathbb{C}_\mathbb{Q} = \{a + i \cdot b | a, b \in \mathbb{Q}\}$ is the set of complex numbers with rational coordinates, and $\mathbb{Q}^+$ is the set of all non-negative rational numbers.

2.3.1 The logic $LCOMP_B$

In logic $LCOMP_B$ we make statements about probability using operators $B_{z,\rho}$. Therefore, basic probability formulas are of the form $B_{z,\rho}\alpha$, where $\alpha \in \mathrm{For}_C$ is a classical formula, $z \in \mathbb{C}_\mathbb{Q}$ and $\rho \in \mathbb{Q}^+$. Probability formulas are Boolean combinations of basic formulas.

Satisfiability for the logic $LCOMP_B$ is defined in measurable Kripke models $M = \langle W, H, \mu, v \rangle$ where the range of the probability measure μ is the field of complex numbers, i.e., $\mu : H \longrightarrow \mathbf{C}$. In an $LCOMP_B$-model satisfiability of the basic probability formulas is defined as follows:

- $M \models B_{z,\rho}\alpha$ iff $||\mu([\alpha]) - z|| \leq \rho$.

The axiom system Ax_{LCOMP_B} of the logic $LCOMP_B$ contains:
Axioms

A1. Substitutional instances of tautologies.
A2. $B_{z,\rho}\alpha \Rightarrow B_{z,\rho'}\alpha$, whenever $\rho' \geq \rho$
A3. $B_{z_1,\rho_1}\alpha \wedge B_{z_2,\rho_2}\beta \wedge B_{0,0}(\alpha \wedge \beta) \Rightarrow B_{z_1+z_2,\rho_1+\rho_2}(\alpha \vee \beta)$
A4. $B_{z_1,\rho_1}\alpha \Rightarrow \neg B_{z_2,\rho_2}\alpha$, if $||z_1 - z_2|| > \rho_1 + \rho_2$
A5. $B_{z_1,\rho_1}\alpha \Rightarrow B_{z_2,\rho_1+\rho_2}\alpha$, if $||z_1 - z_2|| \leq \rho_2$

Inference rules

R1. From A and $A \Rightarrow B$ infer B. A and B are either both propositional, or both probabilistic formulas.
R2. From α infer $B_{1,0}\alpha$.
R3. If $n \in \mathbb{N}$, from $\varphi \Rightarrow \neg B_{z,\frac{1}{n}}\alpha$ for every $z \in \mathbb{C}_{\mathbb{Q}}$, infer $\varphi \Rightarrow \perp$.
R4. From $\alpha \Rightarrow \perp$, infer $B_{0,0}\alpha$.
R5. If $z \in \mathbb{C}_{\mathbb{Q}}$, from $\varphi \Rightarrow B_{z,\rho+\frac{1}{n}}\alpha$ for every $n \in \mathbb{N}$, infer $\varphi \Rightarrow B_{z,\rho}\alpha$.
R6. From $(\alpha \Leftrightarrow \beta)$, infer $(B_{z,\rho}\alpha \Rightarrow B_{z,\rho}\beta)$.

We shortly discuss axioms and inference rules that reflect properties of complex balls. Axiom $A2$ says that a ball of smaller radius is contained in a ball of larger radius. Axiom $A3$ corresponds to the additivity of measures. Axiom $A4$ guarantees that the measure of a formula cannot belong to two disjoint balls. Axiom $A5$ says that whenever the measure of the formula α belongs to the ball B then it belongs to every ball that contains B. Rule $R3$ says that for every classical formula α and every arbitrary small $\frac{1}{n}$, there must be some $z \in \mathbb{C}_{\mathbb{Q}}$ such that the measure of α belongs to the ball $B[z, \frac{1}{n}]$. Rule $R4$ provides that a contradiction has the measure 0. Rule $R5$ says that whenever the measure of α is arbitrary close to some number $z \in \mathbb{C}_{\mathbb{Q}}$, then the measure of α is equal to z. Finally, Rule $R6$ says that equivalent classical formulas have the same measure.

To prove soundness, the Deduction theorem, Lindenbaum's theorem and strong completeness we can combine the procedures introduced in Chapter 1 and earlier in this chapter for p-adic logics. Additionally, here for every maximal consistent set T^* the following holds: for every $\alpha \in \text{For}_C$, $n \in \mathbb{N}$ there is at least one $z \in \mathbb{C}_{\mathbb{Q}}$ such that $B_{z,\frac{1}{n}}\alpha \in T^*$. Actually, there are countably many numbers $z \in \mathbb{C}_{\mathbb{Q}}$ such that $B_{z,\frac{1}{n}}\alpha \in T^*$. Using this property for each formula $\alpha \in \text{For}_C$ we make a sequence z_n in the following way: for every $n \in \mathbb{N}$ we arbitrarily chose any number z such that $B_{z,\frac{1}{n}}\alpha \in T^*$ and this z will be the n-th number of the sequence, i.e., $z_n = z$. In this way we obtain the sequence $z_0, z_1, z_2, \ldots$ (denoted $z(\alpha)$) where $B_{z_n,\frac{1}{n}}\alpha \in T^*$. The sequence $z(\alpha)$ is a Cauchy sequence (with respect to the norm $||\cdot||$). We can prove that if $(z_n)_{n\in\mathbb{N}}$ and $(z'_n)_{n\in\mathbb{N}}$ are two different sequences obtained by the above given construction (i.e., for at least one m, $z_m \neq z'_m$), then $\lim_{n\to\infty} z_n = \lim_{n\to\infty} z'_n$. Using these properties of the sequence $z(\alpha)$ we construct a canonical model $M_{T^*} = \langle W, H, \mu, v\rangle$, where $\mu : H \to C$ is defined as follows:

- Let $z(\alpha) = (z_n)_{n\in\mathbb{N}}$. Then

$$\mu([\alpha]) = \begin{cases} z & \text{if } B_{z,0}\alpha \in T^* \\ \lim_{n\to\infty} z_n & \text{otherwise} \end{cases}$$

We prove strong completeness by showing that every formula holds in M_{T^*} iff it belongs to T^*. To illustrate the idea we consider formulas of the form $B_{z,\rho}\alpha$:

- First assume that $B_{z,\rho}\alpha \in T^*$. Let $\rho > 0$, $z(\alpha) = (z_n)_{n\in\mathbb{N}}$ and $\mu([\alpha]) = \lim^{p}_{n\to\infty} z_n$. Then $(\forall\varepsilon)(\exists n_0)(\forall n)(n \geq n_0 \to ||\mu([\alpha]) - z_n|| \leq \varepsilon)$. Observing the area around the boundary of the ball of the radius ρ we can distinguish the following cases:

  1. There is $\rho_1 < \rho$ such that $B_{z,\rho_1}\alpha \in T^*$. Therefore $\rho - \rho_1 > 0 > \frac{1}{m}$, for some $m \in \mathbb{N}$. Let $\varepsilon = \rho - \rho_1 - \frac{1}{m} > 0$ and choose n_0 such that for every $n \geq n_0$, $||\mu([\alpha]) - z_n|| \leq \rho - \rho_1 - \frac{1}{m}$. If $n \geq max\{n_0, m\}$ then from $T^* \vdash B_{z,\rho_1}$ and $T^* \vdash B_{z_n,\frac{1}{n}\alpha}$, according to Axiom $A4$ we have $||z - z_n|| \leq \rho_1 + \frac{1}{n}$. Therefore $||z - \mu([\alpha])|| = ||z - z_n + z_n - \mu([\alpha])|| \leq ||z - z_n|| + ||z_n - \mu([\alpha])|| \leq \rho_1 + \frac{1}{n} + \rho - \rho_1 - \frac{1}{m} \leq \rho$ and hence $M_{T^*} \models B_{z,\rho}\alpha$.
  2. There is no $\rho_1 < \rho$ such that $B_{z,\rho_1}\alpha \in T^*$. Since $T^* \vdash B_{z,\rho}\alpha$ from Axiom $A2$, we obtain $T^* \vdash B_{z,\rho+\frac{1}{n}}\alpha$, for every $n \in \mathbb{N}$. Thus, $||z - \mu([\alpha])|| \leq \rho + \frac{1}{n}$, for every $n \in \mathbb{N}$, and hence $||z - \mu([\alpha])|| \leq \rho$.

  If $\rho = 0$, then $T^* \vdash B_{z,0}\alpha$ and $\mu([\alpha]) = z$, $||\mu([\alpha]) - z|| = 0$. Thus, $M_{T^*} \models B_{z,0}\alpha$.
- Suppose that $M_{T^*} \models B_{z,\rho}\alpha$. If $\rho > 0$ then we reason as follows:

  1. Let $||\mu([\alpha]) - z|| < \rho$. There is $\rho_1 < \rho$: $M_{T^*} \models B_{z,\rho_1}\alpha$. Let $k \in \mathbb{N}$ be such that $\rho - \rho_1 > \frac{1}{k}$. If $\mu([\alpha]) = \lim^{p}_{n\to\infty} z_n$, then there is n_0 such that $(\forall n)(n \geq n_0 \to ||\mu([\alpha]) - z_n|| \leq \frac{1}{k})$. Let $n \geq n_0$ and $\frac{1}{n} + \frac{1}{k} + \rho_1 \leq \rho$. Then from $||z_n - \mu([\alpha])|| \leq \frac{1}{k}$ and $||z - \mu([\alpha])|| \leq \rho_1$, we obtain $||z - z_n|| \leq \rho_1 + \frac{1}{k}$. Thus: $T^* \vdash B_{z_n,\frac{1}{n}}\alpha, T^* \vdash B_{z_n,\frac{1}{n}}\alpha \Rightarrow B_{z,\frac{1}{n}+\rho_1+\frac{1}{k}}\alpha$, $T^* \vdash B_{z,\frac{1}{n}+\rho_1+\frac{1}{k}}\alpha$, and $T^* \vdash B_{z,\rho}\alpha$.
  2. If there is no $\rho_1 < \rho$ such that $||\mu([\alpha]) - z|| < \rho_1$, then from $||\mu([\alpha]) - z|| < \rho$, we have $||\mu([\alpha]) - z|| < \rho + \frac{1}{n}$, for every $n \in \mathbb{N}$. Hence, $T^* \vdash B_{z,\rho+\frac{1}{n}}\alpha$ for every $n \in \mathbb{N}$, and using Rule $R5$ we conclude that $T^* \vdash B_{z,\rho}\alpha$.

  Finally suppose that $\rho = 0$. Then $||\mu([\alpha]) - z|| = 0$ and hence $||\mu([\alpha]) - z|| < \frac{1}{n}$ for every $n \in \mathbb{N}$. Therefore $T^* \vdash B_{z,0+\frac{1}{n}}\alpha$ for every $n \in \mathbb{N}$ and so $T^* \vdash B_{z,0}\alpha$.

Finally, we briefly discuss decidability of PSAT, the satisfiability problem for the logic $LCOMP_B$. As in the previous sections we consider only probabilistic part and reduce the problem of satisfiability to considering formula of the form:

$$\bigwedge_{j=1,k_i} \pm B_{z_{i,j},\frac{1}{n_{i,j}}} FDNF(\alpha_{i,j})$$

Similarly as above, this formula is satisfiable iff the following system is satisfiable:

$$\sum_{t=1}^{2^n} z_t = 1$$

$$\Upsilon_1 = \begin{cases} (\sum_{a_t \in \alpha_{i,1}} x_t - a_1)^2 + (\sum_{a_t \in \alpha_{i,1}} y_t - b_1)^2 \leq \rho_1^2 & \text{if } \pm B_{z_1,\rho_1}\alpha_{i,1} = B_{z_1,\rho_1}\alpha_{i,1} \\ (\sum_{a_t \in \alpha_{i,1}} x_t - a_1)^2 + (\sum_{a_t \in \alpha_{i,1}} y_t - b_1)^2 > \rho_1^2 & \text{if } \pm B_{z_1,\rho_1}\alpha_{i,1} = \neg B_{z_1,\rho_1}\alpha_{i,1} \end{cases}$$

$$\ldots$$

$$\Upsilon_{k_i} = \begin{cases} (\sum_{a_t \in \alpha_{i,k_i}} x_t - a_{k_i})^2 + (\sum_{a_t \in \alpha_{i,k_i}} y_t - b_{k_i})^2 \leq \rho_{k_i}^2 & \text{if } \pm B_{z_{k_i},\rho_{k_i}}\alpha_{i,k_i} = B_{z_{k_i},\rho_{k_i}}\alpha_{i,k_i} \\ (\sum_{a_t \in \alpha_{i,k_i}} x_t - a_{k_i})^2 + (\sum_{a_t \in \alpha_{i,k_i}} y_t - b_{k_i})^2 > \rho_{k_i}^2 & \text{if } \pm B_{z_{k_i},\rho_{k_i}}\alpha_{i,k_i} = \neg B_{z_{k_i},\rho_{k_i}}\alpha_{i,k_i} \end{cases}$$

where $a_t \in \alpha_{i,j}$ denotes that the atom a_t appears in $FDNF(\alpha_{i,j})$, $z_t = \mu([a_t]) = x_t + i \cdot y_t$ and $z_j = a_j + i \cdot b_j, 1 \leq j \leq k_i$. It is shown in [6] that $LCOMP_B$-formulas can be coded in the existential fragment of RCF (the theory of real closed fields). Since the SAT problem for this fragment is PSPACE-complete, satisfiability for $LCOMP_B$ is in PSPACE. .

2.4 Measure logic

Here we discuss logics for reasoning about uncertainty formalized in terms of a measure $m : H \to G$, where H is a given set of events and G is a partially ordered monoid [3]. More precisely, $(G, \leq, *, 0)$ is a partially ordered countable commutative monoid with the least element 0, which is also the neutral element for $*$.

Definition 2.4. A function $m : H \to G$ is called G-measure or generalized measure on H if the following conditions are satisfied:

1. $m(\emptyset) = 0$,
2. if $A, B \in H$ and $A \cap B = \emptyset$, then $m(A \cup B) = m(A) * m(B)$. ■

Possible structures for G are:

- Additive monoid of nonnegative rational numbers $(\mathbb{Q}^+, \leq, +, 0)$;
- $(\{0, \frac{1}{n}, \ldots, \frac{n-1}{n}, 1\}, \leq, \oplus, 0)$, $x \oplus y = \min\{1, x+y\}$,
- $([0,1]_{\mathbb{Q}}, \leq, \oplus, 0)$,
- $(\mathbb{Q}^+(\varepsilon), \leq, +, 0)$, where $\mathbb{Q}^+(\varepsilon)$ is the set of nonnegative elements of the domain of the nonarchimedean field $\mathbb{Q}(\varepsilon)$ which is the smallest field obtained by adding a positive infinitesimal ε to rational numbers,

- $(\mathbb{Q}^+(\varepsilon)^n, \leq, +, (0,\ldots,0))$, $\mathbb{Q}^+(\varepsilon)^n = \underbrace{\mathbb{Q}^+ \times \ldots \times \mathbb{Q}^+}_{n}$, $\leq$ is lexicographic order,
- $(\mathbb{N}, \leq, +, 0)$, $(\mathbb{N}, \leq, \max, 0)$, $(\mathbb{Q}^+, \leq, \max, 0)$, etc.

2.4.1 The logic L_{PG}

The basic logic for reasoning about G-measure is the logic L_{PG} that enlarges classical propositional logic with formulas of the form $M_{=a}\varphi$, $a \in G$, with the intended meaning "the measure of φ is a."

The language of the logic L_{PG} consists of a countable set $\mathcal{V} = \{p, q, r, p_1, \ldots\}$ of propositional variables, connectives $\wedge$, $\neg$, and the set $\{M_{=a} \mid a \in G\}$ of (unary) measure operators.

The set For_M of M-formulas is the set of all Boolean combinations of the basic M-formulas: $M_{=a}\varphi$, $a \in G$. The set of all L_{PG}-formulas For is a union of For_M and the set of classical propositional formulas For_C. Classical propositional formulas will be denoted by small Greek letters φ, ψ while for For_M-formulas we use designations Φ, Ψ, ... Formulas from For will be denoted by A, B, C, ... etc.

Definition 2.5. An L_{PG}-model is a structure $\mathbf{M} = \langle W, H, m, v\rangle$ where:
- W is a nonempty set of objects called worlds,
- H is an algebra of subsets of W,
- m is a G-measure on H,
- $v : W \times \mathcal{V} \to \{0,1\}$ provides for each world $w \in W$ a two-valued evaluation of the propositional letters; a truth-evaluation $v(w,\cdot)$ is extended to classical propositional formulas as usual. ■

Let $\mathbf{M} = \langle W, H, m, v\rangle$ be a measurable L_{PG}-model, i.e., $H = \{[\alpha] : \alpha \in \mathrm{For}_C\}$. The satisfiability relation for the basic For_M-formulas is defined as follows:
- if $\varphi \in \mathrm{For}_C$, $\mathbf{M} \models M_{=a}\varphi$ iff $m([\varphi]) = a$,

which is extended to all For_M-formulas in an expected way.

The axiom system $Ax_{L_{PG}}$ for L_{PG} contains the following axiom schemata and inference rules:

A1. all instances of classical propositional tautologies,
A2. $\neg M_{=0}\top$,
A3. $M_{=a}\varphi \to \neg M_{=b}\varphi$, $a \neq b$,
A4. $M_{=0}\neg(\varphi \leftrightarrow \psi) \to (M_{=a}\varphi \to M_{=a}\psi)$,
A5. $(M_{=a}\varphi \wedge M_{=b}\psi \wedge M_{=0}(\varphi \wedge \psi)) \to M_{=a*b}(\varphi \vee \psi)$,

and inference rules:

R1. From A and $A \to B$ infer B, where A and B are both classical or both measure formulas.
R2. From φ infer $M_{=0}\neg\varphi$,

R3. From $\Phi \to \neg M_{=a}\varphi$, for every $a \in G$, infer $\neg\Phi$.

Note that Axiom A3 says that the measure of each formula is unique, while A5 corresponds to additivity of measure. Rule R3 define the range of the measure.

Soundness and the Deduction Theorem can be shown using the technique from previous sections. Also, in order to prove the completeness theorem we continue in the standard manner. We show how to extend a consistent set of formulas T to a maximal consistent set: let $\varphi_1, \varphi_2, \ldots$ and $\Phi_1, \Phi_2, \ldots$ be enumerations of For_C and For_M-formulas, respectively. We define a sequence of sets T_i in the following way:

1. $T_0 = T \cup Con^C(T) \cup \{M_{=0}\neg\varphi \mid \varphi \in Con^C(T)\}$, where $Con^C(T)$ is the set of all classical consequences of the set T.
2. for every $i \geq 0$, if $T_{2i} \cup \{\Phi_i\}$ is consistent, then $T_{2i+1} = T_{2i} \cup \{\Phi_i\}$; otherwise $T_{2i+1} = T_{2i} \cup \{\neg\Phi_i\}$,
3. for every $i \geq 0$, $T_{2i+2} = T_{2i+1} \cup \{M_{=a}\varphi_i\}$, for some $a \in G$, so that T_{2i+2} is consistent.

If $\overline{T} = \cup_{n\geq 0} T_n$ then using standard considerations we prove that $\overline{T}$ is deductively closed and maximally consistent. Being maximal $\overline{T}$ meets the following requirements:

1. If $A \in \overline{T}$, then $\neg A \notin \overline{T}$.
2. $A \wedge B \in \overline{T}$ iff $A \in \overline{T}$ and $B \in \overline{T}$.
3. If $\vdash A$, then $A \in \overline{T}$.
4. If $A \in \overline{T}$ and $A \to B \in \overline{T}$, then $B \in \overline{T}$.

Starting from a consistent set T and its maximal consistent extension $\overline{T}$ we construct the canonical model $\mathbf{M}_{\overline{T}} = \langle W, H, m, v\rangle$ such that:

- W consists of all classical propositional interpretations that satisfy the set $Con^C(T)$, i.e., $W = \{w \mid w \models Con^C(T)\}$,
- $H = \{[\varphi] \mid \varphi \in \mathrm{For}_C\}$,
- $v : W \times \mathscr{V} \to \{0,1\}$ is an assignment such that for every world $w \in W$ and every propositional letter $p \in \mathscr{V}$, $v(w,p) = 1$ iff $w \models p$,
- $m : H \to G$ is defined by: $m([\varphi]) = a$ iff $M_{=a}\varphi \in \overline{T}$.

Strong completeness of the logic L_{PG} is a consequence of the following lemma:

Lemma 2.4. *Let $\mathbf{M}_{\overline{T}}$ be the canonical model over a consistent set T.*

1. $\mathbf{M}_{\overline{T}}$ is an L_{PG}-model.
2. $\mathbf{M}_{\overline{T}} \models A$ iff $A \in \overline{T}$.
3. $\mathbf{M}_{\overline{T}} \models T$.

Proof. 1. H is an algebra of subsets of W. Indeed, let φ be arbitrary formula. Then $W = [\varphi \vee \neg\varphi] \in H$. Also, if $[\varphi] \in H$, then the $[\varphi]^c = [\neg\varphi] \in H$. Finally, if $[\varphi_1], \ldots [\varphi_n] \in H$, then the union $[\varphi_1] \cup \ldots \cup [\varphi_n] \in H$ because $[\varphi_1] \cup \ldots \cup [\varphi_n] = [\varphi_1 \vee \ldots \vee \varphi_n]$. Thus, H is an algebra of subsets of W. We need to show that m is a G-measure on H. Assume that $[\varphi] = [\psi]$. Then, by the completeness theorem for classical propositional logic, $Con^C(T) \vdash \varphi \leftrightarrow \psi$, and $M_{=0}\neg(\varphi \leftrightarrow \psi) \in \overline{T}$, by

Rule (R2). It follows that m is well-defined. Let $[\varphi]\cap[\psi]=\emptyset$, $M_{=a}\varphi, M_{=b}\psi \in \overline{T}$. By the completeness theorem for classical propositional logic and Rule R2, we have $\overline{T}\vdash M_{=0}(\varphi\wedge\psi)$. Thus, by Axiom A5, it follows $M_{=a*b}(\varphi\vee\psi)\in\overline{T}$, i.e., $m([\varphi]\cup[\psi])=a*b$. Hence, $\mathbf{M}_{\overline{T}}$ is an L_{PG}-model.
2. If A is a classical formula, the statement obviously holds. Let $A=M_{=a}\varphi$. Then $\mathbf{M}_{\overline{T}}\models M_{=a}\varphi$ iff $m([\varphi])=a$ iff $M_{=a}\in\overline{T}$. By induction, one can show that for every M-formula Φ, $\mathbf{M}\models\Phi$ iff $\Phi\in\overline{T}$. ■

Similarly as for other probability logics, the problem of satisfiability of a L_{PG}-formula, PSAT, is equivalent to satisfiability in G of a set of equalities and negations of equalities of the forms:

- $x_1 * x_2 * \ldots * x_m = a$,
- $x_1 * x_2 * \ldots * x_m \neq b$,

for some a and $b\in G$. For example, PSAT for L_{PG} is decidable if the set G is finite. The paper [3] also points out some infinite sets G, where the corresponding logics are decidable.

2.4.2 Extensions of the logic L_{PG}

The logic L_{PG} can be extended in several ways, and in this section we discuss two of them.

First, we extend the logic L_{PG} with unary operators of the forms $M_{\leq a}$ and $M_{\geq a}$, for every $a\in G$, that mean "the measure is less than (greater than) or equal to a". In that way we obtain the logic $L_{PG^\natural}$. The same class of L_{PG}-models is used here, and satisfiability of new probability formulas is defined in an obvious way:

- $\mathbf{M}\models M_{\leq a}\varphi$ iff $m([\varphi])\leq a$, and
- $\mathbf{M}\models M_{\geq a}\varphi$ iff $m([\varphi])\geq a$.

Note that G need not be linearly ordered, so the operators $M_{\leq a}$ and $M_{\geq a}$ are treated independently of each other. Furthermore, the axiom system $Ax_{L_{PG^\natural}}$ is an extension of $Ax_{L_{PG}}$ obtained by adding the following axioms:

A6. $M_{\geq 0}\varphi$
A7. $M_{\leq a}\varphi\wedge M_{\geq a}\varphi\to M_{=a}\varphi$
A8. $M_{=a}\varphi\to M_{\geq b}\varphi$, for each $b\leq a$
A9. $M_{=a}\varphi\to M_{\leq b}\varphi$, for each $b\geq a$
A10. $M_{=a}(\varphi\vee\psi)\to M_{\leq a}\varphi$,

and rules:

R4. $\dfrac{M_{=b}\varphi\to\Psi \text{ for each } b\geq a}{M_{\geq a}\varphi\to\Psi}$, and

R5. $\dfrac{M_{=b}\varphi\to\Psi \text{ for each } b\leq a}{M_{\leq a}\varphi\to\Psi}$.

The second extension of L_{PG}, the logic $L_{PG^{\natural\natural}}$, concerns qualitative reasoning about G-valued measures. Qualitative reasoning can be based on a binary relation $\leq$ on events (denoted by formulas) instead of exact numerical representations of realization of events.

To formalize qualitative reasoning about G-valued measures[11] we further extend the L_{PG}-language with the binary operators: $\leq$, $\geq$, $<$, $>$, $=$, and $\|$. The intended meaning of the formula $\varphi \leq \psi$ is "the measure of φ is smaller than or equal to the measure of ψ", and similarly for $\varphi \geq \psi$, $\varphi < \psi$, $\varphi > \psi$, and $\varphi = \psi$, while $\varphi \parallel \psi$ says that the measures of φ and ψ are incomparable.

We use L_{PG}-models as the corresponding semantics, while the satisfiability relation fulfils also:

- $\mathbf{M} \models \varphi \Diamond \psi$ iff $m([\varphi]) \Diamond m([\psi])$, for each $\Diamond \in \{\leq, \geq, <, >, =, \|\}$.

The axiom system $Ax_{L_{PG^{\natural\natural}}}$ extends $Ax_{L_{PG^{\natural}}}$ with the following new axioms:

A$\leq$. $\varphi \leq \psi \rightarrow (M_{=a}\varphi \rightarrow M_{\geq a}\psi)$, $a \in G$,
A$\geq$. $\varphi \geq \psi \rightarrow (M_{=a}\varphi \rightarrow M_{\leq a}\psi)$, $a \in G$,
A$<$. $\varphi < \psi \rightarrow (M_{=a}\varphi \rightarrow M_{>a}\psi)$, $a \in G$,
A$>$. $\varphi > \psi \rightarrow (M_{=a}\varphi \rightarrow M_{<a}\psi)$, $a \in G$,
A$=$. $\varphi = \psi \rightarrow (M_{=a}\varphi \rightarrow M_{=a}\psi)$, $a \in G$,
A$\|$. $\varphi \parallel \psi \rightarrow (M_{=a}\varphi \rightarrow M_{\|a}\psi)$, $a \in G$,

and rules

R$\leq$. $\dfrac{\Phi \rightarrow (M_{=a}\varphi \rightarrow M_{\geq a}\psi), \text{ for every } a \in G}{\Phi \rightarrow \varphi \leq \psi}$

R$\geq$. $\dfrac{\Phi \rightarrow (M_{=a}\varphi \rightarrow M_{\leq a}\psi), \text{ for every } a \in G}{\Phi \rightarrow \varphi \geq \psi}$

R$<$. $\dfrac{\Phi \rightarrow (M_{=a}\varphi \rightarrow M_{>a}\psi), \text{ for every } a \in G}{\Phi \rightarrow \varphi < \psi}$

R$>$. $\dfrac{\Phi \rightarrow (M_{=a}\varphi \rightarrow M_{<a}\psi), \text{ for every } a \in G}{\Phi \rightarrow \varphi > \psi}$

R$=$. $\dfrac{\Phi \rightarrow (M_{=a}\varphi \rightarrow M_{=a}\psi), \text{ for every } a \in G}{\Phi \rightarrow \varphi = \psi}$

R$\|$. $\dfrac{\Phi \rightarrow (M_{=a}\varphi \rightarrow M_{\|a}\psi), \text{ for every } a \in G}{\Phi \rightarrow \varphi \parallel \psi}$.

Strong completeness for $Ax_{L_{PG^{\natural}}}$ and $Ax_{L_{PG^{\natural\natural}}}$ can be proved following the ideas from Section 2.4.1.

References

[1] G. Bachman. *Introduction to p-adic numbers and valuation theory*. Academic Press, 1964.

[11] Reasoning about qualitative probabilities is described in Chapter 5. The main difference is that the range of the standard, real-valued, probabilities is linearly ordered.

[2] B. Bidabad and B. Bidabad. Complex probability and Markov stochastic processes. In *Proc. First Iranian Statistics Conference, Isfahan University of Technology*, 1992.
[3] N. Ikodinović, M. Rašković, Z. Marković, and Z. Ognjanović. Logics with Generalized Measure Operators. *Journal of Multiple-Valued Logic and Soft Computing*, 20(5-6):527–555, 2013.
[4] A. Ilić-Stepić. A Logic for Reasoning About Qualitative Probability. *Publications de l'Institut Mathématique*, Ns. 87(101):97–108, 2010.
[5] A. Ilić-Stepić, Z. Ognjanović, N. Ikodinović, and A. Perović. A *p*-adic probability logic. *Mathematical Logic Quarterly*, 58(4-5):263–280, 2012.
[6] A. Ilić-Stepić and Z. Ognjanović. Complex valued probability logics. *Publications de l'Institut Mathématique*, Ns. 95(109):73–86, 2014.
[7] A. Ilić-Stepić and Z. Ognjanović. Logics for reasoning about processes of thinking with information coded by *p*-adic numbers. *Studia Logica*, 103:145–174, 2015.
[8] A. Ilić-Stepić and Z. Ognjanović. Logics to formalise *p*-adic valued probability and their applications. *International Journal of Parallel, Emergent and Distributed Systems*, 33:257–275, 2018.
[9] A. Ilić-Stepić, Z. Ognjanović, and N. Ikodinović. Conditional *p*-adic probability logic. *International Journal of Approximate Reasoning*, 55(9):1843–1865, 2014.
[10] A. Ilić-Stepić, Z. Ognjanović, N. Ikodinović, and A. Perović. *p*-Adic probability logics. *p-Adic Numbers, Ultrametric Analysis and Applications*, 8(3):177–203, 2016.
[11] A.A. Jaoude. The complex statistic paradigm and the law of large numbers. *Journal of Mathematics and Statistics*, 9:289–304, 2013.
[12] A.A. Jaoude. The Paradigm of Complex Probability and Ludwig Boltzmann's Entropy. *Systems Science and Control Engineering*, 6:108–149, 2018.
[13] J.M. Keynes. *Treatise on Probability*. Macmillan & Co, London, 1921.
[14] A. Khrennikov. *Non-Archimedean analysis: quantum paradoxes, dynamical systems and biological models*. Kluwer Academic Publishers, Dordrecht, The Netherlands, 1977.
[15] A. Khrennikov. *p*-adic quantum mehanics with *p*-adic valued functions. *J. Math. Phys*, 32:932–937, 1991.
[16] A. Khrennikov. *p-adic valued distibutions in mathematical physic*. Kluwer Academic Publishers, Dordrecht, The Netherlands, 1994.
[17] A. Khrennikov. *Interpretations of probability*. Walter de Gruyter, Berlin, Germany, 1999.
[18] A. Khrennikov. Toward theory of *p*-adic valued probabilities. *Studies in logic, grammar and rhetoric*, 14:137–154, 2008.
[19] A. Khrennikov, S. Yamada, and A. Van Rooij. The measure-theoretical approach to *p*-adic probability theory. *Annales mathematiques Blaise Pascal*, 6:21–32, 1999.
[20] N. Koblitz. *p-adic numbers, p-adic analysis and Zeta-Functions*, volume 58 of *Graduate Texts in Mathematics*. Springer, New York, 1977.

[21] Z. Ognjanović, A. Perović, and M. Rašković. Logics with the Qualitative Probability Operator. *Logic Journal of IGPL*, 16(2):105–120, 2008.
[22] N.S. Yanofsky and M.A. Mannucci. *Quantum computing for computer scientists*. Cambridge University Press, 2008.
[23] A. Zadehgol. Generalizing the Boltzmann equation in complex phase space. *Physical Review E*, 94(2):1–17, 2016.

Chapter 3
Probabilistic Temporal Logics

Dragan Doder and Aleksandar Perović

Abstract This chapter presents the proof-theoretical and model-theoretical approaches to reasoning about time and probability. Three different ways of combining probabilistic and temporal modalities are presented, and well defined syntax and corresponding semantics is provided for every formalism. Hilbert-style axiomatizations of the logics are presented, and they are proved to be sound and strongly complete with respect to the introduced semantics.

3.1 Introduction

Temporal logics are formal systems that allow reasoning about sentences referred to time. Their modern roots can be traced back to Prior's work in the late fifties of the twentieth century [38]. Temporal logics have found many applications, particularly in the computer science. For instance, in formal description of an agent's knowledge base, concurrent program design, verification and specification of programs, manual program composition, protocols in distributed systems, hardware etc., see [10, 11].

The main question in the modeling of a particular discrete-time temporal logic is how to describe the time flow. In terms of ordering, time flow could be linear or branching, and both choices might be represented by modal-like semantics. The main idea is to add a temporal connection between possible worlds – accessibility relation is expressed in terms of the past (preceding worlds) and the future (succeeding worlds). Similarly as in modal logic, information about any particular world is expressed by means of classical logic, and connections between possible worlds are

---

Dragan Doder
Utrecht University, Princetonplein 5, 3584 CC Utrecht, The Netherlands, e-mail: \email{d.doder@uu.nl}

Aleksandar Perović
Faculty of Transport and Traffic Engineering, University of Belgrade, Vojvode Stepe 305, 11000 Belgrade, Serbia, e-mail: pera@sf.bg.ac.rs

Z. Ognjanović (ed.), *Probabilistic Extensions of Various Logical Systems*,
https://doi.org/10.1007/978-3-030-52954-3_3

described by temporal operators. Linear time flow usually assumes that the structure of time is isomorphic to the ordering of natural numbers, i.e., each time instance (possible world) has a unique time instance reachable from it (*next* time instance). On the other hand, branching time flow assumes non-deterministic future in which a time instance might have more than one successor.

In this chapter we study propositional languages with probabilistic and temporal operators. The time flow will be either linear, with the unary next operator $\bigcirc$, and the until operator U, or branching, with the addition of the unary universal path operator (path quantifier) A. Furthermore, probabilities will be represented either by modal like operators $P_{\geq r}$, or probabilistic terms similar to [14]. The particular choice of operators is determined by the type (nature) of the analyzed problem.

One of the main challenges with such languages is proper formalization of the interplay between probabilistic and temporal operators. We can combine those two types of modalities in at least three ways:

1. *Probabilistic reasoning about temporal information.* In this approach, the basic statements are probability assessments of temporal sentences. An example of this kind is the proposition: "the probability of the overcast in the next two days is at least 50%". The corresponding semantics contains a probability measure defined on sets of temporal models.
2. *Temporal reasoning about probabilities.* This approach is convenient for modeling the changes of probabilities over time. An example of this kind is the proposition "if the probability that the coin lands tails is at least 50%, then it will always be greater than zero". The semantics consist of temporal structures in which each possible world contains a probabilistic model.
3. *Non-restricted modal approach*, where all modalities can be combined in an arbitrary way. In this case, the possible worlds of temporal models are equipped with probability measures.

In this chapter, we present an example of probabilistic temporal logic for each of the three approaches. We start by proposing strongly complete axiomatizations for LTL and CTL* which use infinitary rules of inference, paving the road to temporal extensions of axiomatization techniques presented in Chapter 1. Then we present a very expressive probabilistic temporal logic, which extends CTL* with two types of probability operators: one measuring sets of possible worlds of a path, and one which measures sets of paths passing through a given state. Probabilistic and temporal operators can be combined in an arbitrary way. After that, we apply probabilistic terms to LTL. Finally, we model change of probabilities by applying LTL formulas to a probability language containing observations and hypotheses. For all the considered logics we propose strongly complete axiom systems.

This chapter covers some results from [7–9, 32–35].

3.2 Temporal logics

In this section, we recall the logics LTL and CTL*, which will be the basis for the probabilistic extensions in the rest of the chapter. We first introduce the syntax of linear temporal logic LTL, which is a propositional temporal language built recursively from the atomic propositions using the next and until operators, present the corresponding path semantics and an axiom system which is strongly complete for the semantics. Then we extend the logic to CTL*, whose language is obtained by adding the path switching operators to the language of LTL. We consider Kripke semantics for the logic, with a set of states, a total accessibility relation between them and a set of paths which arise by moving from state to state in accordance with the accessibility relation.

3.2.1 *Linear temporal logic*

Let *Var* be at most countable set of propositional letters.

Definition 3.1 (LTL formula). The set of formulas For_{LTL} of LTL is the smallest set which satisfies the following conditions:

- $Var \subseteq For_{LTL}$,
- if $\alpha, \beta \in For_{LTL}$, then $\alpha \wedge \beta, \neg\alpha \in For_{LTL}$,
- if $\alpha, \beta \in For$, then $\bigcirc\alpha, \alpha U \beta \in For_{LTL}$. ■

Intuitively, the operators mean:

- $\bigcirc\alpha$: α holds in the next time instant
- $\alpha U \beta$: α holds in every time instant until β becomes true.

The other Boolean connectives ($\vee$, $\rightarrow$, $\leftrightarrow$) and constants ($\top$, $\bot$) are introduced as abbreviations in the usual way. The temporal operators F (sometime in future) and G (always in future) are defined as follows:

- $F\alpha$ is $\top U \alpha$,
- $G\alpha$ is $\neg F \neg \alpha$.

If T is a set of LTL formulas, then $\bigcirc T$ denotes the set $\{\bigcirc\alpha \mid \alpha \in T\}$. Further, for $n \in \mathbb{N}$, we define $\bigcirc^{n+1}\alpha$ as $\bigcirc(\bigcirc^n\alpha)$, assuming that $\bigcirc^0\alpha = \alpha$.

Now we turn to the semantics of LTL. We assume that S is a nonempty set of states, which are abstract objects equipped with truth labellings. i.e., there is a function $v : S \times Var \longrightarrow \{0,1\}$ which assigns a truth labeling to every state. Then semantics for LTL formulas consists of the set of *paths*, where a path is an ω-sequence of states of the form $\sigma = s_0, s_1, s_2, \ldots$ Here s_i, called *the i-th time instance of* σ, and $v(s_i, p) = 1$ means that the propositional letter p is true at time i in σ. In the rest of the chapter, we use the following abbreviations:

- $\sigma_{\geq i}$ is the path $s_i, s_{i+1}, s_{i+2}, \ldots$

- σ_i is the state s_i.

We denote the set of all paths by $\overline{\Sigma}$. Now we define satisfiability of an LTL formula on a path.

Definition 3.2. Let α be an LTL formula and σ a path. The satisfiability relation $\models$ is defined recursively as follows:

- if $p \in Var$, then $\sigma \models p$ iff $v(s_0, p) = 1$,
- $\sigma \models \neg\alpha$ iff $\sigma \not\models \alpha$,
- $\sigma \models \alpha \wedge \beta$ iff $\sigma \models \alpha$ and $\sigma \models \beta$,
- $\sigma \models \bigcirc\alpha$ iff $\sigma_{\geq 1} \models \alpha$,
- $\sigma \models \alpha U \beta$ iff there is some $i \in \mathbb{N}$ such that $\sigma_{\geq i} \models \beta$ and for each $j \in \mathbb{N}$, if $0 \leq j < i$ then $\sigma_{\geq j} \models \alpha$. ■

The notions of satisfiability, validity and semantical sequence are defined in a standard way.

Similarly as in the case of the real-valued probability logic, non-compactness is present in LTL as well. Indeed, suppose that α is neither a valid formula nor a contradiction. Then, the theory

$$T = \{F\,\neg\alpha\} \cup \{\alpha, \bigcirc\alpha, \bigcirc^2\alpha, \bigcirc^3\alpha, \ldots\}$$

is obviously finitely satisfiable, but it is not satisfiable. Consequently, there is no strongly complete finitary axiomatization for LTL, and the standard axiom systems focus on weak completeness.

Since the aim of this section is to extend the probability logics from Chapter 1 with temporal operators, we follow that axiomatization approach and introduce an infinitary inference rule in order to achieve strong completeness. The main idea behind the rule is to properly capture the semantic relationship between the operators $\bigcirc$ and U. For that purpose, we found convenient to introduce the binary operators $\overline{U}_n$ ($n \in \mathbb{N}$) in the following way:

$$\alpha \overline{U}_n \beta := (\bigwedge_{k=0}^{n-1} \bigcirc^k \alpha) \wedge \bigcirc^n \beta,$$

for every $\alpha, \beta \in For_{LTL}$. Note that the formula $\alpha \overline{U}_n \beta$ holds on a path iff β holds in n–th time instant, and α holds in every time instant before n. Obviously, if $\sigma \models \alpha U \beta$, then there is $n \in \mathbb{N}$ such that $\sigma \models \alpha \overline{U}_n \beta$. In the infinitary L_{ω_1}–logic we can express this as

$$\alpha U \beta \to \bigvee_{n \in \mathbb{N}} \alpha \overline{U}_n \beta.$$

Using the contrapositive, we obtain an equivalent reformulation:

$$\bigwedge_{n \in \mathbb{N}} \neg(\alpha \overline{U}_n \beta) \to \neg(\alpha U \beta).$$

Since we allow only finitary formulas, the previous formula can be reformulated as an infinitary inference rule

$$\frac{\{\neg(\alpha \overline{U}_n \beta) \mid n \in \mathbb{N}\}}{\neg(\alpha U \beta)}. \tag{3.1}$$

The actual form of the rule (3.1) will vary throughout this section, due to the technical reasons, in order to allow easy proofs of the Deduction theorem and the strong necessitation theorem of a logic under consideration.

For the axiomatization of LTL, we will use four standard axioms [39]:

A1. all the tautologies of the classical propositional logic
A2. $\bigcirc(\alpha \rightarrow \beta) \rightarrow (\bigcirc\alpha \rightarrow \bigcirc\beta)$
A3. $\neg\bigcirc\alpha \leftrightarrow \bigcirc\neg\alpha$
A4. $\alpha U \beta \leftrightarrow \beta \vee (\alpha \wedge \bigcirc(\alpha U \beta))$.

The set of inference rules contains Modus Ponens and the Necessitation rule for "next" operator:

R1. from $\{\alpha, \alpha \rightarrow \beta\}$ infer β
R2. from α infer $\bigcirc\alpha$

Note that we use the temporal necessitation R2 with the next operator, while the standard axiomatizations contains the necessitation for the operator G. However, it can be shown that the latter is a derived inference rule of our system.

In addition, we use the following form of the infinitary rule (3.1):

R3. From the set of premises

$$\{\gamma \rightarrow \neg(\alpha \overline{U}_n \beta) \mid n \in \mathbb{N}\}$$

infer $\gamma \rightarrow \neg(\alpha U \beta)$.

We will denote the axiom system consisting of the axioms A1–A4 and the rules R1–R3 by $Ax(LTL)$.

Now we define the notions of derivation from the inference system introduced above. Since the axiomatization contains the infinitary rule R3, the length of derivation can be any countable successor ordinal. We need to restrict applications of the rule R2 to theorems only, since otherwise any change during the time would be impossible. For that reason, we define the notions of theorems and derivations from a set of formulas separately.

- A formula α is a *theorem* of the logic, ($\vdash \alpha$), if there is an at most countable sequence of formulas $\alpha_0, \alpha_1, \ldots, \alpha$, such that every α_i is an axiom, or it is derived from the preceding formulas by an inference rule.
- A formula α is *deducible* from a set of formulas T ($T \vdash \alpha$) if there is an at most countable sequence of formulas $\alpha_0, \alpha_1, \ldots, \alpha$, such that every α_i is a theorem or a formula from T, or it is derived from the preceding formulas by one of the

inference rules, excluding R2. The corresponding sequence $\alpha_0, \alpha_1, \ldots, \alpha$ is the *proof* of α from T.

We say that a set of formulas is

- *consistent*, iff $T \nvdash \bot$ (otherwise T is *inconsistent*),
- *deductively closed*, iff $T \vdash \alpha$ implies $\alpha \in T$, and
- *maximal consistent*, if T is consistent and every T' such that $T \subset T'$ is inconsistent.

Obviously, a set T is maximal consistent if it is consistent and for every formula α, either $\alpha \in T$ or $\neg\alpha \in T$. It is easy to check that $Ax(LTL)$ is sound. Now we prove strong completeness of the axiomatization. We start with the Deduction theorem.

Theorem 3.1 (Deduction theorem). *If T is a set of formulas, α is a formula, and $T, \alpha \vdash \beta$, then $T \vdash \alpha \to \beta$.*

Proof. The proof is on the transfinite induction on the length of the inference. Here we consider the case when we apply the infinitary rule. Suppose that $T, \delta \vdash \gamma \to \neg(\alpha \mathrm{U} \beta)$ is obtained by R3. Then:

$T, \delta \vdash \gamma \to \neg(\alpha \overline{U}_n \beta)$, for $n \in \mathbb{N}$,
$T \vdash \delta \to (\gamma \to \neg(\alpha \overline{U}_n \beta))$, for $n \in \mathbb{N}$, by the induction hypothesis,
$T \vdash (\delta \wedge \gamma) \to \neg(\alpha \overline{U}_n \beta)$, $n \in \mathbb{N}$,
$T \vdash (\delta \wedge \gamma) \to \neg(\alpha \mathrm{U} \beta)$, by R3,
$T \vdash \delta \to (\gamma \to \neg(\alpha \mathrm{U} \beta))$. ■

Lemma 3.1. *Let α, β, γ be formulas.*

1. $\vdash (\bigcirc\alpha \to \bigcirc\beta) \to \bigcirc(\alpha \to \beta)$,
2. $\vdash \bigcirc(\alpha \wedge \beta) \leftrightarrow (\bigcirc\alpha \wedge \bigcirc\beta)$,
3. *(Strong necessitation) if $T \vdash \alpha$, where T is a set of formulas, then $\bigcirc T \vdash \bigcirc\alpha$,*
4. *for $j \geq 0$, $\bigcirc^j \beta, \bigcirc^0 \alpha, \ldots, \bigcirc^{j-1}\alpha \vdash \alpha \mathrm{U} \beta$.*

Proof. Since the proofs of the first two statements are straightforward, we only prove the third statement. We use the induction on the length of the proof of α from T. Assume that $T \vdash \gamma \to \neg(\alpha \mathrm{U} \beta)$ is obtained by R3. Then:

$T \vdash \gamma \to \neg(\alpha \overline{U}_n \beta)$, for $n \in \mathbb{N}$,
$\bigcirc T \vdash \bigcirc(\gamma \to \neg(\alpha \overline{U}_n \beta))$, for $n \in \mathbb{N}$, by the induction hypothesis,
$\bigcirc T \vdash \bigcirc\gamma \to \neg\bigcirc(\alpha \overline{U}_n \beta)))$, $n \in \mathbb{N}$, by A2 and A3,
$\bigcirc T \vdash \bigcirc\gamma \to \neg(\bigcirc\alpha \overline{U}_n \bigcirc\beta)$, $n \in \mathbb{N}$, by Lemma 3.1(2)
(since $\alpha \overline{U}_n \beta := (\bigwedge_{k=0}^{n-1} \bigcirc^k \alpha) \wedge \bigcirc^n \beta$),
$\bigcirc T \vdash \bigcirc\gamma \to \neg(\bigcirc\alpha \mathrm{U} \bigcirc\beta)$, by R3,
$\bigcirc T \vdash \bigcirc\gamma \to \neg\bigcirc(\alpha \mathrm{U} \beta)$, by Lemma 3.1(2),
$\bigcirc T \vdash \bigcirc(\gamma \to \neg(\alpha \mathrm{U} \beta))$, by A3 and Lemma 3.1(1). ■

Theorem 3.2 (Lindenbaum's theorem). *Every consistent set T of LTL formulas can be extended to a maximal consistent set T^*.*

Proof. Let us assume an enumeration $\{\alpha_i \mid i \in \mathbb{N}\}$ of all LTL formulas. The maximal consistent set T^* is then defined recursively, as follows:

1. $T_0 = T$.
2. If α_i is consistent with T_i, then $T_{i+1} = T_i \cup \{\alpha_i\}$.
3. If α_i is not consistent with T_i, then:

   a. If α_i is of the form $\gamma \to \neg(\alpha \mathrm{U} \beta)$, then

   $$T_{i+1} = T_i \cup \{\gamma \to (\alpha \overline{U}_{n_0} \beta)\},$$

   where n_0 is a positive integer such that T_{i+1} is consistent.

   b. Otherwise, $T_{i+1} = T_i$.

4. $T^* = \bigcup_{n \in \mathbb{N}} T_n$.

Let us prove the existence of the number n_0 in 3(a). If we suppose that $\gamma \to (\alpha \overline{U}_n \beta)$ is not consistent with T_i, for every $n \in \mathbb{N}$, then, by Theorem 3.1, $T_i \vdash \neg(\gamma \to (\alpha \overline{U}_n \beta)$, for every $n \in \mathbb{N}$. By A1 we obtain $T_i \vdash \gamma \to \neg(\alpha \overline{U}_n \beta)$, for every $n \in \mathbb{N}$. By R3 we have $T_i \vdash \gamma \to \neg(\alpha \mathrm{U} \beta)$, which contradicts the assumption.

It is easy to show that T_i is consistent for every i, and that for each $\alpha \in For$, either $\alpha \in T^*$ or $\neg\alpha \in T^*$.

Let us prove that T^* is deductively closed. It is sufficient to prove that it is closed under the inference rules, since all instances of axioms are obviously in T^*. We will only consider the case for the inference rule R3, while the other cases follows straightforwardly.

Suppose that $\gamma \to \neg(\alpha \mathrm{U} \beta) \notin T^*$, while $\gamma \to \neg(\alpha \overline{U}_n \beta) \in T^*$ for every $n \in \mathbb{N}$. By maximality of T^*, $\neg(\gamma \to \neg(\alpha \mathrm{U} \beta)) \in T^*$, or, equivalently, $\gamma \wedge (\alpha \mathrm{U} \beta) \in T^*$. Consequently, $\gamma \in T^*$ and $\alpha \mathrm{U} \beta \in T^*$, so there are $m, n \in \mathbb{N}$ such that $\gamma \in T_m$ and $\alpha \mathrm{U} \beta \in T_n$. If $\gamma \to \neg(\alpha \mathrm{U} \beta) = \alpha_l$, then, by the construction of T^*, there is n_0 such that $\gamma \to (\alpha \overline{U}_{n_0} \beta) \in T_l$. By Lemma 3.1(4), $T_l \vdash \alpha \mathrm{U} \beta$. Consequently, $\alpha \mathrm{U} \beta \in T_{\max\{l,m,n\}}$, which is in contradiction with consistency of $T_{\max\{l,m,n\}}$.

Note that deductive closeness of T^* implies its consistency. Indeed, $T^* \vdash \bot$ implies $\bot \in T^*$, thus there exists i such that $\bot \in T_i$, which is impossible. ■

For any maximal consistent set, we define the corresponding path as follows.

Definition 3.3. Let T^* be a maximal consistent set. Then the path $\sigma(T^*) = s_0, s_1, \ldots$ is defined as follows:

- $s_0 = T^*$,
- $s_{i+1} = \{\alpha \mid \bigcirc \alpha \in s_i\}$, for every $i \in \mathbb{N}$,
- for every propositional letter p, and every s_i, $v(s_i, p) = 1$ iff $p \in s_i$. ■

Lemma 3.2. *Let T^* be a maximal consistent set and $\sigma(T^*) = s_0, s_1, \ldots$ the corresponding path. Then the following hold:*

1. s_i is a maximal consistent set for every $i \in \mathbb{N}$.

2. $\sigma(T^*)_k = \{\alpha \mid \bigcirc^k \alpha \in T^*\} = \sigma(s_k)$.
3. $\sigma(T^*) \models \gamma$ *iff* $T^* \vdash \gamma$, *for every formula* γ.

Proof. (1) By definition, s_0 is a maximal consistent set. We use induction on i to prove that for every $i \in \mathbb{N}$ and every α, either $\alpha \in s_i$ or $\neg\alpha \in s_i$. Suppose that there is a formula α such that $\{\alpha, \neg\alpha\} \cap s_{i+1} = \emptyset$. Consequently, $\{\bigcirc\alpha, \bigcirc\neg\alpha\} \cap s_i = \emptyset$. We obtain that $\{\bigcirc\alpha, \neg\bigcirc\alpha\} \cap s_i = \emptyset$ which contradicts the induction hypothesis. Next we use induction on i to prove that each s_i is consistent. Suppose that s_{i+1} is not consistent. Then $s_{i+1} \vdash \alpha \wedge \neg\alpha$, for any formula α. Using Lemma 3.1(3) we obtain $\bigcirc s_{i+1} \vdash \bigcirc(\alpha \wedge \neg\alpha)$. Consequently, $s_i \vdash \bigcirc(\alpha \wedge \neg\alpha)$, so, by A3 and Lemma 3.1(2), $s_i \vdash \bigcirc\alpha \wedge \neg\bigcirc\alpha$, which is in contradiction with consistency of s_i.
(2) This is a direct consequence of Definition 3.3.
(3) The proof is by induction on the complexity of γ. For example, let γ be a formula of the form $\alpha U \beta$. Then

$\sigma(T^*) \not\models \gamma$ iff $\sigma(T^*) \models \neg(\alpha U \beta))$
iff for all $n \in \mathbb{N}$, it is not the case that $\sigma(T^*)_{\geq n} \models \beta$ and for all $k < n$, $\sigma(T^*)_{\geq k} \models \alpha$
iff for all $n \in \mathbb{N}$, it is not the case that $\models \sigma(s_n), \beta$ and for all $k < n$, $\sigma(s_k) \models \alpha$, by Lemma 3.2(2)
iff for all $n \in \mathbb{N}$, it is not the case that $s_n \vdash \beta$ and for all $k < n$, $s_k \vdash \alpha$ (by induction hypothesis)
iff for all $n \in \mathbb{N}$, it is not the case that $T^* \vdash \bigcirc^n \beta$ and for all $k < n$, $T^* \vdash \bigcirc^k \alpha$
iff for all $n \in \mathbb{N}$, $T^* \vdash \neg(\alpha \overline{U}_n \beta)$ (by the maximal consistency of T^*)
iff $T^* \vdash \neg(\alpha U \beta)$ (by R3). ■

Theorem 3.3 (Strong completeness theorem). *Every consistent set of LTL formulas is satisfiable.*

Proof. Let T be a consistent set of formulas. By Theorem 3.2, T can be extended to a maximal consistent set T^*. Let us consider the corresponding path $\sigma(T^*)$. For every $\alpha \in T$ we have that $\alpha \in T^*$, so, by Lemma 3.2(3), $\sigma(T^*) \models \alpha$. Therefore, $\sigma(T^*) \models T$. ■

Using the above theorem and soundness of $Ax(LTL)$, one can easily obtain the usual formulation of soundness and completeness, stated in the following result.

Theorem 3.4. *If* α *is a formula and* T *is a set of formulas, then*

$$T \vdash \alpha \text{ iff } T \models \alpha.$$

3.2.2 Branching time logic

The language of CTL* is obtained by adding the universal path operator A to the language of LTL. The corresponding language For_{CTL^*} is defined as the smallest set which contain given set *Var* of propositional letters, such that if $\alpha, \beta \in For_{CTL^*}$, then $\alpha \wedge \beta, \neg\alpha, \bigcirc\alpha, \alpha U \beta, A\alpha \in For_{CTL^*}$. This language is used to describe several different types of models in which one state might have more than one successor state, leading to branching structure of time. We read the formula $A\alpha$ in the following way: "α holds on every path which passes through the current state".

The dual operator E (existential path switching operator) is introduced as an abbreviation: $E\alpha$ is $\neg A \neg \alpha$. Similarly as in the case of next operator, for any set of formulas T we denote by AT the set $\{A\alpha \mid \alpha \in T\}$. As we mentioned, there are several classes of models for the language of CTL*. In this section, we focus on a semantics with minimum of requirements

Definition 3.4 (CTL* model). A model $\mathscr{M}$ is any tuple $\langle S, v, R, \Sigma \rangle$ such that:

- S is a nonempty set of states (time instants),
- $v : S \times Var \longrightarrow \{0,1\}$ assigns a truth labeling to every state.
- R is a binary relation on S, which is total (for every $s \in S$ there is $t \in S$ such that sRt),
- Σ is a set of paths $\sigma = s_0, s_1, s_2, \ldots$, where each path is an ω−sequence of states from S, such that $s_i R s_{i+1}$, for all $i \in \mathbb{N}$. We assume that Σ is suffix-closed, i.e., if $\sigma = s_0, s_1, s_2, \ldots$ is a path and $i \in \mathbb{N}$, then the sequence $s_i, s_{i+1}, s_{i+2}, \ldots$ is also a path. ■

We evaluate the formulas of CTL* on a path of a model, i.e., we define $\mathscr{M}, \sigma \models \alpha$ by extending the definition for LTL with

- $\mathscr{M}, \sigma \models A\alpha$ iff for every path π, if $\sigma_0 = \pi_0$ then $\mathscr{M}, \pi \models \alpha$.

A formula is a *state formula* if it is a Boolean combination of propositional letters and formulas of the form $A\alpha$. Note that evaluation of a state formula α does not depend on a path, but only on its initial state, i.e., if $\sigma_0 = \sigma'_0$, then $\mathscr{M}, \sigma \models \alpha$ iff $\mathscr{M}, \sigma' \models \alpha$.

Next we extend the axiomatization for LTL by adding the standard axioms for the operator A ([42]):

A5. $p \rightarrow Ap,\ p \in Var$
A6. $Ep \rightarrow p,\ p \in Var$
A7. $A\alpha \rightarrow \alpha$
A8. $A(\alpha \rightarrow \beta) \rightarrow (A\alpha \rightarrow A\beta)$
A9. $A\alpha \rightarrow AA\alpha$
A10. $E\alpha \rightarrow AE\alpha$

and the necessitation rule for the operator A:

R4. from α infer $A\alpha$.

Similarly as in the case of R2, applications of the rule R4 are restricted to theorems only. Due to the presence of an infinitary rule, the proof of Strong necessitation theorem (if $T \vdash \alpha$ then $AT \vdash A\alpha$) is non-trivial and requires a modification of the rule R3 using the formulas in the form of k-nested implications.[1]

Definition 3.5 (k-nested implication). A k-nested implication $\Phi_k(\tau, \overline{\gamma}, \overline{X})$ for the formula τ, based on the sequence $\overline{\gamma} = (\gamma_0, ..., \gamma_k)$ of formulas and the sequence $\overline{X} = (X_1, ..., X_k)$ of operators, where every $X_i \in \{\bigcirc, A\}$, is defined recursively as follows:

$$\Phi_0(\tau, \overline{\gamma}) = \gamma_0 \to \tau$$

$$\Phi_k(\tau, \gamma) = \gamma_k \to X_k(\Phi_{k-1}(\tau, (\gamma_0, ..., \gamma_{k-1})).$$

■

Example 3.1. For example, $\Phi_2(\alpha, (\gamma_0, \gamma_1, \gamma_2), (\bigcirc, A))$ is the formula

$$\gamma_2 \to A(\gamma_1 \to \bigcirc(\gamma_0 \to \alpha)),$$

while $\Phi_3(\alpha, (\gamma_0, \gamma_1, \gamma_2, \gamma_3), (\bigcirc, A, A))$ is the formula

$$\gamma_3 \to A(\gamma_2 \to A(\gamma_1 \to \bigcirc(\gamma_0 \to \alpha))).$$

■

Now we propose an axiomatization, that we denote simply by $Ax(CTL^*)$, for the class of models introduced in Definition 3.4. It contains the axioms A1–A10, the inference rules R1, R2 and R4, and the following modification of the infinitary rule R3:

R3'. From the set of premises

$$\{\Phi_k(\neg(\alpha \overline{U}_n \beta), \overline{\gamma}, \overline{X}) \mid n \in \mathbb{N}\}$$

infer $\Phi_k(\neg(\alpha U \beta), \overline{\gamma}, \overline{X})$.

This axiom system is sound and strongly complete.

Theorem 3.5. *If α is a formula and T is a set of CTL* formulas, then*

$$T \vdash \alpha \text{ iff } T \models \alpha.$$

■

We omit the proof of this theorem, since it follows from completeness of a probabilistic extension of CTL*, presented in the following section.

[1] The k-nested implications are also used in the infinitary axiomatizations of epistemic logics and epistemic probabilistic logic [5, 31, 43, 44].

3.3 Probabilistic branching time logic pBTL

This section is devoted to a probabilistic extension of CTL*. We extend the language of CTL* with two types of probability operators, $P^p_{\geqslant r}$ and $P^s_{\geqslant r}$, the former ranging over future states on a given path, and the latter ranging over paths starting from a given state.

3.3.1 Syntax and semantics

For given set of propositional letters *Var*, the set of formulas is defined as follows.

Definition 3.6 (pBTL formula). The set of pBTL formulas For_{pBTL} is the smallest set which contains *Var*, such that if $\alpha, \beta \in For_{\text{pBTL}}$, then

- $\alpha \wedge \beta, \neg\alpha, \bigcirc\alpha, A\alpha, \alpha U \beta \in For_{\text{pBTL}}$,
- for every $r \in \mathbb{Q} \cap [0,1]$, $P^p_{\geqslant r}\alpha, P^s_{\geqslant r}\alpha \in For_{\text{pBTL}}$. ■

Intuitively, the operators mean:

- $P^p_{\geqslant r}\alpha$: probability that α holds at the randomly chosen time instant on a particular path is at least r, and
- $P^s_{\geqslant r}\alpha$: probability that α holds at the randomly chosen path (with a particular initial state) is at least r.

Similarly as in the previous chapters, we introduce the other types of probability operators as abbreviations:

- $P^p_{<r}\alpha$ is $\neg P^p_{\geqslant r}\alpha$, $P^p_{\leqslant r}\alpha$ is $P^p_{\geqslant 1-r}\neg\alpha$, $P^p_{>r}\alpha$ is $\neg P^p_{\leqslant r}\alpha$ and $P^p_{=r}\alpha$ is $P^p_{\geqslant r}\alpha \wedge P^p_{\leqslant r}\alpha$,
- $P^s_{<r}\alpha$, $P^s_{\leqslant r}\alpha$, $P^s_{>r}\alpha$ and $P^s_{=r}\alpha$ are defined in a similar way.

An example of a formula is

$$EG\alpha \rightarrow P^s_{\geqslant\frac{1}{2}} P^p_{\geqslant\frac{1}{3}} \alpha,$$

which can be read as: "if there exists a path on which the formula α always holds, then on at least a half of paths α holds in at least $\frac{1}{3}$ of time instants".

Example 3.2. In order to illustrate the different nature of the two types of probability operators, let us code the following two statements in the language of pBTL, assuming that a time unit is a day:

i. "In this region, there are more than 10 hours of daylight at least 80% of the year."
ii. "Whenever you send a shipment to Utrecht, the chance that our company will deliver it within five days is more than 98 %."

If *daylight* denotes "more than 10 hours of daylight", then the first statement can be formalized by

$$P^p_{\geqslant 0.8} daylight.$$

On the other hand, the second statement can be coded by the following formula:

$$G(sent \rightarrow P^s_{\geqslant 0.98}(\bigvee_{n=0}^{5} \bigcirc^n delivered)).$$

■

Definition 3.7. A pBTL model $\mathscr{M}$ is any tuple $\langle S, v, R, \Sigma, Prob^{state}, Prob^{path} \rangle$ such that:

- the set of states S, truth labelling v, accessibility relation R and the set of paths Σ are defined in the same way as in Definition 3.4,
- $Prob^{state}$ associates to every $s \in S$, a probability space $Prob_s = \langle H_s, \mu_s \rangle$ such that:
  - H_s is an algebra of subsets of $\Sigma_s = \{\sigma \in \Sigma \mid \sigma_0 = s\}$, i.e., it contains Σ_s and it is closed under complements and finite union,
  - $\mu_s : H_s \longrightarrow [0,1]$ is a finitely additive probability measure, i.e.,
    - $\mu_s(H_s) = 1$, and
    - $\mu_s(X \cup Y) = \mu_s(X) + \mu_s(Y)$, whenever X and Y are disjoint.
- $Prob^{path}$ associates to every $\sigma \in \Sigma$ a probability space $Prob_\sigma = \langle H_\sigma, \mu_\sigma \rangle$ such that:
  - H_σ is an algebra of subsets of $S_\sigma = \{\sigma_{\geq i} \mid i \in \mathbb{N}\}$,
  - $\mu_\sigma : H_\sigma \longrightarrow [0,1]$ is a finitely additive probability measure. ■

Definition 3.8. Let $\mathscr{M} = \langle S, v, R, \Sigma, Prob^{state}, Prob^{path} \rangle$ be any model. The satisfiability of a formula α on a path σ in a model $\mathscr{M}$, denoted by $\mathscr{M}, \sigma \models \alpha$, is defined recursively as follows:

- if $p \in Var$, then $\mathscr{M}, \sigma \models p$ iff $v(s_0, p) = 1$,
- $\mathscr{M}, \sigma \models \neg\alpha$ iff $\mathscr{M}, \sigma \not\models \alpha$,
- $\mathscr{M}, \sigma \models \alpha \wedge \beta$ iff $\mathscr{M}, \sigma \models \alpha$ and $\mathscr{M}, \sigma \models \beta$,
- $\mathscr{M}, \sigma \models \bigcirc\alpha$ iff $\mathscr{M}, \sigma_{\geq 1} \models \alpha$,
- $\mathscr{M}, \sigma \models A\alpha$ iff for every path π, if $\sigma_0 = \pi_0$ then $\mathscr{M}, \pi \models \alpha$.
- $\mathscr{M}, \sigma \models \alpha U \beta$ iff there is some $i \in \mathbb{N}$ such that $\mathscr{M}, \sigma_{\geq i} \models \beta$ and for each $j \in \mathbb{N}$, if $0 \leq j < i$ then $\mathscr{M}, \sigma_{\geq j} \models \alpha$,
- $\mathscr{M}, \sigma \models P^s_{\geqslant r}\alpha$ iff $\mu_{\sigma_0}(\{\pi \in \Sigma_{\sigma_0} \mid \mathscr{M}, \pi \models \alpha\}) \geqslant r$,
- $\mathscr{M}, \sigma \models P^p_{\geqslant r}\alpha$ iff $\mu_\sigma(\{\pi \in S_\sigma \mid \mathscr{M}, \pi \models \alpha\}) \geqslant r$. ■

Note that the satisfiability of the formulas of the form $P^s_{\geqslant r}\alpha$ depends only on the initial state of the path. For that reason, we need to extend the definition of state formulas.

Definition 3.9 (State formula). A formula is a *state formula* if it is a Boolean combination of propositional letters, formulas of the form $P^s_{\geqslant r}\alpha$ and formulas of the form $A\alpha$. We denote the set of all state formulas by St. ■

Similarly as in the previous chapters, we will restrict our attention to the class of measurable models. For a model $\mathscr{M} = \langle S, v, R, \Sigma, Prob^{state}, Prob^{path}\rangle$ and a path $\sigma \in \Sigma$, we use the following notation:

- $[\alpha]^{path}_{\mathscr{M},\sigma} = \{\pi \in S_\sigma \mid \mathscr{M}, \pi \models \alpha\}$, and
- $[\alpha]^{state}_{\mathscr{M},s} = \{\pi \in \Sigma_s \mid \mathscr{M}, \pi \models \alpha\}$.

Definition 3.10 (Measurable pBTL-model). A model $\mathscr{M} = \langle S, v, R, \Sigma, Prob^{state}, Prob^{path}\rangle$ is *measurable* if the following conditions are satisfied:

- $[\alpha]^{path}_{\mathscr{M},\sigma} \in A_\sigma$, for every $\alpha \in For$,
- $[\alpha]^{state}_{\mathscr{M},s} \in H_s$, for every $\alpha \in For$. ■

We will denote the probabilistic branching-time temporal logic characterized by the class of all measurable models by pBTL_{Meas}.

3.3.2 Axiomatization

Now we present the axiom system for the logic pBTL, denoted by $Ax(\text{pBTL})$. It includes the axiomatization $Ax(CTL^*)$, presented in the previous section, and probabilistic axioms for the logic LPP_1 (see Chapter 1). Since we have two types of probability operators, we have two variants of each probabilistic axiom. In addition, we have four axioms that capture the interplay between probabilistic and temporal modalities. The Archimedean rule has a modified form, that uses the k-nested implications.

The axiomatization $Ax(\text{pBTL})$:

Temporal axioms A1–A10.

Probabilistic axioms ($x \in \{p, s\}$)

A11. $P^x_{\geq 0}\alpha$
A12. $P^x_{\leq s}\alpha \rightarrow P^x_{<t}\alpha,\ t > s$
A13. $P^x_{<s}\alpha \rightarrow P^x_{\leq s}\alpha$
A14. $(P^x_{\geq s}\alpha \wedge P^x_{\geq r}\beta \wedge P^x_{\geq 1}(\neg\alpha \vee \neg\beta)) \rightarrow P^x_{\geq \min(1,s+r)}(\alpha \vee \beta)$
A15. $(P^x_{\leq s}\alpha \wedge P^x_{<r}\beta) \rightarrow P^x_{<s+r}(\alpha \vee \beta),\ s + r \leq 1$

Axioms about probability and time

A16. $G\alpha \rightarrow P^p_{\geq 1}\alpha$
A17. $A\alpha \rightarrow P^s_{\geq 1}\alpha$
A18. $P^s_{\geq r}\alpha \rightarrow AP^s_{\geq r}\alpha$
A19. $EP^s_{\geq r}\alpha \rightarrow P^s_{\geq r}\alpha$

Inference rules

The rules R1, R2, R3' and R4 (the inference rules of $Ax(CTL^*)$)

R5. From the set of premises

$$\{\Phi_k(P^x_{\geq r-\frac{1}{k}}\alpha,\overline{\gamma},\overline{X}) \mid k\in\mathbb{N}, k\geq \frac{1}{r}\}$$

infer $\Phi_k(P^x_{\geq r}\alpha,\overline{\gamma},\overline{X})$.

By the axioms A1–A10 and the inference rules R1, pBTL extends the logic CTL*.

3.3.3 Completeness

Following the proof strategy for the completeness of $Ax(LTL)$, we start with the Deduction theorem and strong necessitation, and then we proceed with Lindenbaum's theorem.

Theorem 3.6 (Deduction theorem). *If T is a set of* pBTL *formulas, α is a formula, and $T,\alpha\vdash\beta$, then $T\vdash\alpha\rightarrow\beta$.*

Proof. We use the transfinite induction on the length of the inference. Let us consider the case when β is obtained by the rule R5, i.e $\beta=\Phi_k(P^x_{\geq r}\alpha,\overline{\gamma},\overline{X})$, where $x\in\{p,s\}$, $\overline{\gamma}=(\gamma_0,\dots,\gamma_k)$ and $\overline{X}=(X_1,\dots,X_k)$, $X_i\in\{\bigcirc,A\}$. Then we have

$T,\alpha\vdash\Phi_k(P^x_{\geq r-\frac{1}{k}}\alpha,\overline{\gamma},\overline{X})$, for all $k\in\mathbb{N},k\geq\frac{1}{r}$
$T,\alpha\vdash\gamma_k\rightarrow X_k\Phi_{k-1}((P^x_{\geq r-\frac{1}{k}}\alpha,(\gamma_0,\dots,\gamma_{k-1}),(X_1,\dots,X_{k-1})))$, for all $k\in\mathbb{N},k\geq\frac{1}{r}$
$T\vdash(\alpha\wedge\gamma_k)\rightarrow X_k\Phi_{k-1}((P^x_{\geq r-\frac{1}{k}}\alpha,(\gamma_0,\dots,\gamma_{k-1}),(X_1,\dots,X_{k-1})))$, for all $k\in\mathbb{N},k\geq\frac{1}{r}$ (by (A1))
$T\vdash\Phi_k(P^x_{\geq r-\frac{1}{k}}\alpha,(\gamma_0,\dots,\gamma_{k-1},\alpha\wedge\gamma_k),\overline{X})$, for all $k\in\mathbb{N},k\geq\frac{1}{r}$
$T\vdash\Phi_k(P^x_{\geq r}\alpha,(\gamma_0,\dots,\gamma_{k-1},\alpha\wedge\gamma_k),\overline{X})$ (by R5)
$T\vdash(\alpha\wedge\gamma_k)\rightarrow X_k\Phi_{k-1}((P^x_{\geq r}\alpha,(\gamma_0,\dots,\gamma_{k-1}),(X_1,\dots,X_{k-1})))$
$T\vdash\alpha\rightarrow(\gamma_k\rightarrow X_k\Phi_{k-1}((P^x_{\geq r}\alpha,(\gamma_0,\dots,\gamma_{k-1}),(X_1,\dots,X_{k-1}))))$
$T\vdash\alpha\rightarrow\Phi_k(P^x_{\geq r}\alpha,\overline{\gamma},\overline{X})$
$T\vdash\alpha\rightarrow\beta$.

In the case when we apply the infinitary rule R3', the proof is similar as the one for R3, presented in the proof of Theorem 3.1. ■

Theorem 3.7 (Strong necessitation). *For any formula α and a set of formulas T such that $T\vdash\alpha$, the following hold.*

1. $\bigcirc T\vdash\bigcirc\alpha$.
2. $AT\vdash A\alpha$.

Proof. (1) We use the induction on the depth of the derivation of α from T.[2] The case when we apply R1 is an easy consequence of A2, while the cases when we apply R2 and R4 are trivial, since those two rules can be applied to theorems only. Suppose that, for some $x \in \{p,s\}$, $\overline{\gamma} = (\gamma_0, \ldots, \gamma_k)$ and $\overline{X} = (X_1, \ldots, X_k)$, $X_i \in \{\bigcirc, A\}$. , $T \vdash \Phi_k(P^x_{\geq r}\alpha, \overline{\gamma}, \overline{X}))$ iz obtained from $T \vdash \Phi_k(P^x_{\geq r - \frac{1}{k}}\alpha, \overline{\gamma}, \overline{X})$, for all $k \in \mathbb{N}$ such that $k \geq \frac{1}{r}$, by the inference rule R5. Then we have

$$\bigcirc T \vdash \bigcirc \Phi_k(P^x_{\geq r - \frac{1}{k}}\alpha, \overline{\gamma}, \overline{X})) \text{ for all } k \in \mathbb{N}, k \geq \tfrac{1}{r} \text{ (by the induction hypothesis),}$$
$$\bigcirc T \vdash \top \to \bigcirc(\Phi_k(P^x_{\geq r - \frac{1}{k}}\alpha, \overline{\gamma}, \overline{X})) \text{ for all } k \in \mathbb{N}, k \geq \tfrac{1}{r}$$
$$\bigcirc T \vdash \Phi_{k+1}(P^x_{\geq r - \frac{1}{k}}\alpha, (\gamma_0, \ldots, \gamma_k, \top), (X_1, \ldots, X_k, \bigcirc)) \text{ for all } k \in \mathbb{N}, k \geq \tfrac{1}{r}$$
$$\bigcirc T \vdash \Phi_{k+1}(P^x_{\geq r}\alpha, (\gamma_0, \ldots, \gamma_k, \top), (X_1, \ldots, X_k, \bigcirc)) \text{ (by R5)}$$
$$\bigcirc T \vdash \top \to \bigcirc(\Phi_k(P^x_{\geq r}\alpha, (\gamma_0, \ldots, \gamma_k), (X_1, \ldots, X_k)))$$
$$\bigcirc T \vdash \bigcirc(\Phi_k(P^x_{\geq r}\alpha, \overline{\gamma}, \overline{X}))$$

The case when we consider infinitary rule R3' is similar.

(2) The prof is similar to the proof of (1). ■

Theorem 3.8 (Lindenbaum's theorem). *Every consistent set T of CTL* formulas can be extended to a maximal consistent set T^*.*

Proof. We follow the proof of Theorem 3.2. We assume an enumeration $\{\alpha_i \mid i \in \mathbb{N}\}$ of all pBTLformulas, and define the maximal consistent set T^* in the following way:

1. $T_0 = T$.
2. If α_i is consistent with T_i, then $T_{i+1} = T_i \cup \{\alpha_i\}$.
3. If α_i is not consistent with T_i, then:
   a. If α_i is of the form $\Phi_k(\neg(\alpha U \beta), \overline{\gamma}, \overline{X})$, then
   $$T_{i+1} = T_i \cup \{\neg\Phi_k(\neg(\alpha \overline{U}_{n_0} \beta), \overline{\gamma}, \overline{X})\},$$
   where n_0 is a positive integer such that T_{i+1} is consistent.
   b. If α_i is of the form $\Phi_k(P^x_{\geq r}\alpha, \overline{\gamma}, \overline{X})$, then
   $$T_{i+1} = T_i \cup \{\neg\Phi_k(P^x_{\geq r - \frac{1}{k_0}}\alpha, \overline{\gamma}, \overline{X})\},$$
   where k_0 is a positive integer such that $k_0 \geq \frac{1}{r}$ and T_{i+1} is consistent.
   c. Otherwise, $T_{i+1} = T_i$.
4. $T^* = \bigcup_{n \in \mathbb{N}} T_n$.

The proof that T^* is a maximal consistent set is based on the proof of Theorem 3.2. The only difference is in the proof that T^* is deductively closed, since we consider

[2] Note that the proof of this statement is essentially different from the proof of Lemma 3.1(3), due to different forms of the infinitary inference rules.

different inference rules. As an illustration, we will prove that T^* is closed under the inference rule R5.

Suppose $T^* \vdash \Phi_k(P^x_{\geq r}\alpha, \overline{\gamma}, \overline{X})$ is obtained by R5. Then $\Phi_k(P^x_{\geq r-\frac{1}{k}}\alpha, \overline{\gamma}, \overline{X}) \in T^*$ for all $k \in \mathbb{N}$ such that $k \geq \frac{1}{r}$. Assume that $\Phi_k(P^x_{\geq r}\alpha, \overline{\gamma}, \overline{X}) \notin T^*$. Let i be the positive integer such that $\alpha_i = \Phi_k(P^x_{\geq r}\alpha, \overline{\gamma}, \overline{X})$. Then $T_i \cup \{\alpha_i\}$ is inconsistent, since otherwise $\alpha_i = \Phi_k(P^x_{\geq r}\alpha, \overline{\gamma}, \overline{X}) \in T_{i+1} \subset T^*$. Therefore $T_{i+1} = T_i \cup \{\neg\Phi_k(P^x_{\geq r-\frac{1}{k_0}}\alpha, \overline{\gamma}, \overline{X})\}$ for some k_0, by the construction of T^*. Let j be the positive integer such that $\alpha_j = \Phi_k(P^x_{\geq r-\frac{1}{k_0}}\alpha, \overline{\gamma}, \overline{X})$. Then $T_{\max\{i,j\}+1}$ contains both $\Phi_k(P^x_{\geq r-\frac{1}{k_0}}\alpha, \overline{\gamma}, \overline{X})$ and $\neg\Phi_k(P^x_{\geq r-\frac{1}{k_0}}\alpha, \overline{\gamma}, \overline{X})$, which is in contradiction with consistency of $T_{\max\{i,j\}+1}$. ■

Next we define the equivalence relation $\sim$ on the set of maximal consistent sets of formulas as follows (recall that St is the set of state formulas):

$$T_1^* \sim T_2^* \quad \text{iff} \quad T_1^* \cap St = T_2^* \cap St.$$

The equivalence class of T^* is $[T^*] = \{T_1^* \mid T_1^* \sim T^*\}$.

Lemma 3.3. *If T^* is a maximal consistent set of formulas such that $A\alpha \notin T^*$, then there exists $T_1^* \in [T^*]$ such that $\alpha \notin T_1^*$.*

Proof. Let $T = T^* \cap St$. If the set $T \cup \{\neg\alpha\}$ is consistent, then it can be extended to a maximal consistent set T_1^*, by Theorem 3.8. Obviously $\alpha \notin T_1^*$. Moreover, from $T_1^* \cap St = T$ we obtain $T_1^* \in [T^*]$. On the other hand, if $T \cup \{\neg\alpha\}$ is inconsistent, then $T \vdash \alpha$, by Theorem 3.6. Using Theorem 3.7(2), we obtain $AT \vdash A\alpha$. Since $T \subseteq St$, by temporal reasoning and A18 and A19, we obtain $T \vdash A\alpha$. Since T^* is a maximal consistent set which contains T, we have $A\alpha \in T^*$, a contradiction. ■

Theorem 3.9 (Strong completeness theorem). *Every consistent set of* pBTL *formulas is satisfiable.*

Proof. First we construct the canonical model $\mathscr{M}^* = \langle S, v, R, \Sigma, Prob^{state}, Prob^{path}\rangle$ in the following way:

- $S = \{[T^*] \mid T^* \text{ is maximal consistent set of formulas}\}$,
- $v([T^*], p) = 1$ iff $T^* \vdash p$, $p \in Var$,
- $[T_1^*]R[T_2^*]$ if there exist $T_3^* \sim T_1^*$, $T_4^* \sim T_2^*$ such that $T_4^* = \{\alpha \mid \bigcirc\alpha \in T_3^*\}$,
- Σ is the set of all paths $[T_0^*]$, $[T_1^*]$, $[T_2^*]$,... such that $T_{i+1}^* = \{\alpha \mid \bigcirc\alpha \in T_i^*\}$, for all $i \in \mathbb{N}$. If the sequence $\{T_i^*\}_{i\in\mathbb{N}}$ determines a path σ, we write $\sigma(i)$ for T_i^*,
- $Prob^{path}$ is defined as follows: for every $\sigma = [T_0^*]$, $[T_1^*]$, $[T_2^*]$,..., $Prob_\sigma = \langle H_\sigma, \mu_\sigma\rangle$ is a probability space such that:
  - $A_\sigma = \{[\alpha]_\sigma \mid \alpha \in For\}$, where $[\alpha]_\sigma = \{\sigma_{\geqslant} i \mid T_i^* \vdash \alpha, i \in \mathbb{N}\}$,
  - $\mu_\sigma([\alpha]_\sigma) = \sup\{r \in \mathbb{Q} \cap [0,1] \mid T_0^* \vdash P^p_{\geqslant r}\alpha\}$,
- $Prob^{state}$ is defined as follows: for every $\sigma = [T_0^*]$, $[T_1^*]$, $[T_2^*]$,..., the probability space $Prob_\sigma = \langle H_\sigma, \mu_\sigma\rangle$ is determined by the following conditions: that:
  - $H_s = \{[\alpha]_s \mid \alpha \in For\}$, where $[\alpha]_s = \{\pi \mid \pi(0) \sim T_0^*, \pi(0) \vdash \alpha\}$,

– $\mu_s([\alpha]_s) = \sup\{r \in \mathbb{Q} \cap [0,1] \mid T_0^* \vdash P^s_{\geqslant r}\alpha\}$.

It is easy to verify that $\mathscr{M}^*$ is well defined pBTL-model:

- v is well defined, since $Var \subseteq St$, so $T_1^* \vdash p$ iff $T_2^* \vdash p$, whenever $T_1^* \sim T_2^*$, $p \in Var$.
- The definition of R is correct – using Theorem 3.7(2), one can show, similarly as in the proof of Theorem 3.2(1), that $\{\alpha \mid \bigcirc\alpha \in T^*\}$ is a maximal consistent set, whenever T^* is a maximal consistent set. Moreover, R is obviously a total relation.
- H_σ is an algebra of sets. It is easy to show that $S_\sigma = [\top]_\sigma$, $[\alpha]^c_\sigma = [\neg\alpha]_\sigma$ and $[\alpha]_\sigma \cup [\beta]_\sigma = [\alpha \vee \beta]_\sigma$. Similarly, H_s is an algebra of sets.
- The function μ_s is well defined, since any formula of the form $P^s_{\geqslant r}\alpha$ is a state formula, so it belongs to a maximal consistent set T_1^* if and only if it belongs to any other maximal consistent set $T_2^* \in [T_1^*]$. Consequently, $\sup\{r \in \mathbb{Q} \cap [0,1] \mid T_1^* \vdash P^s_{\geqslant r}\alpha\} = \sup\{r \in \mathbb{Q} \cap [0,1] \mid T_2^* \vdash P^s_{\geqslant r}\alpha\}$.
- μ_s and μ_σ are finitely additive probability measures- the proof is essentially the same as the corresponding proof presented in Chapter 1.

Finally, we need to prove that for every $\alpha \in For$, $\mathscr{M}^*, \sigma \models \alpha$ iff $\alpha \in \sigma(0)$. The proof is on the complexity of α, the cases when α is a propositional letter or of the form $\beta \wedge \gamma$, $\neg\beta$ or $\bigcirc\beta$ are trivial. The case when α is of the form $\beta U \gamma$ follows the ideas from the proof of Theorem 3.2(3). The cases when α is of the form $P^s_{\geqslant r}\beta$ or $P^p_{\geqslant r}\beta$ are the same as the corresponding proofs presented in Chapter 1.

Finally, let $\alpha = A\beta$. If $\mathscr{M}^*, \sigma \not\models A\beta$, then there exists $\pi \in \Sigma_{\sigma_0}$ such that $\mathscr{M}^*, \pi \models \neg\beta$. By the induction hypothesis we obtain $\neg\beta \in \pi(0)$, so $\beta \notin \pi(0)$. By Axiom A7, $A\beta \notin \pi(0)$. From $\pi(0) \sim \sigma(0)$ and $A\beta \in St$, we conclude $A\beta \notin \sigma(0)$. For the other direction, suppose that $\mathscr{M}^*, \sigma \models A\beta$. Then for all $\pi \in \Sigma_{\sigma_0}$, $\mathscr{M}^*, \sigma \models \beta$. Consequently, by the induction hypothesis, for all $\pi \in \Sigma_{\sigma_0}$, $\beta \in \pi(0)$. If $A\beta \notin \sigma(0)$, by Lemma 3.3 there exists $\rho \in \Sigma_{\sigma_0}$ such that $\beta \notin \rho(0)$, which contradicts the assumption.

Note that, by the proof above, $[\alpha]_\sigma = \{\sigma_{\geqslant i} \mid T_i^* \vdash \alpha, i \in \mathbb{N}\} = \{\pi \in S_\sigma \mid \mathscr{M}^*, \pi \models \alpha\} = [\alpha]^{path}_{\mathscr{M},\sigma}$. Similarly, $[\alpha]_s = [\alpha]^{state}_{\mathscr{M}^*,\sigma}$, so $\mathscr{M}^*$ is a measurable model. ■

3.4 Probabilistic reasoning about temporal information

We present the syntax and semantics of the logic for probabilistic reasoning about linear time formulas, that we denote by PL_{LTL}. The logic contains two types of formulas: formulas of LTL without probabilities, and the linear weight formulas in the style of [14], with weights applied to temporal formulas.

3.4.1 Syntax and semantics

In this section, probability operators are applied to the formulas of LTL. Following the notation introduced in Section 3.2.1, we denote by For_{LTL} the set of all LTL formulas, and by α, β and γ, possibly subscripted, the elements of For_{LTL}.

Now we introduce the probabilistic terms.

Definition 3.11 (Probabilistic term). A *probabilistic term* is any expression of the form

$$r_1 w(\alpha_1) + \cdots + r_k w(\alpha_k) + r_{k+1},$$

where k is a positive integer, and for all $i \leq k+1$, $\alpha_i \in For_{LTL}$ and $r_i \in \mathbb{Q}$.[3] ■

We use f and g, possibly subscripted, to denote probabilistic terms.

Definition 3.12 (Probabilistic formula). A *basic* probabilistic formula is any formula of the form $f \geq r$, where f is a probabilistic term and $r \in \mathbb{Q}$. The set For_P of probabilistic formulas is the smallest set containing all basic probabilistic formulas, closed under Boolean connectives. ■

We denote elements of For_P by ϕ, ψ and θ (possibly with indices). To simplify notation, we define the following abbreviations: $f \geq g$ is $f - g \geq 0$, $f \leq g$ is $g \geq f$, $f < g$ is $\neg f \geq g$, , $f > g$ is $\neg f \leq g$ and $f = g$ is $f \geq g \wedge f \leq g$.

Example 3.3. The expression

$$w(p \vee q) = w(\bigcirc p) \rightarrow w(Gq) \leq \frac{1}{2}$$

is a probabilistic formula. Its meaning is "if the probability that either p or q hold in this moment is equal to the probability that p will hold in the next moment, then the probability that q will always hold is at most one half". ■

Definition 3.13 (PL_{LTL} Formula). The set $For_{PL_{LTL}}$ of all formulas of the logic PL_{LTL} is $For_{PL_{LTL}} = For_{LTL} \cup For_P$. ■

We denote arbitrary formulas by Φ and Ψ (possibly with subscripts). We denote by $\perp$ both $\phi \wedge \neg\phi$ and $\alpha \wedge \neg\alpha$, letting the context determines the meaning. Similarly, we use $\top$ for both LTL and probabilistic formulas.

Example 3.4. The expression $Gp \rightarrow w(\bigcirc p) > 0$ is not a formula, since mixing LTL formulas and probabilistic formulas is not allowed, by Definition 3.13. ■

Example 3.5. In order to illustrate the expressive power of the introduced formal language, let us code the following statement:

---

[3] In [14], r_{k+1} does not appear in the definition of terms. We introduce it for the simpler presentation, when we introduce other formulas as abbreviations.

"The chance of a success within ten days is at least twice bigger then the chance of a success later on."

If we denote the success by p, then the formal representation is given by the formula

$$w(\bigvee_{k=0}^{10} \bigcirc^k p) \geq 2w(\bigcirc^{11} Fp).$$

■

Now we introduce the semantics of PL_{LTL}. The models contain $\sigma-$additive probabilities. Recall that $\mu : H \longrightarrow [0,1]$ is a $\sigma-$additive probability measure on an algebra H if the finite additivity condition is strengthen to the following condition:

$$\mu(\bigcup_{i\in\mathbb{N}} A_i) = \sum_{i\in\mathbb{N}} \mu(A_i), \tag{3.2}$$

whenever $A, A_i \in H$ and $A_i \cap A_j = \emptyset$ for all $i \neq j$. We also say that an algebra H is a σ-algebra, if $\bigcup_{i\in\mathbb{N}} A_i \in H$ whenever $A_i \in H$ for every $i \in \mathbb{N}$.

For a finitely additive μ, the condition (3.2) is equivalent to the condition

$$\mu(\bigcup_{i\in\mathbb{N}} A_i) = \lim_{n\to+\infty} \mu(\bigcup_{i=0}^{n} A_i). \tag{3.3}$$

We use (3.3) in the axiomatization of the logic PL_{LTL} (see the inference rule R6).

The logic PL_{LTL} contains two types of formula; for that reason, in this section we use the symbol $\models$ for satisfiability relation of the logic PL_{LTL}. Since the satisfiability of an LTL formula on a path is defined as a straightforward extension of a truth labeling $v : S \times Var \longrightarrow \{0,1\}$ (see Definition 3.2,) in this section we write $v(\pi, \alpha) = 1$, instead of $\pi \models \alpha$, to denote that α is true on the path π, and $v(\pi, \alpha) = 0$ instead of $\pi \not\models \alpha$. Recall that $\overline{\Sigma}$ denotes the set of all paths.

Definition 3.14 (PL_{LTL} structure). A PL_{LTL} structure is a tuple $M = \langle W, H, \mu, \pi \rangle$ where:

- W is a nonempty set of worlds,
- H is an algebra of subsets of W,
- μ is a $\sigma-$additive probability measure on H, and
- $\pi : W \longrightarrow \overline{\Sigma}$ provides for each world $w \in W$ a path $\pi(w)$. ■

For a PL_{LTL} structure $M = \langle W, H, \mu, \pi \rangle$, we define $[\alpha]_M = \{w \in W \mid v(\pi(w), \alpha) = 1\}$. We say that M is *measurable*, if $[\alpha]_M \in H$ for every $\alpha \in For_{LTL}$. We denote the class of all measurable PL_{LTL} structures with PL_{LTL}^{Meas}.

Now we define the satisfiability of a formula from For in a structure from PL_{LTL}^{Meas}.

Definition 3.15 (Satisfiability). Let $M = \langle W, H, \mu, \pi \rangle$ be a PL_{LTL} structure. We define the satisfiability relation $\models \subseteq PL_{LTL}^{Meas} \times For$ recursively as follows:

- $M \models \alpha$ iff $v(\pi(w), \alpha) = 1$ for every $w \in W$,
- $M \models r_1 w(\alpha_1) + \cdots + r_k w(\alpha_k) \geq r$ iff $r_1 \mu([\alpha_1]_M) + \cdots + r_k \mu([\alpha_k]_M) \geq r$,
- $M \models \neg\phi$ iff $M \not\models \phi$; and $M \models \phi \wedge \psi$ iff $M \models \phi$ and $M \models \psi$. ■

We say that $M \in PL_{LTL}^{Meas}$ is a model of Φ, if $M \models \Phi$. The notions of validity, entailment and semantical consequence are defined in the usual way.

3.4.2 The axiomatization of PL_{LTL}

Recall from Section 3.2.1 that we denote by $\alpha\overline{U}_n\beta$ the formula $(\bigwedge_{k=0}^{n-1} \bigcirc^k \alpha) \wedge \bigcirc^n \beta$. In addition, let us introduce the operator U_n as follows:

$$\alpha U_n \beta := \bigvee_{k=0}^{n} \alpha \overline{U}_n \beta.$$

Obviously

$$[\alpha U \beta]_M = \bigcup_{n \in \mathbb{N}} [\alpha U_n \beta]_M. \tag{3.4}$$

We will use (3.4) in the axiomatization of the logic for capturing σ-additivity.

Now we present the axiomatization $Ax(PL_{LTL})$.

Tautologies

A1. All instances of classical propositional tautologies for both LTL and probabilistic formulas.

LTL axioms

A2 – A4 from the axiomatization $Ax(LTL)$

Axioms for reasoning about linear inequalities

A5. All instances of valid formulas about linear inequalities.

Probabilistic axioms

A6. $w(\alpha) \geq 0$.
A7. $w(\alpha \wedge \beta) + w(\alpha \wedge \neg\beta) = w(\alpha)$.
A8. $w(\alpha \to \beta) = 1 \to w(\alpha) \leq w(\beta)$.

Inference rules

R1. From Φ and $\Phi \to \Psi$ infer Ψ (where either $\Phi, \Psi \in For_{LTL}$ or $\Phi, \Psi \in For_P$).
R2. and R3. from the axiomatization $Ax(LTL)$
R4. From α infer $w(\alpha) = 1$.
R5. From the set of premises $\{\phi \to f \geq r - \frac{1}{n} \mid n \in \mathbb{N} \setminus \{0\}\}$ infer $\phi \to f \geq r$.

R6. From the set of premises $\{\phi \rightarrow w(\alpha U_n \beta) \leq r \mid n \in \mathbb{N}\}$ infer $\phi \rightarrow w(\alpha U \beta) \leq r$.

The axiom A5 includes all valid formulas about linear inequalities. For example, $f+1 \leq f+2$ and $f+g = g+f$ are instances of A5. The infinitary rule R5 is a variant of Archimedean rule. The rule R6 are crucial for the proof of σ-additivity. The notions of proof, theorem and deducibility are introduced as in Section 3.2.1, with the constraint that application of the rule R2 is restricted to theorems only.

In $Ax(PL_{LTL})$ a set of formulas T is maximal consistent if it is consistent and for all $\Phi \notin T$, $T \cup \{\Phi\}$ is inconsistent. Note that in this logic maximal consistency of T doesn't imply that for every $\alpha \in For_{LTL}$ either $T \vdash \alpha$ or $T \vdash \neg\alpha$. Indeed, suppose that $w(\alpha) = \frac{1}{2} \in T$ for some α. If $T \vdash \alpha$ or $T \vdash \neg\alpha$, then by R4 (and some probabilistic reasoning) we have $T \vdash w(\alpha) = 1$ or $T \vdash w(\alpha) = 0$, which would make T inconsistent. On the other hand, for a $\phi \in For_P$ we have either $T \vdash \phi$ or $T \vdash \neg\phi$.

3.4.3 Completeness of $Ax(PL_{LTL})$

Since $Ax(PL_{LTL})$ contains the axiomatization $Ax(LTL)$, we extend the proof strategy from Section 3.2.1. We start with the Deduction theorem.

Theorem 3.10 (Deduction theorem). *Let T be a set of formulas and let Φ and Ψ be two formulas such that either $\Phi, \Psi \in For_{LTL}$ or $\Phi, \Psi \in For_{LTL}$. Then $T \cup \{\Phi\} \vdash \Psi$ iff $T \vdash \Phi \rightarrow \Psi$.*

Proof. As before, we use induction on the length of the inference. Here we consider the case when R6 is applied. Suppose that $T \cup \{\phi\} \vdash \psi \rightarrow w(\alpha U \beta) \leq r$ is obtained by R6. Then $T \cup \{\phi\} \vdash \psi \rightarrow w(\alpha U_n \beta) \leq r$ holds, by assumption, for every $n \in \mathbb{N}$. Using induction hypothesis, we have:

$T \vdash \phi \rightarrow (\psi \rightarrow w(\alpha U_n \beta) \leq r)$, for for every $n \in \mathbb{N}$
$T \vdash (\phi \wedge \psi) \rightarrow w(\alpha U_n \beta) \leq r$, for every $n \in \mathbb{N}$;
$T \vdash (\phi \wedge \psi) \rightarrow w(\alpha U \beta) \leq r$, by R6
$T \vdash \phi \rightarrow (\psi \rightarrow w(\alpha U \beta) \leq r)$. ∎

Theorem 3.11 (Lindenbaum's theorem). *Every consistent set of formulas can be extended to a maximal consistent set.*

Proof. Let T be a consistent set and let $\Phi_0, \Phi_1, \ldots$ be an enumeration of all formulas. We define the sequence of sets T_i, $i = 0, 1, 2, \ldots$ and the set T^* recursively as follows:

1. $T_0 = T$,
2. for every $i \geq 0$,

   a. if $T_i \cup \{\Phi_i\}$ is consistent, then $T_{i+1} = T_i \cup \{\Phi_i\}$, otherwise
   b. if Φ_i is of the form $\gamma \rightarrow \neg(\alpha U \beta)$, then $T_{i+1} = T_i \cup \{\gamma \rightarrow (\alpha \overline{U}_n \beta)\}$, where n is the smallest nonnegative integer such that T_{i+1} is consistent, otherwise

c. if Φ_i is of the form $\phi \to f \geq r$, then $T_{i+1} = T_i \cup \{\phi \to f < r - \frac{1}{n}\}$, where n is the smallest positive integer such that T_{i+1} is consistent, otherwise
d. if Φ_i is of the form $\phi \to w(\alpha U \beta) \leq r$, then $T_{i+1} = T_i \cup \{\phi \to w(\alpha U_n \beta) > r\}$, where n is the smallest nonnegative integer such that T_{i+1} is consistent, otherwise
e. $T_{i+1} = T_i$.

3. $T^\star = \bigcup_{i=0}^{\infty} T_i$.

First, using Theorem 3.10 one can prove that the set T^* is correctly defined, i.e., there exist n from the parts 2(b)–2(d) of the construction. Each T_i, $i > 0$ is consistent. The steps (1) and (2) of the construction ensure that $T^\star$ is maximal. Also, $T^\star$ obviously doesn't contain all formulas. Finally, one can show that $T^\star$ is deductively closed set, and as a consequence we obtain that $T^\star$ is consistent. ■

Definition 3.16 (Canonical model). For a maximal consistent set T^*, we define a PL_{LTL} structure as a tuple $M_{T^*} = \langle W, H, \mu, \pi \rangle$, such that:

1. $W = \{\sigma \in \overline{\Sigma} \mid v(\sigma, \alpha) = 1 \text{ for all } \alpha \in T^* \cap For_{LTL}\}$,
2. $H = \{[\alpha] \mid \alpha \in For_{LTL}\}$, where $[\alpha] = \{w \in W \mid v(w, \alpha) = 1\}$,
3. $\mu([\alpha]) = \sup\{r \in \mathbb{Q} \mid T^* \vdash w(\alpha) \geq r\}$, for every $\alpha \in For_{LTL}$,
4. $\pi(w) = w$ for every $w \in W$. ■

Now we show that M_{T^*} is a measurable PL_{LTL} structure.

Theorem 3.12. *For every maximal consistent set T^*, $M_{T^*} \in PL_{LTL}^{Meas}$.*

Proof. First we need to show that the definition is correct. The set $\{[\alpha] \mid \alpha \in For_{LTL}\}$ is an algebra of subsets of W, since $W = [\top]$, $W \setminus [\alpha] = [\neg\alpha]$ and $[\alpha] \cup [\beta] = [\alpha \vee \beta]$. We also need to check that μ is correctly defined, i.e., that if $[\alpha] = [\beta]$ then $\mu([\alpha]) = \mu([\beta])$. From $[\alpha] = [\beta]$ we conclude that if σ is a path such that $v(\sigma, \gamma) = 1$ for all $\gamma \in T^* \cap For_{LTL}$, then $v(\sigma, \alpha \leftrightarrow \beta) = 1$. From Theorem 3.4 we obtain $T^* \vdash \alpha \leftrightarrow \beta$. Consequently, $T^* \vdash w(\alpha) = w(\beta)$ by R4 and A8, so $\mu([\alpha]) = \mu([\beta])$. Obviously $\mu(W) = \mu([\top]) = 1$ by R4. Similarly, using A6 we conclude that μ is nonnegative, and using A7 we conclude that μ is a finitely additive probability measure on A. We need to prove that μ is σ-additive.
Let $H_{\overline{\Sigma}} = \{[\alpha]_{\overline{\Sigma}} \mid \alpha \in For_{LTL}\}$, where $[\alpha]_{\overline{\Sigma}} = \{\sigma \in \overline{\Sigma} \mid v(w, \alpha) = 1\}$. By $For_{LTL}^{\bigcirc}$ we denote the set of all LTL formulas in which $\bigcirc$ is the only temporal operator (i.e. there are no appearances of U). We also introduce the set $A = \{[\alpha] \mid \alpha \in For_{LTL}^{\bigcirc}\}$. Using the same argument as above, we can show that the sets $H_{\overline{\Sigma}}$ and A are two algebras of subsets of $\overline{\Sigma}$. Similarly as in the definition of M_{T^*}, we define μ^* on $H_{\overline{\Sigma}}$ by

$$\mu^*([\alpha]_{\overline{\Sigma}}) = \sup\{r \in \mathbb{Q} \mid T^* \vdash w(\alpha) \geq r\}.$$

Reasoning as above, we conclude that μ^* is a finitely additive measure. We also use the same symbol μ^* to denote the restriction of μ^* to A. We actually want to show that μ^* is σ-additive on A. It is sufficient to show that if $B = \bigcup_{n \in \mathbb{N}} B_i$, where $B, B_i \in A$, then there is n such that $B = \bigcup_{n=0}^{\mathbb{N}} B_i$.

If 2^{Var} denotes the set of subsets of Var, note that $\overline{\Sigma} = 2^{Var} \times 2^{Var} \times 2^{Var} \times \dots$ If we assume discrete topology on the finite set 2^{Var} and the induced product topology on $\overline{\Sigma}$, then $\overline{\Sigma}$ is a compact space as a product of compact spaces. By definition of evaluation function v, we obtain that for every $\alpha \in For^{\bigcirc}_{LTL}$ there exist $n \in \mathbb{N}$ (for example n is the number of appearances of $\bigcirc$) and $S \subseteq (2^{Var})^n$ such that $[\alpha]_{\overline{\Sigma}} = S \times 2^{Var} \times 2^{Var} \times \dots$ are Note that the sets of the form $S \times 2^{Var} \times 2^{Var} \times \dots$, where $S \subseteq (2^{Var})^n$ for some $n \in \mathbb{N}$, are clopen (both closed and open) sets in product topology. Thus, each $[\alpha]_{\overline{\Sigma}} \in A$ is a clopen set in $\overline{\Sigma}$. Now assume $[\alpha]_{\overline{\Sigma}} = \bigcup_{n \in \mathbb{N}} [\alpha_n]_{\overline{\Sigma}}$, where $\alpha \in For^{\bigcirc}_{LTL}$ and $\alpha_n \in For^{\bigcirc}_{LTL}$ for every $n \in \mathbb{N}$. The set $\{[\alpha_n]_{\overline{\Sigma}} \mid n \in \mathbb{N}\}$ is an open cover of the closed subset $[\alpha]_{\overline{\Sigma}}$ of the compact space $\overline{\Sigma}$, so there is a finite subcover $\{[\alpha_{n_1}]_{\overline{\Sigma}}, \dots, [\alpha_{n_1}]_{\overline{\Sigma}}\}$ of $[\alpha]_{\overline{\Sigma}}$. Thus, μ^* is σ-additive on A.

Let F be the σ-algebra generated by A. Since $[\alpha U \beta]_{\overline{\Sigma}} = \bigcup_{n \in \mathbb{N}} [\alpha U_n \beta]_{\overline{\Sigma}}$, we can show that $[\alpha]_{\overline{\Sigma}} \in F$ for every $\alpha \in For_{LTL}$, using the induction on the number of appearances of U in α. Thus, $H_{\overline{\Sigma}} \subseteq F$. By Caratheodory's extension theorem, there is a unique σ-additive probability measure v on F which coincide with μ^* on A. We will actually show that μ^* is the restriction of v to $H_{\overline{\Sigma}}$, i.e., that $\mu^*([\alpha]_{\overline{\Sigma}}) = v([\alpha]_{\overline{\Sigma}})$ for all $\alpha \in For_{LTL}$, using the induction on the number of appearances of U in α. Indeed, $v([\alpha]_{\overline{\Sigma}}) = v(\bigcup_{n \in \mathbb{N}} [\alpha U_n \beta]_{\overline{\Sigma}}) = \lim_{k \to +\infty} v(\bigcup_{n=1}^{k} [\alpha U_n \beta]_{\overline{\Sigma}}) = \lim_{k \to +\infty} \mu^*(\bigcup_{n=1}^{k} [\alpha U_n \beta]_{\overline{\Sigma}}) = \mu^*([\alpha U \beta]_{\overline{\Sigma}})$. Here we used σ-additivity of v, the induction hypothesis and, in the last step, the definition of μ^* and R6. Thus, μ^* is a σ-additive probability measure on $H_{\overline{\Sigma}}$. Note that we have that $\mu^*([\alpha]_{\overline{\Sigma}}) = 1$ whenever $T^* \vdash \alpha$, by R4. Thus, $\mu^*(W) = \mu^*(\bigcap_{\alpha : T^* \vdash \alpha} [\alpha]_{\overline{\Sigma}}) = 1$, by σ-additivity of μ^*. Note that $[\alpha] = [\alpha]_{\overline{\Sigma}} \cap W$, so $H \subseteq F$. Let $\overline{\mu}$ be the σ-additive probability measure on H induced by μ^* by

$$\overline{\mu}([\alpha]) = \overline{\mu}([\alpha]_{\overline{\Sigma}} \cap W) = \mu^*([\alpha]_{\overline{\Sigma}}).$$

Note that $\mu^*(W) = 1$ implies $\mu^*([\alpha]_{\overline{\Sigma}}) = \mu^*([\alpha]_{\overline{\Sigma}} \cap W)$, so $\mu^*([\alpha]) = v([\alpha])$. By definitions of μ and μ^* it follows that μ and v coincide. Thus, μ is σ-additive. We showed that M_{T^*} is a PL_{LTL} structure. Finally, note that $[\alpha] = [\alpha]_{M_{T^*}}$, by the choice of π, so $M_{T^*} \in PL^{Meas}_{LTL}$. ■

Theorem 3.13 (Strong completeness theorem). *A set of formulas $T \subseteq For$ is consistent iff it is satisfiable.*

Proof. The direction from right to left is straightforward. For the other direction, we extend a consistent set of formulas T to a maximal consistent set T^*, we construct the canonical model M_{T^*}, and we show that M_{T^*} is a model of T^*. It is sufficient to prove that for all $\Phi \in For$, $T^* \vdash \Phi$ iff $M_{T^*} \models \Phi$.

If $\Phi = \alpha \in For_{LTL}$. If $\alpha \in T^*$, then by the definition of W from M_{T^*}, $M_{T^*} \models \alpha$. Conversely, if $M_{T^*} \models \alpha$, by completeness of $Ax(LTL)$, we have $\alpha \in T^*$.

If $\Phi \in For_P$, we use the induction on the complexity of Φ. We only consider the case when $\Phi = f \geq r$. If $f = r_1 w(\alpha_1) + \dots + r_k w(\alpha_k) + r_{k+1}$. Then we can show, using properties of supremum, that $r_1 \mu([\alpha_1]) + \dots + r_k \mu([\alpha_k]) + r_{k+1} = \sup\{s \mid T^* \vdash f \geq s\}$. If we suppose that $f \geq r \in T^*$, then $r \leq \sup\{s \mid T^* \vdash f \geq s\}$, so $M_{T^*} \models f \geq r$. For the other direction, assume that $M_{T^*} \models f \geq r$. Then $M_{T^*} \not\models f < r$. If $f < r \in T^*$, then, reasoning as above, we conclude $M_{T^*} \models f < r$, a contradiction. By Maximality of T^*, we obtain $f \geq r \in T^*$. ■

3.4.4 Decidability

It is well known that the logic LTL is decidable. Indeed, [41] proved that the problem of deciding whether an LTL formula is satisfiable in a path is *PSPACE*-complete. Note that if α is not satisfiable in any path, then by Definition 3.15 it is not satisfiable in the logic PL_{LTL}. On the other hand, if there is a path σ such that $v(\sigma, \alpha) = 1$, then we can define a measurable structure $M = \langle W, H, \mu, \pi \rangle$, such that $W = \{w\}$ is a singleton and $\pi(w) = \sigma$ (note that in that case the range of μ is $\{0, 1\}$). Obviously, $v(\pi(w), \alpha) = 1$ for every $w \in W$, so $M \models \alpha$. Thus, we proved that the satisfiability problem of LTL formulas for the logic PL_{LTL} is *PSPACE*-complete.

Furthermore, the problem of satisfiability of a formula $\varphi \in For_P$ can be translated to the problem of solving a system of linear equations and inequalities, which lead to a procedure which decides satisfiability of the formula in *PSPACE*. For the proof we refer the reader to [7]. Thus, the following result holds.

Theorem 3.14. *The problem of deciding whether a formula of the logic PL_{LTL} is satisfiable in a measurable structure from PL_{LTL}^{Meas} is* PSPACE*-complete.* ■

3.5 Temporal reasoning about evidence

This section is devoted to logical formalization of the notion of evidence. Halpern and Fagin [20] have suggested that evidence can be seen as a function from prior beliefs to beliefs after making an observation (posterior beliefs). A formalism based on temporal logic for reasoning about evidence is proposed in [21]. In this section we show how the methods presented in this chapter can be adapted for development of a logic which captures how probabilities change over time in the presence of new evidence.

In this section we will deal with formulas over different sets of propositional letters. For any set of propositions V, by $For(V)$ we denote the set of propositional formulas over V. Since our models are linear time structures, where time instances are equipped with probabilities, for simplicity we define probability measures directly on formulas (instead of the corresponding measurable sets), i.e., we define them as functions $\mu : For(V) \longrightarrow [0, 1]$ which satisfy such that $\mu(\phi) = 1$, whenever ϕ is a tautology, and $\mu(\alpha \vee \beta) = \mu(\alpha) + \mu(\beta)$, whenever $\neg(\alpha \wedge \beta)$ is a tautology.

3.5.1 Weight of evidence

We start with several definitions from [21], which give mathematical formalization of the notion of *weight of evidence*. Intuitively, the probability of a hypothesis after making an observation depends on:

- the prior probabilities of the hypothesis
- to what extent the observations support the hypothesis.

The second item is formalized by the weight of evidence – the function which assigns a number from the unit interval to every observation and hypothesis.

Let $H = \{h_1, \ldots, h_m\}$ be the set that represents mutually exclusive and exhaustive hypotheses and $O = \{o_1, \ldots, o_n\}$ be the set of possible observations. For the hypothesis h_i, let μ_i be a likelihood function on O, i.e., the function which satisfies

- $\mu_i : O \longrightarrow [0,1]$;
- $\mu_i(o_1) + \cdots + \mu_i(o_n) = 1$.

We assume that for every observation $o \in O$ there is $i \in \{1, \ldots, m\}$ such that $\mu_i(o) > 0$. An *evidence space* is a tuple $\mathrm{E} = \langle H, O, \mu_1, \ldots, \mu_m \rangle$. For an evidence space E, we define the weight function $w_{\mathrm{E}} : O \times H \longrightarrow [0,1]$ as follows:

$$w_{\mathrm{E}}(o_i, h_j) = \frac{\mu_j(o_i)}{\mu_1(o_i) + \ldots + \mu_m(o_i)}.$$

The weight function can be seen as a qualitative assessment of the evidence in favor of one of the hypotheses, even if the prior probabilities are unknown, so one cannot determine the probability of the hypotheses after an observation. Intuitively, $w_{\mathrm{E}}(o,h)$ is the likelihood that h holds, if o is observed. Specially, $w_{\mathrm{E}}(o,h) = 0$ means that h is certainly false if o is observed, while $w_{\mathrm{E}}(o,h) = 1$ means that o fully confirms h.

Halpern and Pucella provided a characterization of weight functions.

Theorem 3.15 ([21]). *Let $H = \{h_1, \ldots, h_m\}$ and $O = \{o_1, \ldots, o_n\}$, and let f be a real-valued function, $f : O \times H \longrightarrow [0,1]$. Then there exists an evidence space $\mathrm{E} = \langle H, O, \mu_1, \ldots, \mu_m \rangle$ such that $f = w_{\mathrm{E}}$ iff f satisfies the following properties:*

1. *$f(o_i, h_1) + \ldots + f(o_i, h_m) = 1$, for every $i \in \{1, \ldots, n\}$.*
2. *There exist $x_1, \ldots, x_n > 0$ such that, for all $j \in \{1, \ldots, m\}$, $x_1 f(o_1, h_j) + \ldots + x_n f(o_n, h_j) = 1$.* ∎

Moreover, if 1. and 2. are satisfied, then the likelihood functions μ_i, $i \in \{1, \ldots, m\}$ are defined by

$$\mu_j(o_i) = \frac{f(o_i, h_j)}{x_j}.$$

If we know the prior probabilities, then we can use the Dempster's rule of combination to calculate the posterior probabilities. For given finite set $\mathscr{H}$ and two arbitrary probability distributions ν_1 and ν_2 on $\mathscr{H}$, the rule combines ν_1 and ν_2 in the following way: for every $H \subseteq \mathscr{H}$

$$(v_1 \oplus v_2)(H) = \frac{\sum_{h \in H} v_1(h) v_2(h)}{\sum_{h \in \mathcal{H}} v_1(h) v_2(h)}.$$

Let μ be a probability measure on the set $For(H)$ of propositional formulas over H, which satisfies $\mu(h_1) + \ldots + \mu(h_m) = 1$. Since the hypotheses are mutually exclusive, μ should satisfy $\mu(h_i \wedge h_j) = 0$, for $i \neq j$. Then $\mu(h_1 \vee \ldots \vee h_m) = \mu(h_1) + \ldots + \mu(h_m)$, so the fact that hypotheses are exhaustive may be expressed by the equality $\mu(h_1) + \ldots + \mu(h_m) = 1$.

Note that for any $\phi \in For(H)$, there exists $\phi' \in For(H)$ of the form $\bigvee_{i \in I} h_i$, for some $I \subseteq \{1, \ldots, m\}$,[4] such that $\mu(\phi) = \mu(\phi')$. Indeed, it is obvious if $\phi \in H$; suppose that $\mu(\phi_1) = \mu(\phi_1')$ and $\mu(\phi_2) = \mu(\phi_2')$, where ϕ_1' is of the form $\bigvee_{i \in I_1} h_i$ and ϕ_2' is of the form $\bigvee_{i \in I_2} h_i$. Then, $\mu(\phi_1 \wedge \phi_2) = \mu((\phi_1 \wedge \phi_2)')$ and $\mu(\neg\phi_1) = \mu((\neg\phi_1)')$ for $(\phi_1 \wedge \phi_2)' = \bigvee_{i \in I_1 \cap I_2} h_i$ and $(\neg\phi_1)' = \bigvee_{i \in \{1,\ldots,m\} \setminus I_1} h_i$.

In [21] Halpern and Pucella noticed that, for each observation o for which $\mu_1(o) + \cdots + \mu_m(o) > 0$, $w_{\mathrm{E}}(o, h_1) + \ldots + w_{\mathrm{E}}(o, h_m) = 1$ holds, so there is a unique probability measure on $For(H)$ which is an extension of $w_{\mathrm{E}}(o, \cdot)$, such that hypotheses are mutually exclusive. Hence, we will also denote that measure with $w_{\mathrm{E}}(o, \cdot)$. Informally, we may assume that elements of $For(H)$ are subsets of H.

If we let μ be a prior probability on hypotheses, then we can calculate the probability of hypotheses after observing o in the following way:

$$\mu_o = \mu \oplus w_{\mathrm{E}}(o, \cdot).$$

3.5.2 Reasoning about prior and posterior probabilities

Let $H = \{h_1, \ldots, h_m\}$, $O = \{o_1, \ldots, o_n\}$ and $C = \{c_1, \ldots, c_n\}$.

Definition 3.17. We define the set *Term* of all probabilistic terms as follows:

- $Term(0) = \{P_0(\alpha), P_1(\alpha) \mid \alpha \in For(H)\} \cup \{w(o,h) \mid o \in O, h \in H\} \cup C \cup \{0, 1\}$.
- $Term(n+1) = Term(n) \cup \{(\mathtt{f} + \mathtt{g}), (\mathtt{f} \cdot \mathtt{g}), (-\mathtt{f}) \mid \mathtt{f}, \mathtt{g} \in Term(n)\}$.
- $Term = \bigcup_{n=0}^{\infty} Term(n)$. ■

We denote probabilistic terms by $\mathtt{f}, \mathtt{g}$ and $\mathtt{h}$, possibly with indices. Furthermore, we introduce the usual abbreviations: $\mathtt{f} + \mathtt{g}$ is $(\mathtt{f} + \mathtt{g})$, $\mathtt{f} + \mathtt{g} + \mathtt{h}$ is $((\mathtt{f} + \mathtt{g}) + \mathtt{h})$, $\mathtt{f} \cdot \mathtt{g} \cdot \mathtt{h}$ is $((f \cdot \mathtt{g}) \cdot \mathtt{h})$, $-\mathtt{f}$ is $(-\mathtt{f})$, $\mathtt{f} - \mathtt{g}$ is $(\mathtt{f} + (-\mathtt{g}))$, 2 is $1 + 1$ and so on.

Definition 3.18. A basic probabilistic formula is any formula of the form $\mathtt{f} \geqslant 0$. The set *For* of formulas is the smallest set containing basic probabilistic formulas, observations and hypotheses that is closed under $\neg$ and $\wedge$. ■

Formulas will be denoted by ϕ, ψ and θ, possibly with indices. To simplify notation, we define the other types of inequalities as abbreviations, eg., $\mathtt{f} > 0$ is $\neg(\mathtt{f} \leqslant 0)$.

[4] Note that we allow the possibility that I is $\emptyset$, in which case we adopt the convention $\bigvee_{i \in \emptyset} h_i = \bot$.

Without loss of generality, we may assume that rational numbers are also terms, since they can be eliminated from a formula by clearing the denominator. For example, the formula $\frac{1}{3}\mathtt{f} \geqslant \frac{1}{2}\mathtt{g}$ is an abbreviation for $2\mathtt{f} - 3\mathtt{g} \geqslant 0$.

Definition 3.19. A *model* $\mathcal{M}$ is any tuple $\langle \mathtt{E}, \mu, o, h, d_1, \ldots, d_n \rangle$ such that

- $\mathtt{E} = \langle H, O, \mu_1, \ldots, \mu_m \rangle$ is an evidence space.
- μ is a finitely additive probability measure on $For(H)$, such that $\mu(h_1) + \ldots + \mu(h_m) = 1$.
- $o \in O$ is an observation.
- $h \in H$ is a hypothesis.
- $d_1, \ldots, d_n$ are positive real numbers such that $d_1 w_{\mathtt{E}}(o_1, \mathtt{h}_j) + \ldots + d_n w_{\mathtt{E}}(o_n, h_j) = 1$ for all $j \in \{1, \ldots, m\}$.[5] ■

We mentioned before that we may assume that elements of $For(H)$ are subsets of H. This allows us to apply Dempster's Rule of Combination in the following definition.

Definition 3.20. Let $\mathcal{M} = \langle \mathtt{E}, \mu, o, h, d_1, \ldots, d_n \rangle$ be any model. We define the satisfiability relation $\models$ recursively as follows:

1. For $h' \in H$, $\mathcal{M} \models h'$ if $h' = h$.
2. For $o' \in O$, $\mathcal{M} \models o'$ if $o' = o$.
3. $\mathcal{M} \models \mathtt{f} \geqslant 0$ if $\mathtt{f}^{\mathcal{M}} \geqslant 0$, where $\mathtt{f}^{\mathcal{M}}$ is recursively defined in the following way:

   a. $0^{\mathcal{M}} = 0$, $1^{\mathcal{M}} = 1$.
   b. $c_i^{\mathcal{M}} = d_i$.
   c. $P_0(\phi)^{\mathcal{M}} = \mu(\phi)$, $\phi \in For(H)$.
   d. $w(o,h)^{\mathcal{M}} = w_{\mathtt{E}}(o,h)$.
   e. $P_1(\phi)^{\mathcal{M}} = (\mu \oplus w_{\mathtt{E}}(o,\cdot))(\phi)$, $\phi \in For(H)$.
   f. $(\mathtt{f}+\mathtt{g})^{\mathcal{M}} = \mathtt{f}^{\mathcal{M}} + \mathtt{g}^{\mathcal{M}}$.
   g. $(\mathtt{f} \cdot \mathtt{g})^{\mathcal{M}} = \mathtt{f}^{\mathcal{M}} \cdot \mathtt{g}^{\mathcal{M}}$.
   h. $(-\mathtt{f})^{\mathcal{M}} = -(\mathtt{f}^{\mathcal{M}})$.

4. $\mathcal{M} \models \neg\phi$ if $\mathcal{M} \not\models \phi$; and $\mathcal{M} \models \phi \wedge \psi$ if $\mathcal{M} \models \phi$ and $\mathcal{M} \models \psi$. ■

Let us now introduce the axiom system for the logic.

Propositional axioms

A1. $\tau(\phi_1, \ldots, \phi_n)$, where $\tau(p_1, \ldots, p_n) \in For_C$ is any propositional tautology.
A2. $\mathtt{f} = \mathtt{g} \to (\phi(\ldots, \mathtt{f}, \ldots) \to \phi(\ldots, \mathtt{g}, \ldots))$

Probabilistic axioms ($i \in \{0,1\}$)

A3. $P_i(\alpha) \geqslant 0$.
A4. $P_i(\top) = 1$.
A5. $P_i(\alpha) = P_i(\beta)$, whenever $\alpha \leftrightarrow \beta$ is a propositional tautology.

---

[5] The existence of those numbers is provided by Theorem 3.15.

A6. $P_i(\alpha \vee \beta) = P_i(\alpha) + P_i(\beta) - P_i(\alpha \wedge \beta)$.

Axioms about hypotheses

A7. $h_1 \vee \ldots \vee h_m$.
A8. $h_i \rightarrow \neg h_j$, for all $i, j \in \{1, \ldots, m\}$, $i \neq j$.

Axioms about observations

A7. $o_1 \vee \ldots \vee o_n$.
A8. $o_i \rightarrow \neg o_j$, for all $i, j \in \{1, \ldots, n\}$, $i \neq j$.

Axioms of a commutative ordered ring

A10. All the valid formulas about commutative ordered rings

Axioms about evidence

A26. $w(o,h) \geqslant 0$, $o \in O$, $h \in H$.
A27. $w(o,h_1) + \ldots + w(o,h_m) = 1$, $o \in O$.
A28. $o \rightarrow P_0(h)w(o,h) = P_1(h)(P_0(h_1)w(o,h_1) + \ldots + P_0(h_m)w(o,h_m))$, $o \in O$, $h \in H$.
A29. $c_1 > 0 \wedge \ldots \wedge c_n > 0 \wedge c_1 w(o_1,h_1) + \ldots + c_n w(o_n,h_1) = 1 \wedge \ldots \wedge c_1 w(o_1,h_m) + \ldots + c_n w(o_n,h_m) = 1$.

Inference rules

R1. From ϕ and $\phi \rightarrow \psi$ infer ψ.
R2. From the set of premises $\{\phi \rightarrow \mathtt{f} \geqslant -n^{-1} \mid n = 1,2,3,\ldots\}$ infer $\phi \rightarrow \mathtt{f} \geqslant 0$.

The axioms A27 and A29 are counterparts of the items 1 and 2 of Theorem 3.15 respectively. Axiom A28 is a reformulation of Dempster's rule of combination and asserts that prior probability μ (syntactical counterpart in A28 is P_0), posterior probability μ_o (P_1 in A28) and evidence w_{E} (w in A28) are connected via $\mu_o(h) = \frac{\mu(h)w_{\mathrm{E}}(o,h)}{\sum_{i=1}^{m}\mu(h_i)w_{\mathrm{E}}(o,h_i)}$, or equivalently $\mu(h)w_{\mathrm{E}}(o,h) = \mu_o(h) \cdot \sum_{i=1}^{m}\mu(h_i)w_{\mathrm{E}}(o,h_i)$.

Lemma 3.4. *For any term t* $\mathtt{f}$*, the theory* $T = \{\mathtt{f} \geqslant n \mid n \in \mathbb{N}\}$ *is inconsistent.*

Proof. It is sufficient to prove that constants c_i are bounded since all terms are polynomials in variables $P_0(\alpha)$, $P_1(\alpha)$, $w(o,h)$ and c_i, while the probability axioms guarantee that $\vdash 0 \leqslant P_j(\alpha) \leqslant 1$ and the axioms 26 and 27 imply the same for $w(o,h)$. It is easy to see (A26 and A27) that $\vdash \bigvee_{i=1}^{m} w(o_k,h_i) \geqslant \frac{1}{m}$ for all $k \in \{1,\ldots,n\}$. Moreover, by Axiom A29 $\vdash c_k w(o_k,h_i) \leqslant 1$ for all $k \in \{1,\ldots,n\}$ and $i \in \{1,\ldots,m\}$. Since $\{c_k \geqslant n \mid n \in \mathbb{N}\} \vdash c_k \geqslant m+1$, we have that $\{c_k \geqslant n \mid n \in \mathbb{N}\} \vdash (m+1)w(o_k,h_i) \leqslant 1$ for all $k \in \{1,\ldots,n\}$ and $i \in \{1,\ldots,m\}$, so $\{c_k \geqslant n \mid n \in \mathbb{N}\} \vdash \bigwedge_{i=1}^{m}(m+1)w(o_k,h_i) \leqslant 1$. Finally, $\vdash \bigvee_{i=1}^{m} w(o,h_i) \geqslant \frac{1}{m} \rightarrow \neg(\bigwedge_{i=1}^{m}(m+1)w(o_k,h_i) \leqslant 1)$, so $\{c_k \geqslant n \mid n \in \mathbb{N}\} \vdash \perp$ for all $k \in \{1,\ldots,n\}$. ■

Theorem 3.16 (Strong completeness theorem). *Every consistent set T of formulas is satisfiable.*

Proof. Similarly as in the previous sections, we can prove that the Deduction theorem holds and that T can be extended to a maximal consistent set T^*. Then we define the canonical model $\mathcal{M}^* = \langle \mathrm{E}, \mu, o, h, d_1, \ldots, d_n \rangle$ as follows:

- $d_i = \sup\{r \in \mathbb{Q} \mid T^* \vdash c_i \geqslant r\}$. By Lemma 3.4 the set $\{r \in \mathbb{Q} \mid T^* \vdash c_i \geqslant r\}$ has an upper bound in $\mathbb{R}$, so this definition is correct.
- Evidence space $\mathrm{E} = \langle H, O, \mu_1, \ldots, \mu_m \rangle$ is defined by sets H and O and likelihood functions $\mu_j : O \longrightarrow [0,1]$ defined by
$$\mu_j(o_i) = \frac{w_{\mathrm{E}}(o_i, h_j)}{d_j},$$
where $w_{\mathrm{E}}(o_i, h_j) = \sup\{r \in [0,1] \cap \mathbb{Q} \mid T^* \vdash w(o_i, h_j) \geqslant r\}$.
- $\mu(\phi) = \sup\{r \in [0,1] \cap \mathbb{Q} \mid T^* \vdash P_0(\phi) \geqslant r\}$.
- h is the unique hypothesis such that $T^* \vdash h$.
- o is the unique observation such that $T^* \vdash o$.

Let us show that $\mathcal{M}^*$ is a well defined model. First we show that $d_i > 0$. It is sufficient to show that there exist a positive rational number r such that $T^* \vdash c_i \geqslant r$. Since T^* is complete, for each positive $s \in \mathbb{Q}$, either $T^* \vdash c_i \geqslant s$ or $T^* \vdash c_i < s$. Suppose that $T^* \vdash c_i < s$ for all positive $s \in \mathbb{Q}$. Then, $T^* \vdash c_i = c_i \rightarrow c_i < s$ for all positive $s \in \mathbb{Q}$. By R2, $T^* \vdash c_i = c_i \rightarrow c_i \leqslant 0$. Since $T^* \vdash c_i = c_i$, by modus ponens $T^* \vdash c_i \leqslant 0$. However, $T^* \vdash c_i > 0$ (axiom A29), so T^* is inconsistent: a contradiction. Hence, there exist a positive $r \in \mathbb{Q}$ such that $T^* \vdash c_i > r$.

The proof that μ is a probability measure is similar to the corresponding proofs for the other probability logics in the previous chapters.

Let us prove the equality $d_1 w_{\mathrm{E}}(o_1, h_1) + \ldots + d_n w_{\mathrm{E}}(o_n, h_1) = 1$. We can chose increasing sequences of rational numbers $(\underline{a}^i_k)_{k \in \mathbb{N}}$ and $(\underline{b}^i_k)_{k \in \mathbb{N}}$, and decreasing sequences of rational numbers $(\overline{a}^i_k)_{k \in \mathbb{N}}$ and $(\overline{b}^i_k)_{k \in \mathbb{N}}$, $i \in \{1, \ldots, n\}$, such that $\lim \underline{a}^i_k = \lim \overline{a}^i_k = d_i$ and $\lim \underline{b}^i_k = \lim \overline{b}^i_k = w_{\mathrm{E}}(o_i, h_1)$ $(i \in \{1, \ldots, n\})$. Then, the equality is immediate consequence of the following facts:

- $T^* \vdash \underline{a}^1_k \underline{b}^1_k + \ldots + \underline{a}^n_k \underline{b}^n_k \leqslant 1$, for all k.
- $T^* \vdash \overline{a}^1_k \overline{b}^1_k + \ldots + \overline{a}^n_k \overline{b}^n_k > 1$, for all k.
- $\vdash c_1 w(o_1, h_1) + \ldots + c_n w(o_n, h_1) = 1$.
- $\lim \underline{a}^1_k \underline{b}^1_k + \ldots + \underline{a}^n_k \underline{b}^n_k = \lim \overline{a}^1_k \overline{b}^1_k + \ldots + \overline{a}^n_k \overline{b}^n_k = d_1 w_{\mathrm{E}}(o_1, h_1) + \ldots + d_n w_{\mathrm{E}}(o_n, h_1)$.

Similarly, we can show that the other conditions of Theorem 3.15 hold, so w_{E} is a weight function and it defines functions μ_i. Finally, it is obvious from axioms about hypotheses and observations (and completeness of T^*) that there is the unique h such that $T^* \vdash h$ and the unique o such that $T^* \vdash o$, so $\mathcal{M}^*$ is a well defined model.

Finally, we prove that for every formula ϕ, $\mathcal{M}^* \models \phi$ iff $\phi \in T^*$ by the induction on the complexity of formulas. We only consider the case $\phi = \mathtt{f} \geqslant 0$.

Let $\mathtt{f} \geqslant 0 \in T^*$. Using the axioms for ordered commutative rings, we can prove $\mathtt{g}_i^{\mathcal{M}^*} = \sup\{r \in \mathbb{Q} \mid T^* \vdash \mathtt{g}_i \geqslant r\}$. Without the loss of generality, suppose that $T^* \vdash r_i \geqslant 0$, for $1 \leqslant i \leqslant m_{\mathtt{f}}$, and $T^* \vdash r_i < 0$, for $m_{\mathtt{f}} < i \leqslant n_{\mathtt{f}}$. Using increasing sequences of rational numbers $(\underline{b}_k^i)_{k\in\mathbb{N}}$, and decreasing sequences of rational numbers $(\overline{b}_k^i)_{k\in\mathbb{N}}$, such that $\lim \underline{b}_k^i = \lim \overline{a}_k^i = \mathrm{g}_i^{\mathscr{M}^*}$, we have $\vdash r_1\underline{b}_k^1 + \ldots + r_{m_\mathtt{f}}\underline{b}_k^{m_\mathtt{f}} + r_{m_\mathtt{f}+1}\overline{b}_k^{m_\mathtt{f}+1} + \ldots + r_{n_\mathtt{f}}\overline{b}_k^{n_\mathtt{f}} \leqslant \mathtt{f}$, and $\vdash \mathtt{f} < r_1\overline{b}_k^1 + \ldots + r_{m_\mathtt{f}}\overline{b}_k^{m_\mathtt{f}} + r_{m_\mathtt{f}+1}\underline{b}_k^{m_\mathtt{f}+1} + \ldots + r_{n_\mathtt{f}}\underline{b}_k^{n_\mathtt{f}}$, for all $k \in \mathbb{N}$. Finally,

$$\mathtt{f}^{\mathscr{M}^*} = \sup\{r \in \mathbb{Q} \mid T^* \vdash \mathtt{f} \geqslant r\},$$

so $\mathtt{f}^{\mathscr{M}^*} \geqslant 0$ or, equivalently, $\mathscr{M}^* \models \mathtt{f} \geqslant 0$.

For the other direction, let $\mathscr{M}^* \models \mathtt{f} \geqslant 0$. Since T^* is deductively closed, it is sufficient to show that $T^* \vdash \mathtt{f} \geqslant 0$. By definition of $\models$, we have that $\mathtt{f}^{\mathscr{M}^*} \geqslant 0$. By definition of $\mathscr{M}^*$, $\mathtt{f}^{\mathscr{M}^*} = \sup\{r \in \mathbb{Q} \mid T^* \vdash \mathtt{f} \geqslant r\}$. If $f^{\mathscr{M}^*} > 0$, then there exists positive number $r \in \mathbb{Q}$ such that $T^* \vdash \mathtt{f} > r$, hence $T^* \vdash \mathtt{f} > 0$. Let $\mathtt{f}^{\mathscr{M}^*} = 0$. Then, for each negative $r \in \mathbb{Q}$, $T^* \vdash \mathtt{f} \geqslant r$. Since $T^* \vdash \mathtt{f} = \mathtt{f}$, we have that $T^* \vdash \mathtt{f} = \mathtt{f} \to \mathtt{f} \geqslant r$ for all negative $r \in \mathbb{Q}$. By archimedean rule, $T^* \vdash \mathtt{f} = \mathtt{f} \to \mathtt{f} \geqslant 0$, so $T^* \vdash \mathtt{f} \geqslant 0$ (modus ponens + $\vdash \mathtt{f} = \mathtt{f}$). On the other hand, $T^* \vdash \mathtt{f} \leqslant s$ for all positive $s \in \mathbb{Q}$, so similarly as above, $T^* \vdash \mathtt{f} = \mathtt{f} \to \mathtt{f} \leqslant s$ for all positive $s \in \mathbb{Q}$. By R2, $T^* \vdash \mathtt{f} = \mathtt{f} \to \mathtt{f} \leqslant 0$, and consequently $T^* \vdash \mathtt{f} \leqslant 0$. Hence, $T^* \vdash \mathtt{f} = 0$. ■

3.5.3 Adding temporal operators

In this section we introduce a linear time temporal logic that can deal with sequences of observations made over the time, assuming that the flow of the time is isomorphic to natural numbers. Here we only outline the main differences with respect to the previously developed logic. For an evidence space $\mathtt{E} = \langle H, O, \mu_1, \ldots, \mu_m \rangle$, we define $\mathtt{E}^* = \langle H, O^*, \mu_1^*, \ldots, \mu_m^* \rangle$ as follows:

- $O^* = \{\langle o^1 \ldots, o^k \rangle \mid k \in \mathbb{N}, o^i \in O\}$.
- $\mu_i^* : O^* \longrightarrow [0,1]$ is defined by $\mu_i^*(\langle o^1, \ldots, o^k \rangle) = \mu_i(o^1) \cdots \mu_i(o^k)$.

The sequence $\langle o^1, \ldots, o^k \rangle$ can be seen as a conjunction of its members, so the previous formula implicitly reflects independence of observations $o^1, \ldots, o^k$. It is shown in [21] that $w_{\mathtt{E}^*}(\langle o^1, \ldots, o^k \rangle, \cdot) = w_\mathtt{E}(o^1, \cdot) \oplus \cdots \oplus w_\mathtt{E}(o^k, \cdot)$. Informally, $w_{\mathtt{E}^*}(\langle o^1, \ldots, o^k \rangle, h)$ is the weight that hypothesis h is true, after observing $o^1 \ldots, o^k$. We will also use the following equality in an axiomatization of our temporal logic:

$$w_{\mathtt{E}^*}(\langle o^1, \ldots, o^k \rangle, h_i) = \frac{w_{\mathtt{E}^*}(o^1, h_i) \cdots w_{\mathtt{E}^*}(o^k, h_i)}{w_{\mathtt{E}^*}(o^1, h_1) \cdots w_{\mathtt{E}^*}(o^k, h_1) + \cdots + w_{\mathtt{E}^*}(o^1, h_m) \cdots w_{\mathtt{E}^*}(o^k, h_m)}.$$

If the prior probability μ on the set of hypotheses H is known, then we can calculate the probability of hypotheses after observing $\langle o^1, \ldots, o^k \rangle$ in the following way: $\mu_{\langle o^1, \ldots, o^k \rangle} = \mu \oplus w_\mathtt{E}(\langle o^1, \ldots, o^k \rangle, \cdot)$.

The set of terms *Term* is defined similarly as in Definition 3.17, the difference is that now we don't need two probability operators, since the posterior probability P_1 is now represented as the probability in the next time step. Therefore, we have one operator P, and $Term(0)$ contains the set $\{P(\alpha) \mid \alpha \in For(H)\}$. Then the set *For* of formulas is defined recursively as the smallest set that satisfies the conditions:

- Expressions of the form $\mathtt{f} \geqslant 0$, observations and hypotheses are formulas.
- If ϕ and ψ are formulas, then $\neg\phi$, $\phi \wedge \psi$, $\bigcirc\phi$ and $\phi U \psi$ are formulas.

Example 3.6. Let us illustrate the expressivity of the introduced formal language by coding the following statement:

> "The probability of hypothesis h_1 is twice larger than the probability of hypothesis h_2. However, if o is constantly observed in the next ten days, this ratio changes at a constant rate, until the odds are reversed (after 10 days)."

The formal representation of this statement is given by the formula

$$P(h_1) = 2P(h_2) \wedge (\bigwedge_{k=0}^{10} o \rightarrow \bigwedge_{k=0}^{10} \bigcirc^k ((1+k/10)P(h_1) = (2-k/10)P(h_2)).$$

■

We adapt the semantics from the previous section, and we define a model $\overline{\mathcal{M}}$ as an infinite sequence $\langle \mathcal{M}_0, \mathcal{M}_1, \mathcal{M}_2, \ldots \rangle$ such that

$$\mathcal{M}_k = \langle \mathrm{E}^*, \mu, h, d_1, \ldots, d_n, o^1, o^2, \ldots, o^k \rangle$$

for all $k \geqslant 1$ and $\mathcal{M}_0 = \langle \mathrm{E}^*, \mu, h, d_1, \ldots, d_n \rangle$. Notice that E^*, μ, $d_1, \ldots, d_n$ are the same in all $\mathcal{M}_k$'s and that with the increment of k we just add one new observation.

Then we define the satisfiability of a formula in a time instance of a model, denoted by $\mathcal{M}_k \models \phi$, modifying Definition 3.20 in the following way:

- The conditions 3.c, 3.d and 3.e are replaced with
  - $P(\phi)^{\mathcal{M}_k} = \mu \oplus w_{\mathrm{E}^*}(\langle o^1, \ldots, o^k \rangle, \cdot)(\phi)$, $\phi \in For(H)$.
  - $w(\langle o^{i_1}, \ldots, o^{i_k} \rangle, h')^{\mathcal{M}_k} = w_{\mathrm{E}^*}(\langle o^{i_1}, \ldots, o^{i_k} \rangle, h')$.
- The cases when we apply the temporal operators are added:
  - $\mathcal{M}_k \models \bigcirc\phi$ if $\mathcal{M}_{k+1} \models \phi$.
  - $\mathcal{M}_k \models \phi U \psi$ if there is $l \in \mathbb{N}$ such that $\mathcal{M}_{k+l} \models \psi$, and for every $l' \in \mathbb{N}$ such that $l' < l$, $\mathcal{M}_{k+l'} \models \phi$.

Now we modify axiomatization from the previous section: it includes propositional axioms and axioms about observations, hypotheses and commutative ordered rings, as well as probabilistic axioms (with the difference that we drop indices, since there is only one probabilistic operator P).

Axioms about evidence

Axioms A26, A27, and A29 are the same as before. Axiom A28 is replaced with:

A30. $\bigcirc(P(h) \geqslant r) \rightarrow P(h)w(o,h) \geqslant r(P(h_1)w(o,h_1) + \ldots + P(h_m)w(o,h_m))$, $o \in O$, $h \in H$, $r \in [0,1] \cap \mathbb{Q}$.

The following axiom is formal representation of $w_{\mathtt{E}^*}(\langle o^1,\dots,o^k\rangle,\cdot) = w_{\mathtt{E}}(o^1,\cdot)\oplus\cdots\oplus w_{\mathtt{E}}(o^k,\cdot)$, or, in expanded form

$$w_{\mathtt{E}^*}(\langle o^1,\dots,o^k\rangle,h_i) = \frac{w_{\mathtt{E}^*}(o^1,h_i)\cdots w_{\mathtt{E}^*}(o^k,h_i)}{w_{\mathtt{E}^*}(o^1,h_1)\cdots w_{\mathtt{E}^*}(o^k,h_1)+\cdots+w_{\mathtt{E}^*}(o^1,h_m)\cdots w_{\mathtt{E}^*}(o^k,h_m)}.$$

A31. $w(o^1,h)\cdots w(o^k,h) =$
$w(\langle o^1,\dots,o^k\rangle,h)(w(o^1,h_1)\cdots w(o^k,h_1)+\cdots+w(o^1,h_m)\cdots w(o^k,h_m))$.

Temporal axioms

The axioms from $Ax(LTL)$ and, in addition, the two axioms which state that only probabilities and observations might change over time:

A32. $\phi \leftrightarrow \bigcirc\phi$, $\phi \in For(H)$.
A33. $\mathtt{f} \geqslant 0 \leftrightarrow \bigcirc(\mathtt{f} \geqslant 0)$, if f does not contain an occurrence of P.

Inference rules

The inference rules of $Ax(LTL)$ and the modified Archimedean rule:

From $\{\phi \to \bigcirc^m\mathtt{f} \geqslant -n^{-1} \mid n = 1,2,3,\dots\}$ infer $\phi \to \bigcirc^m\mathtt{f} \geqslant 0$ (for any $m \in \mathbb{N}$).

The modification of the form of the rule is required for the proof of strong necessitation theorem. The proof is similar to the proof of Theorem 3.7, but the form of the rule doesn't include k-nested implications since now we don't use the operator A.

The proof of strong completeness combines the ideas from Section 3.5.2 with the proof of completeness of $Ax(LTL)$. For a given consistent theory T we can build a model $\overline{\mathcal{M}}$ as follows:

- First we extend T to the maximal consistent set T_0^*.
- For any positive integer n, let $T_n^* = \{\phi \mid \bigcirc\phi T_{n-1}^*\}$.
- For any $n \in \mathbb{N}$, let $\mathcal{M}_n' = \langle \mathtt{E}^*,\mu,h,d_1,\dots,d_n,o^n\rangle$ be the canonical model of T_n^*, defined as in Section 3.5.2.
- $\mathcal{M}_n = \langle \mathtt{E}^*,\mu,h,d_1,\dots,d_n,o^1,\dots,o^n\rangle$, where $\mathcal{M}_k' = \langle \mathtt{E}^*,\mu,h,d_1,\dots,d_n,o^k\rangle$, for all $k \leqslant n$.
- $\overline{\mathcal{M}} = \langle \mathcal{M}_0,\mathcal{M}_1,\mathcal{M}_2,\dots\rangle$.

It can be shown that the model $\overline{\mathcal{M}}$ is well defined and that $\overline{\mathcal{M}} \vdash T$.

3.6 Related work

In the early paper [46], the language of classical logic is enriched with two temporal operators: the "future" operator F and "past" operator P. A more expressive language with the new binary operators "until" (U) and "since" (S) is introduced in [25]; the corresponding completeness result is presented in [3] . Propositional linear time logic with the "next" operator $\bigcirc$ and "until" operator is introduced in [37], and a complete axiomatization is proposed in [16] (see also [28]); its first-order extension is given in [30]. The computation tree logic CTL, introduced in [12], is a branching time logic in which LTL operators are always immediately preceded by quantifiers over paths (A or E). A complete axiomatization can be found in [10]. The logic CTL* with the basic operators $\bigcirc$, U and A, and without restrictions on their application, is introduced in [13]. CTL* generalizes both LTL and CTL. There are a few complete axiomatizations for CTL*, with respect to different classes of models [39, 42].

Some papers that either implicitly or explicitly combine probabilistic and temporal reasoning are motivated by the need to analyze probabilistic programs and stochastic systems. In [23, 27] propositional temporal formulas that do not mention probabilities explicitly are interpreted over Markov systems which can simulate the execution of probabilistic programs. In both papers, the complete axiomatic systems are provided. The propositional logics from [27] extend linear temporal logic. The formulas do not mention probabilities, except the probability one expressed as a unary modal operator ∇ qualifying assertions that are certain, while possibly α can be expressed as $\triangle\alpha = \neg\nabla\neg\alpha$. The logic is suitable to prove the correctness of probabilistic programs that do not depend on specific distributions and are guaranteed to give correct results with probability one. The language can be interpreted on tree structures of branching time logic, with the symbol ∇ be interpreted as a path quantifier. The paper [23] introduces two propositional logics which extend CTL. The first one, PTL_f, is interpreted over finite models and allows to reason about finite state sequential probabilistic programs, while the second one, PTL_b, is interpreted over infinite models with transitions probabilities bounded away from 0 and allows to reason about finite state concurrent probabilistic programs, without any explicit reference to the values of their state-transition probabilities.

The branching-time logic PCTL for reasoning about time and probability is described in [22]. The underlying temporal logic is CTL. An example of a statement expressible in PCTL is: "after a request for service there is at least a 98% probability that the service will be carried out within 2 seconds" are expressible in the language of PCTL. Formulas are interpreted over discrete time Markov chains and algorithms for checking satisfiability of formulas by a given Markov chain are described. The paper does not provide an axiomatization for the logic. Similarly as in CTL, the formulas are classified into state formulas and path formulas, and their mixing is not allowed. The classical propositional language is enriched in the following way:

- $\alpha \mathrm{U}^{\leq n}\beta$ and $\alpha \mathscr{U}^{\leq n}\beta$ are path formulas, if α and β are state formulas, and $n \in \mathbb{N}\cup\{\infty\}$. The formula $\alpha \mathrm{U}^{\leq n}\beta$ corresponds to our formula $\alpha \mathrm{U}_n\beta$, while for $n = \infty$, $\mathrm{U}^{\leq n}$ and U coincide. If $\alpha\mathscr{U}\beta$ is the formula $\alpha \mathrm{U}\beta \vee G\alpha$, then the relationship between $\mathscr{U}^{\leq n}$ and $\alpha\mathscr{U}\beta$ is is the same as the relationship between U_n and U.
- $\alpha \mathrm{U}^{\leq n}_{\geq r}\beta$ and $\alpha \mathscr{U}^{\leq n}_{\geq r}\beta$ are state formulas, if α and β are path formulas, and $n \in \mathbb{N}\cup\{\infty\}$. Those operators are probabilistic variants of the above introduced operators, with probabilities measuring the sets of paths. In particular, $\alpha \mathrm{U}^{\leq n}_{\geq r}\beta$ correspond to the formula $P^s_{\geq r}(\alpha \mathrm{U}_n\beta)$ of pBTL.

On the other hand, the operator $P^p_{\geqslant r}$ is not expressible in PCTL.

A more expressive branching-time logic denoted PCTL* is described in [1]. The underlying temporal logic is CTL* with path quantifiers replaced by probabilities. The propositional language contains

- state formulas: $P_{\geq r}\alpha$ (α is a path formula),
- path formulas are obtained by applying LTL operators to the state formulas.

Similarly as in PCTL, a conjunction of a state formula and a path formula is not a formula. According to their definition of satisfiability, the operator $P_{\geqslant r}$ from [1] corresponds to the operator $P^s_{\geqslant r}$ of pBTL, while the operator $P^p_{\geqslant r}$ is not expressible in PCTL*. No axiomatization for PCTL* is given. The paper [2] presents model-checking algorithms for extensions of PCTL and PCTL* that involve non-determinism.

A probabilistic extension of CTL* presented in [4] contains actions attached to elements of accessibility relation. Semantically, it represents a probabilistic generalization of PAL logic from [47, 49], which is used for modeling revision of beliefs and intentions [48]. The formulas of the logic has probability operators applied to temporal formulas, similarly as in PL_{LTL}.

A probabilistic modal logic PPL is introduced in [45]. It allows applying probabilities to sequences of formulas (giving so called path expressiveness). A Gentzen-style axiom system is presented and proved to be sound and complete. The language allows linear combinations of terms (similarly as in [14] and PL_{LTL}) of the form $P(\alpha_1,\dots,\alpha_n)$ which means “the probability of the sequence of formulas.” Iteration of probabilities in a term is allowed. The formula $P(\alpha_1,\dots,\alpha_n) \geq r$ is expressible in the logic pBTL as $P^s_{\geqslant r}(\bigwedge_{i=1}^n \bigcirc^i \alpha_i)$.

The papers [24, 29] introduce real-time interval logics and calculus (Probabilistic duration calculus) which support calculation of the probability of a composed formula. They can be used in design of an embedded real-time systems to calculate whether a requirement holds with a sufficiently high probability based on the given probability of failures of components. The model of such a system is a finite (probabilistic timed) automaton with fixed transition probabilities (Markov process). Some examples of specification in Probabilistic duration calculus and valid formulas are given, but a proof system is not presented. The paper [18] gives a proof system for Probabilistic neighborhood logic, interval based temporal logic with probabilities. Neighborhood logic is a first order modal logic which is an extension of Probabilistic duration calculus that possesses modal operators that refer to intervals outside the current one. The infinite intervals are considered in [19].

Several probabilistic logics for reasoning about objects located in space and time are introduced in [6], and strongly complete axiomatization systems are proposed. The probability operators assign an interval to formulas. Those logics don't contain temporal operators. Instead, the formulas contain positive integers which represent moments in time.

The language of the logic presented in [15] is based on the propositional dynamic logic, and the main objects are programs. Probabilistic operators can be applied on a limited class of formulas, and the completeness problem is not solved. A fragment of [15] is considered in [26]. A first order probability dynamic logic is proposed in [36], and strong completeness is obtained using a variant of Archimedean rule. A dynamic generalization of the logic with qualitative probabilities from [40] is presented in [17]. Probabilities of formulas can be compared using a number of $\leq$ operators that correspond to operations in proportional dynamic logic (compositions, union, iterations). The formula $\alpha \leq_t \beta$ of the logic means: "the probability for a transition t to transform the world w into possible world that satisfies α is smaller or equal to the probability for t to transform the world w into possible world that satisfies β". A complete axiomatization is obtained using an infinitary rule, similarly as in the approach in this chapter.

References

[1] A. Aziz, V. Singhal, F. Balarin, R. K. Brayton, and A. L. Sangiovanni-Vincentelli. It usually works: The temporal logic of stochastic systems. In Pierre Wolper, editor, *Computer Aided Verification*, pages 155–165, Berlin, Heidelberg, 1995. Springer.

[2] A. Bianco and L. de Alfaro. Model checking of probabilistic and nondeterministic systems. In P. S. Thiagarajan, editor, *Foundations of Software Technology and Theoretical Computer Science*, pages 499–513, Berlin, Heidelberg, 1995. Springer.

[3] J. P. Burgess. Axioms for tense logic. i. "since" and "until". *Notre Dame Journal of Formal Logic*, 23(4):367–374, 1982.

[4] Š. Dautović and D. Doder. Probabilistic logic for reasoning about actions in time. In G. Kern-Isberner and Z. Ognjanović, editors, *Proceedings of the 15th European Conference on Symbolic and Quantitative Approaches to Reasoning with Uncertainty, ECSQARU 2019, Belgrade, Serbia, 18-20 September 2019*, volume 11726 of *Lecture Notes in Computer Science*. Springer, 2019.

[5] G.R.R. de Lavalette, B. Kooi, and R. Verbrugge. A strongly complete proof system for propositional dynamic logic. In *AiML2002, Advances in Modal Logic (conference proceedings)*, pages 377–393, 2002.

[6] D. Doder, J. Grant, and Z. Ognjanović. Probabilistic logics for objects located in space and time. *Journal of Logic and Computation*, 23(3):487–515, 2013.

[7] D. Doder and Z. Ognjanović. A probabilistic logic for reasoning about uncertain temporal information. In M. Meila and T. Heskes, editors, *Proceedings of the Thirty-First Conference on Uncertainty in Artificial Intelligence, UAI 2015, July 12-16, 2015, Amsterdam, The Netherlands*, pages 248–257. AUAI Press, 2015.

[8] D. Doder, Z. Ognjanović, and Z. Marković. An Axiomatization of a First-order Branching Time Temporal Logic. *Journal of Universal Computer Science*, 16:1439–1451, 2010.

[9] D. Doder, Z. Ognjanović, Z. Marković, A. Perović, and M. Rašković. A probabilistic temporal logic that can model reasoning about evidence. In *Foundations of Information and Knowledge Systems, 6th International Symposium,*

FoIKS 2010, Sofia, Bulgaria, February 15-19, 2010. Proceedings, pages 9–24. Springer, 2010.
[10] E.A. Emerson. Temporal and modal logic. In J. van Leeuwen, editor, *Handbook of theoretical computer science (vol. B): formal models and semantics*, pages 995–1072, Cambridge, MA, USA, 1990. MIT Press.
[11] E.A. Emerson. Automated temporal reasoning about reactive systems. In F. Moller and G. Birtwistle, editors, *Logics for Concurrency: Structure versus Automata*, pages 41–101, Berlin, Heidelberg, 1996. Springer.
[12] E.A. Emerson and E.M. Clarke. Using branching time temporal logic to synthesize synchronization skeletons. *Sci. Comput. Program.*, 2(3):241–266, 1982.
[13] E.A. Emerson and J. Halpern. Decision procedures and expressiveness in the temporal logic of branching time. *Journal of Computer and System Sciences*, 30(1):1–24, 1985.
[14] R. Fagin, J.Y. Halpern, and N. Megiddo. A logic for reasoning about probabilities. *Information and Computation*, 87:78–128, 1990.
[15] Y. Feldman. A decidable propositional dynamic logic with explicit probabilities. *Information and Control*, 63(1):11–38, 1984.
[16] D. Gabbay, A. Pnueli, S. Shelah, and J. Stavi. On the temporal basis of fairness. In *Conference Record of the Seventh Annual ACM Symposium on Principles of Programming Languages, Las Vegas, Nevada, USA, January 1980*, pages 163–173, 1980.
[17] D. Guelev. A propositional dynamic logic with qualitative probabilities. *Journal of Philosophical Logic*, 28:575–605, 1999.
[18] D. Guelev. Probabilistic neighbourhood logic. In J. Mathai, editor, *Formal Techniques in Real-Time and Fault-Tolerant Systems*, pages 264–275, Berlin, Heidelberg, 2000. Springer.
[19] D. Guelev. Probabilistic interval temporal logic and duration calculus with infinite intervals: Complete proof systems. *Logical Methods in Computer Science*, 3(3), 2007.
[20] J.Y. Halpern and R. Fagin. Two views of belief: Belief as generalized probability and belief as evidence. *Artificial Intelligence*, 54(2):275–317, 1992.
[21] J.Y. Halpern and R. Pucella. A logic for reasoning about evidence. *Journal of Artificial Intelligence Research*, 26:1–34, 2006.
[22] H. Hansson and B. Jonsson. A logic for reasoning about time and reliability. *Formal Asp. Comput.*, 6(5):512–535, 1994.
[23] S. Hart and M. Sharir. Probabilistic temporal logics for finite and bounded models. In *16th ACM Symposium on Theory of Computing*, STOC '84, pages 1–13, New York, NY, USA, 1984. ACM.
[24] D.V. Hung and Z. Chaochen. Probabilistic duration calculus for continuous time. *Formal Asp. Comput.*, 11(1):21–44, 1999.
[25] J.A.W. Kamp. *Tense Logic and the Theory of Linear Order*. University Microfilms, 1968.
[26] D. Kozen. A probabilistic PDL. *Journal of Computer and System Sciences*, 30(2):162–178, 1985.

[27] D. Lehmann and S. Shelah. Reasoning with time and chance. *Information and Control*, 53(3):165–198, 1982.

[28] O. Lichtenstein and A. Pnueli. Propositional temporal logics: Decidability and completeness. *Logic Journal of the IGPL*, 8(1):55–85, 2000.

[29] Z. Liu, A. P. Ravn, E. V. Sørensen, and C. Zhou. A probabilistic duration calculus. In H. Kopetz and Y. Kakuda, editors, *Responsive Computer Systems*, pages 29–52, Vienna, 1993. Springer Vienna.

[30] Z. Manna and A. Pnueli. Verification of concurrent programs, part i: The temporal framework. 1981.

[31] B. Marinković, P. Glavan, Z. Ognjanović, and T. Studer. A Temporal Epistemic Logic with a Non-rigid Set of Agents for Analyzing the Blockchain protocol. *Journal of logic and computation*, 29(5):803–830, 2019.

[32] B. Marinković, Z. Ognjanović, D. Doder, and A. Perović. A propositional linear time logic with time flow isomorphic to ω^2. *Journal of Applied Logic*, 12(2):208–229, 2014.

[33] Z. Ognjanović. Discrete Linear-time Probabilistic Logics: Completeness, Decidability and Complexity. *Journal of Logic and Computation*, 16(2):257–285, 2006.

[34] Z. Ognjanović, D. Doder, and Z. Marković. A Branching Time Logic with Two Types of Probability Operators. In S. Benferhat and J. Grant, editors, *Fifth International Conference on Scalable Uncertainty Management SUM-2011, October 10 - 12, 2011, Dayton, Ohio, USA*, volume 6929 of *Lecture Notes in Computer Science*, pages 219–232. Springer, 2011.

[35] Z. Ognjanović, Z. Marković, M. Rašković, D. Doder, and A. Perović. A propositional probabilistic logic with discrete linear time for reasoning about evidence. *Annals of Mathematics and Artificial Intelligence*, 65(2-3):217–243, 2012.

[36] Z. Ognjanović, A. Perović, and D. Doder. A first-order dynamic probability logic. In L.C. van der Gaag, editor, *Proceedings of the 12th European Conference Symbolic and Quantitative Approaches to Reasoning with Uncertainty, ECSQARU 2013, Utrecht, The Netherlands, July 8-10, 2013*, volume 7958 of *Lecture Notes in Computer Science*, pages 461–472. Springer, 2013.

[37] A. Pnueli. The temporal logic of programs. In *18th Annual Symposium on Foundations of Computer Science, Providence, Rhode Island, USA, 31 October - 1 November 1977*, pages 46–57. IEEE Computer Society, 1977.

[38] A.N. Prior. *Time and Modality*. Oxford University Press, Oxford, 1957.

[39] M. Reynolds. An axiomatization of full computation tree logic. *The Journal of Symbolic Logic*, 66(3):1011–1057, 2001.

[40] K. Segerberg. Qualitative Probability in a Modal Setting. In E. Fenstad, editor, *Proceedings 2nd Scandinavian Logic Symposium*, volume 63 of *Studies in Logic and the Foundations of Mathematics*, pages 341–352, Amsterdam, 1971. North-Holland.

[41] A.P. Sistla and E.M. Clarke. The complexity of propositional linear temporal logic. *Journal of the ACM*, 32(3):733–749, 1985.

[42] C. Stirling. Handbook of logic in computer science (vol. 2). chapter Modal and Temporal Logics, pages 477–563. Oxford University Press, Inc., New York, NY, USA, 1992.

[43] S. Tomović, Z. Ognjanović, and D. Doder. Probabilistic Common Knowledge Among Infinite Number of Agents. In S. Destercke and T. Denoeux, editors, *Proceedings of the 13th European Conference Symbolic and Quantitative Approaches to Reasoning with Uncertainty ECSQARU 2015, Compiegne, France, July 15-17, 2015*, volume 9161 of *Lecture Notes in Computer Science*, pages 496–505, 2015.

[44] S. Tomović, Z. Ognjanović, and D. Doder. A First-Order Logic for Reasoning about Knowledge and Probability. *ACM Transactions on Computational Logic*, 21(2):16:1–16:30, 2020.

[45] E. Tzanis and R. Hirsch. Probabilistic logic over paths. *Electronic Notes in Theoretical Computer Science*, 220(3):79–96, 2008. Proceedings of the Sixth Workshop on Quantitative Aspects of Programming Languages (QAPL 2008).

[46] J. van Benthem. *The logic of time - a model-theoretic investigation into the varieties of temporal ontology and temporal discourse, 2nd Edition*, volume 156 of *Synthese library*. Kluwer, 1991.

[47] M. van Zee, M. Dastani, D. Doder, and L.W.N. van der Torre. Consistency conditions for beliefs and intentions. In *2015 AAAI Spring Symposia, Stanford University, Palo Alto, California, USA, March 22-25, 2015*. AAAI Press, 2015.

[48] M. van Zee and D. Doder. AGM-style revision of beliefs and intentions. In *ECAI 2016 - 22nd European Conference on Artificial Intelligence, 29 August-2 September 2016, The Hague, The Netherlands - Including Prestigious Applications of Artificial Intelligence (PAIS 2016)*, pages 1511–1519, 2016.

[49] M. van Zee, D. Doder, M. Dastani, and L.W.N. van der Torre. AGM revision of beliefs about action and time. In Q. Yang and M. J. Wooldridge, editors, *Proceedings of the Twenty-Fourth International Joint Conference on Artificial Intelligence, IJCAI 2015, Buenos Aires, Argentina, July 25-31, 2015*, pages 3250–3256. AAAI Press, 2015.

Chapter 4
Probabilistic Modeling of Default Reasoning

Nebojša Ikodinović and Zoran Ognjanović

Abstract This chapter addresses the main approaches to formalization of default reasoning in first-order framework. The first part of the chapter overviews the historical development of nonmonotonic reasoning systems. After discussing three well-known early formalisms (Default Logic, Nonmonotonic modal logic and Circumscription), we focus on nonmonotonic consequence relations and their close connections with conditional logics and probabilistic reasoning systems. The logic $L_{\omega\omega}^{\mathrm{P},\mathbb{I}}$ occupies the central position in this chapter. This logic extends classical first-order logic with a list of Keisler-style conditional probability quantifiers with the intention to express statistical knowledge as well as approximate probabilities. We provide a sound and complete deductive system with respect to semantics based on first-order structures endowed with suitable probability spaces. In addition, we describe two decidable fragments of $L_{\omega\omega}^{\mathrm{P},\mathbb{I}}$ that seem to be rich enough for many practical purposes. Finally, we discuss how the logic $L_{\omega\omega}^{\mathrm{P},\mathbb{I}}$ and its suitable extensions can be used to model default reasoning and analyze some properties of nonmonotonic consequence relations in the light of probabilistic inference.

4.1 Introduction

Human commonsense reasoning differs from traditional logical reasoning in at least three important respects. People manage to cope with situations when their knowledge is:

Nebojša Ikodinović
Faculty of Mathematics, University of Belgrade, Studentski trg 16, 11000 Belgrade, Serbia, e-mail: ikodinovic@matf.bg.ac.rs

Zoran Ognjanović
Mathematical Institute of the Serbian Academy of Sciences and Arts, Kneza Mihaila 36, 11000 Belgrade, Serbia, e-mail: zorano@mi.sanu.ac.rs

Z. Ognjanović (ed.), *Probabilistic Extensions of Various Logical Systems*,
https://doi.org/10.1007/978-3-030-52954-3_4

- incomplete,
- contradictory and/or
- defeasible.

The first, we usually assume vast numbers of conditions without explicit justification. In the absence of full information (because some pieces of information are unavailable, or because certain decisions must be made quickly), the available information has to be augmented with additional "normal" assumptions in order to overcome the incompleteness. Next, in situations where evidence is contradictory we would like to obtain acceptable conclusions and, at the same time, to block mutually inconsistent consequences that would be derived in classical logic. An example of such situation is the so-called Nixon diamond when we assume that (usually) Quakers are pacifist while Republicans are not, and there are persons (e.g. Richard Nixon) that are both Quakers and Republicans. Finally, commonsense reasoning has to be flexible. For example, in a well-known example, we accept that the bird Tweety flies until we learn that it is a penguin, after which we reject the previous conclusion. More precisely, commonsense conclusions based on information that are not irrefutable, may have to be retracted (withdrawn) at some later time in the light of new evidence. Thus, a key property of such flexible inference is that of nonmonotonicity. Reasoning is nonmonotonic if additional information may invalidate previous conclusions. Of course, classical logic is monotonic: whenever a formula is a consequence of a set of premises T, then that formula is also a consequence of any superset of T. In other words, adding information never invalidates any conclusions.

Assumptions that could be accepted in the above mentioned situations are generally named as defaults. A default might be regarded as (proto)typical information, a defeasible presumption or a rule of thumb, and its main role is to fill in gaps in knowledge. The previous observations capture the essence of default reasoning. Roughly speaking, default reasoning is any kind of inference in which reasoners draw conclusions based on both supporting evidence and defaults, reserving the right to retract conclusions in the face of further information.

The previous observations capture the essence of default reasoning. Roughly speaking, default reasoning is any kind of inference in which reasoners draw conclusions based on both supporting evidence and defaults, reserving the right to retract the conclusions in the face of further information.

In the late 1970's, various problems in Computer Science and AI initiated intensive development of formal frameworks for default reasoning. Database theory was one of the earliest sources of examples, especially as regards the closed world assumption, see [40]. Shortly afterwards, the importance of default reasoning have been recognized in a wide range of application domains such as medical diagnosis, legal reasoning, intelligent systems, natural language understanding etc. In 1980, the journal Artificial Intelligence published an issue (vol. 13 (1-2)) dedicated to three well-known formalisms: Default logic [41], Modal nonmonotonic logic [34], and Circumscription [32]. Today, forty years later, the family of formal systems for default reasoning is extraordinarily rich so that it is almost impossible to make a detailed overview of all approaches. In Section 4.2, we present the basic ideas behind some of the most influential approaches. For a more detailed overview of formalisms we refer the reader to the following excellent surveys: [11, 12, 18, 42].

All above approaches have been intensively studied and further developed. In the late 1980's, there was a growing need for a general framework in which those approaches could be compared and classified. A new line of research was initiated by [19] and [31]. Their main idea was to understand default reasoning in terms of structural properties of nonmonotonic consequence relations. In Section 4.3 we shortly present this "orthogonal" area in default reasoning research, that is we outline an abstract study of nonmonotonic inference relations. Special attention will be paid to System **P**, i.e., to the core properties of nonmonotonic reasoning systems, see [25]. Moreover, we highlight some close connections between consequence relations and conditional logic.

In Section 4.4, we present a probabilistic first-order logic introduced in [23]. That logic is an extension of ordinary first-order logic which enables us to generalize considerations from the previous section. We use $L^{\mathrm{P},\mathbb{I}}_{\omega\omega}$ to denote our logic in order to emphasize that it is an extension of first order classical logic $L_{\omega\omega}$, while P and $\mathbb{I}$ indicate that the values of the probabilistic functions will belong to $\mathbb{I}$. The logic $L^{\mathrm{P},\mathbb{I}}_{\omega\omega}$ is presented in detail: syntax and semantics are specified, and sound and complete deductive system is provided. Moreover, we have found two very expressive fragments of $L^{\mathrm{P},\mathbb{I}}_{\omega\omega}$ being decidable. We also discuss how one could straightforwardly develop some extensions of $L^{\mathrm{P},\mathbb{I}}_{\omega\omega}$ by adding new (qualitative and/or quantitative) quantifiers which might be very interesting for further applications.

In Section 4.6 we describe how our system can be used to model default reasoning and analyze some properties of the corresponding default consequence relation. We conclude in Section 4.7.

4.2 Some formal systems for default reasoning

The large majority of work in default reasoning has centered on propositional systems. There is a general agreement that predicate framework is more adequate for modeling phenomena related to default reasoning. In this chapter, we consider only those formal systems that are based on traditional first-order logic (FOL). The general idea is to augment first-order logic with some mechanism for predicting conclusions on the absence of specific knowledge.

Although we assume the reader to be familiar with first-order logic (cf. [9, 17]), some basic concepts and notations will be outlined (in addition to notations adopted in Chapter 1). A standard (possibly many-sorted) first-order language L will be used for knowledge representation. It is enough to specify only the non-logical part of a language, i.e., constant, function and relation symbols used in a particular application. As usual, constant symbols can be seen as 0-ary function symbols.

First-order derivability will be denoted by $\vdash$ and the corresponding consequence operator by Cn; $\mathrm{Cn}(T) = \{\alpha \mid T \vdash \alpha\}$; $\mathrm{Cn}(T)$ is the deductive closure of a set of formulas T. A theory T is deductively closed if $\mathrm{Cn}(T) = T$. The fundamental computational intractability of first-order derivability is a reason for limiting to a certain more manageable subsets of first-order formulas: atomic formulas, literals (a literal is either an atomic formulas or its negation), clauses (a clause is a disjunction of literals), Horn clauses (a Horn clause is a clause containing at most one positive literal) etc.

Given a language L, an L-structure (L-model) is determined by its universe M and by the interpretations $S^{\mathbf{M}}$ of all constant, function and predicate symbols from the language. The first-order entailment is denoted by $\models$. A variable valuation (assignment) in $\mathbf{M}$ is a map $v : \mathrm{Var} \to M$ that associates an element of the universe with each variable. If $\overline{x} = (x_1, \ldots, x_n)$ is a tuple of pairwise distinct variables and $\overline{a} \in A^n$, $v(\overline{a}/\overline{x})$ is the valuation which maps x_i to a_i, $i = 1, n$, and agrees with v on all variables distinct from x_i's. An L-interpretation is a pair $(\mathbf{M}, v)$ consisting of L-structure $\mathbf{M}$ and a valuation v in $\mathbf{M}$. Using induction on terms, we associate with every interpretation $(\mathbf{M}, v)$ and every term t an element $t^{\mathbf{M},v}$ from M. Next, we define the satisfiability relation $\mathbf{M}, v \models \varphi$ using induction on formulas. Given an interpretation $(\mathbf{M}, v)$, any formula φ and a tuple $\overline{x}$ of pairwise distinct variables determine the set $[\varphi]^{\overline{x}}_{\mathbf{M},v} =_{\mathrm{def}} \{\overline{a} \in M^n \mid \mathbf{M}, v(\overline{a}/\overline{x}) \models \varphi\}$.

The set of ground (i.e., variable-free) terms of a language L is called the Herbrand universe of L. The Herbrand universe is especially important for applications in artificial intelligence because it contains precisely the individuals for which the corresponding language provides names. The Herbrand universe is closely related to the structure produced in the proof of the completeness theorem by standard Henkin procedure.

There are two prominent approaches to formalization of default reasoning in predicate framework:

1. Proof-theoretic formalizations are typically based on fixed-point operators that generates (possible multiple) sets of acceptable believes (called extensions), taking into account certain first-order consistency conditions.
2. Model-theoretic formalizations are based on restriction of the semantics to some special (normal, preferred) models. Nonmonotonicity arises since adding new information changes the set of preferred models: models that were not preferred before may become preferred once we learn new facts.

4.2.1 Default logic

Default Logic is one of the most widely-used formalisms for default reasoning. It was introduced in [41]. A detailed description of Default logic, its properties and variations can be found in [3, 11, 45]. In Default logic, defaults are represented as nonstandard rules of inference which are added to standard first-order logic. A default rule is any expression of the form

$$(*) \qquad \frac{\alpha(\overline{x}) : \beta_1(\overline{x}), \ldots, \beta_m(\overline{x})}{\gamma(\overline{x})},$$

where $\alpha(\bar{x})$ (called *prerequisite*), $\beta_j(\bar{x})$'s (*justifications*), and $\gamma(\bar{x})$ (*consequent*) are first-order formulas whose free variables are among $\bar{x} = (x_1, \ldots, x_n)$, for $m \geq 1$ and $n \geq 0$. If none of the formulas in $(*)$ contains free variables, the default $(*)$ is said to be *closed*, otherwise it is *open*.

The informal interpretation of $(*)$ is that $\gamma(\bar{x})$ should be believed if $\alpha(\bar{x})$ is believed and it is consistent to believe $\beta_j(\bar{x})$'s. That is, if our initial theory contains the prerequisite of $(*)$, and we do not know the justifications, but we can consistently assume that they do not contradict our resulting theory, then we can apply default rule $(*)$ and extend our initial theory with the consequent.

Almost all default rules in the literature have only a single justification. A default rule where the justification and consequent are the same is called a *normal* rule: $\alpha(\bar{x}) : \beta(\bar{x})/\beta(\bar{x})$. Such a rule is by far the most common case, see [30]: if we know $\alpha(\bar{x})$, then we should assume that $\beta(\bar{x})$ if it is consistent to assume $\beta(\bar{x})$. Even normal defaults without prerequisite could be interesting. Intuitively, $: \alpha(x)/\alpha(x)$ expresses that $\alpha(x)$ is assumed to be true whenever possible.

A default theory (T, D) consists of a consistent set of formulas T (representing, generally incomplete, description of what is known to be true), and a set of default rules D (that provide means for filling in the gaps in our initial knowledge). Roughly speaking, default rules are used as mappings from an incomplete theory to a more complete extension of it. An *extension* of a default theory is the crucial concept in Default logic. Reiter first defined the concept of an extension for closed default theories (with closed defaults only), and then generalized it to arbitrary default theories (possibly with open defaults). Extensions are defined by a fixed point construction.

Let (T, D) be a closed default theory. For any set of sentences S, let $\Gamma(S)$ be the smallest deductively closed (under classical logic) set of sentences which includes S, and satisfies the condition:

[Rei] For any default $(*)$ from D, if $\alpha \in \Gamma(S)$ and $\neg\beta_1, \ldots, \neg\beta_m \notin S$ then $\gamma \in \Gamma(S)$.

$\Gamma(S)$ can be considered as the set of all formulas that can be derived from S by means of classical derivability and by using those defaults whose each justification turns out to be consistent with S. In other words, if β_j's turn out to be consistent with S, the default $\alpha : \beta_1, \ldots, \beta_m/\gamma$ starts to function as the inference rule α/γ, other defaults are ignored.

A set of sentences E is said to be an extension for (T, D), if it is a fixpoint of Γ, that is, if $\Gamma(E) = E$. The following pseudo-iterative construction is somewhat more comprehensive characterization of extensions. Let E be a set of formulas. Define $E_0 = T$, and for $i \geq 0$,

$$E_{i+1} = \mathrm{Cn}(E_i) \cup \left\{ \gamma \mid \frac{\alpha : \beta_1, \ldots, \beta_m}{\gamma} \in D, \alpha \in E_i, \neg\beta_1, \ldots, \neg\beta_m \notin E \right\}.$$

Then E is an extension of (T, D) iff $E = \cup_{i=0}^{\infty} E_i$.

Default theories can have an arbitrary number of extensions, including having no extension at all. Each extension E determines an inference relation: a formula is entailed by a default theory if it belongs to the given extension. Further, one can use extensions to define a *skeptical* inference relation (a formula is skeptically entailed by a default theory if it belongs to all of its extensions), or a *credulous* inference relation (a formula is credulously entailed by a default theory if it belongs to at least one of its extensions). The key reasoning problems for default logic are deciding skeptical and credulous inference and finding extensions. For first-order default logic these problems are not even semi-decidable, see [41].

Whereas closed defaults have been quite thoroughly investigated, open ones are much less studied. However, interesting cases of default reasoning usually deal with open defaults, because the intended use of defaults is to determine, whether an object possesses a given property, rather than accepting or rejecting a closed statement.

Example 4.1. Let us consider how Default logic deals with the following informal reasoning ([9]):

- "Hard" premises: Tweety is a bird. Chilly is a bird. Chilly does not fly.
- "Soft" premise: Most birds fly.
- Conclusion: Does Tweety fly? Yes, since there is no reason to think that Tweety does not belong to the majority of birds, i.e., to the flying birds.

We have the following default theory:

$T = \{\mathtt{Bird(tweety)}, \mathtt{Bird(chilly)}, \neg\mathtt{Flies(chilly)}\}$,
$D = \{\mathtt{Bird}(x) : \mathtt{Flies}(x)/\mathtt{Flies}(x)\}$.

Here, the only default rule has one free variable, and it should be understood as the set of all ground instations (formed by replacing the variable with a ground term). Thus, in our case, the default rule stands for the rules:

$$\frac{\mathtt{Bird(tweety)} : \mathtt{Flies(tweety)}}{\mathtt{Flies(tweety)}}, \frac{\mathtt{Bird(chilly)} : \mathtt{Flies(chilly)}}{\mathtt{Flies(chilly)}}$$

We use a short characterization of extensions: A set of sentences E is an extension of our default theory (T, D) iff for every sentence φ,

$$\varphi \in E \text{ iff } T \cup \left\{ \gamma \mid \frac{\alpha : \beta}{\gamma} \in D, \alpha \in E, \neg\beta \notin E \right\} \models \varphi.$$

It is easy to show that the set E of the entailments of $T \cup \{\mathtt{Flies(tweety)}\}$ is the unique extension.

Next, let us consider another similar reasoning.

- "Hard" premises: Chilly is a penguin. All penguins are birds.
- "Soft" premises: Most birds fly. Most penguins do not fly.
- Conclusion: Does Chilly fly? No, since more specific (by set inclusion) information should win.

The following default theory corresponds to this reasoning.

$T_1 = \{\mathtt{Penguin(chilly)}, \forall x(\mathtt{Penguin}(x) \to \mathtt{Bird}(x))\}$,
$D_1 = \{\mathtt{Bird}(x) : \mathtt{Flies}(x)/\mathtt{Flies}(x), \mathtt{Penguin}(x) : \neg\mathtt{Flies}(x)/\neg\mathtt{Flies}(x)\}$.

Now, we have two extensions: one where Chilly is assumed to fly and one where Chilly is assumed not to fly. Arbitrating among conflicting default rules is a severe (challenging) problem of Default logic, and there are many ideas (strategies) how to cope with it ([45]). ■

Reiter treated a default with free variables as the set of all its ground instances. He defines extensions of arbitrary default theories (T,D) (i.e., default theories with open defaults) as follows. First, the formulae of T and the consequents of the defaults from T are Skolemized, i.e., existentially quantified variables are replaced by Skolem functions whose arity is the number of universal quantifiers in whose scope the existential quantifier lies. Second, a (possibly infinite) set D' of closed default rules is generated by taking all ground instances (over the initial signature together with the newly introduced Skolem functions) of the Skolemized versions of the defaults of D. There are many examples that demonstrate that Reiter's treatment of open defaults is problematic, see [4].

[28] proposed a treatment of open defaults which avoid Skolemization by working with classes of models instead of sets of formulas in the definition of default extensions. For any nonempty set U, let L_U be an enrichment of a language L with new constant symbols to serve as names for all the elements of U. For a set of models $\mathfrak{M}$ with the universe U, let $\mathrm{Th}_U(\mathfrak{M})$ be the set of sentences in L_U which are true in all models from $\mathfrak{M}$. For any set of models $\mathfrak{M}$, consider the largest set of models $\Delta(\mathfrak{M})$ which satisfies the condition:

[Lif] For any default $(*)$ from D and any tuple of constant symbols $\overline{c}$, if $\alpha(\overline{c}) \in \mathrm{Th}_U(\Delta(\mathfrak{M}))$ and $\neg\beta_1(\overline{c}),\ldots,\neg\beta_m(\overline{c}) \notin \mathrm{Th}_U(\mathfrak{M})$ then $\gamma(\overline{c}) \in \mathrm{Th}_U(\Delta(\mathfrak{M}))$.

A U-extension for (T,D) is any set of sentences of the form $\mathrm{Th}(\mathfrak{E})$, where $\mathfrak{E}$ is a fixpoint of Δ. Notice that U-extensions, just like extensions in Reiter's logic, consist of sentences in the language of (T,D); they do not contain new names for the elements of U. For every U, there is exactly one U-extension, see [28]. Extensions determine inference relation in a similar way as in Reiter's logic, and so we omit details. It is worth to mention that the definition of a U-extension can be expressed by a higher order logical formula, so that Lifschitz's modification of Default Logic can be viewed as a syntactic transformation of higher-order sentences. In general, Lifschitz's and Reiter's version of Default Logic are not equivalent, even for closed default theories.

4.2.2 Autoepistemic logic

In Reiter's Default logic, defaults are rules whose role is to complete an underlying first-order theory, and we actually reason in first-order framework using additional default rules. We cannot reason about defaults because there is no notion of entailment among them. This is regarded as one serious disadvantage of Default Logic, see [9]. To overcome these shortcomings, many systems have been developed to encode defaults as formulas in an appropriate language, rather than inference rules.

Such a system is so-called modal nonmonotonic logic proposed by [34]. In that approach, a first-order language L is augmented with a unary modal operator M, where a formula $\mathsf{M}\alpha$ is to be read as "α is consistent (with everything believed)", and roughly thought of as $\not\vdash \neg$. Defaults are formulas of this modal extension of standard first-order logic. The operator M is somehow related to Unless operator in [44].

In this approach, default reasoning is based on a fixed-point construction which is similar to that of Reiter's Default logic, but formulated entirely in the modal language. We omit details. Although any default $\alpha : \beta_1, \ldots, \beta_m / \gamma$ could be intuitively approximated by $\alpha \wedge \mathsf{M}\beta_1 \wedge \cdots \wedge \mathsf{M}\beta_m \rightarrow \gamma$, the extensions of default theories and the fixed-points of modal non-monotonic theories are incomparable in general, see [42].

Moore's criticism of modal nonmonotonic logic, see [35, 36], was based on a clear distinction between default reasoning and "autoepistemic" reasoning (about our own beliefs). Moore argues that the possible sets of beliefs based on a consistent set of premises T, are those sets E such that for every sentence α:

$$\alpha \in E \text{ iff } T \cup \{\mathsf{M}\beta \mid \beta \in E\} \cup \{\neg\mathsf{M}\beta \mid \beta \notin E\} \models_{\mathsf{M}} \alpha,$$

where $\models_{\mathsf{M}}$ denotes the corresponding modal entailment relation. These sets are called *stable extensions* of T. A stable extension includes the premises, and accurately determined what is and what is not believed.

Example 4.2. Let us look at the following theory:

$\mathtt{Bird(chilly)}, \mathtt{Bird(tweety)}, \mathtt{tweety} \neq \mathtt{chilly}, \neg\mathtt{Flies(chilly)}$,
$\forall x(\mathtt{Bird}(x) \wedge \neg\mathsf{M}\neg\mathtt{Flies}(x) \rightarrow \mathtt{Flies}(x))$.

We use the simple fact that any stable extension E satisfies the following minimal properties ([9]):

1. Closure under entailment: if $E \models_{\mathsf{M}} \alpha$, then $\alpha \in E$;
2. Positive introspection: if $\alpha \in E$, then $\mathsf{M}\alpha \in E$;
3. Negative introspection: if $\alpha \notin E$, then $\neg\mathsf{M}\alpha \in E$;

Let us informally consider the consequences of the above theory. First, we see that there is no way to conclude $\neg\mathtt{Flies(tweety)}$. Indeed, our theory does not contain $\neg\mathtt{Flies(tweety)}$, and there is no way that would allow us to conclude it (any conclusion of our "default" statement is of the form $\mathtt{Flies}(x)$). Because of the negative introspection property, a stable extension that did not include the fact $\neg\mathtt{Flies(tweety)}$ would include $\neg\mathsf{M}\neg\mathtt{Flies(tweety)}$. Now, using this fact and $\forall x(\mathtt{Bird}(x) \wedge \neg\mathsf{M}\neg\mathtt{Flies}(x) \rightarrow \mathtt{Flies}(x))$, we conclude $\mathtt{Flies(tweety)}$ by ordinary logical entailment. ■

Moore's approach has been turned into many logics for knowledge and belief, and in particular led to autoepistemic logic. This logic responds to many problems with the initial formulation of modal nonmonotonic logic, but in a propositional framework. There are only modest steps in developing a first-order version. One of the main problems arises when quantifiers bind open variables that lie within the scope of a modality.

4.2.3 Circumscription

One of the earliest formalisms for default reasoning was introduced in [32, 33]. His approach makes explicit the intuition that interpretations of an incomplete theory should be minimal in a certain sense.

The simplest way to illustrate the basic idea of circumscription is in terms of abnormalities. That means that defaults are represented as formulas in a first-order language extended with so-called abnormality predicates. If we have a predicate A to talk about the exceptional or abnormal cases where a default should not apply, the strategy for making default conclusions is to consider only those models where the interpretation of A is as small as possible. In other words, the strategy is to minimize abnormality. This is called circumscribing the predicate A, and the technique is called circumscription. The root of this idea is in the closed world reasoning, see [40], one of the simplest way for dealing with incompleteness in AI.

Example 4.3. Suppose we have the following theory:

$$\forall x(\texttt{Bird}(x) \wedge \neg A(x) \rightarrow \texttt{Flies}(x)), \texttt{Bird(tweety)},$$
$$\texttt{Bird(chilly)}, \texttt{Bird(tweety)}, \texttt{tweety} \neq \texttt{chilly}, \neg\texttt{Flies(chilly)}$$

We would like to conclude, by default, that Tweety flies. Our theory does not classically entail `Flies(tweety)`, because there are models of the theory, where `Flies(tweety)` is false. In this example, we know that Chilly is an abnormal bird, but we do not know that about Tweety. The default assumption we make is that the interpretation of A is as large as it has to be given what we know. Hence, Tweety is not abnormal bird, and we conclude: "Tweety flies". ■

Semantically, circumscription forces minimal sets of predicate instances. Let L be a first-order language augmented with abnormality predicates from $\mathcal{A} = \{A_1,\ldots,A_n\}$. Given two models $\mathbf{M}$ and $\mathbf{M}'$ over the same universe M, we are saying that $\mathbf{M}$ is $\mathcal{A}$-less than $\mathbf{M}'$, written $\mathbf{M} \leq_{\mathcal{A}} \mathbf{M}'$, if $A_i^{\mathbf{M}} \subseteq A_i^{\mathbf{M}'}$, and $S^{\mathbf{M}} = S^{\mathbf{M}'}$, for any symbol S different from A_i's. Intuitively, this ordering makes the interpretations of all the abnormality predicates smaller. We can think of an model that is less than another as more normal. With this idea, one can define a new form of entailment $\models_{\leq}$, called minimal entailment, as follows:

> $T \models_{\leq} \alpha$ iff for every model $\mathbf{M}$ such that $\mathbf{M} \models T$, either $\mathbf{M} \models \alpha$ or there is a model $\mathbf{M}'$ such that $\mathbf{M}' <_{\mathcal{A}} \mathbf{M}$ and $\mathbf{M}' \models T$.

That is, a formula α follows from a theory T by circumscription, if α is true in all models of T that are minimal. Notice that α is not required to be true in all models satisfying T, but only in the minimal (most normal) ones satisfying the T. Predicate circumscription could be syntactically realized in second-order logic by adding the following axiom schemata:

$$\left(\alpha(A_1,\dots,A_n)\wedge\bigwedge_{i=1}^{n}\forall\overline{x}(P_i(\overline{x})\rightarrow A_i(\overline{x})\right)\rightarrow\bigwedge_{i=1}^{n}\forall\overline{x}(A_i(\overline{x})\rightarrow P_i(\overline{x})),$$

where P_i's are predicate variables, with the same arities as A_i's, respectively. Lifschitz has investigated many aspects of circumscription, and we refer the reader to his excellent overview in [29].

In general, the relationship between Default logic and Circumscription appears to be complex. As already mentioned, the semantic intuition underlying circumscription is to force minimal sets of predicate instances. Unlike this "minimalist reasoning", Default logic is more flexible in determining which models are considered interesting. Reiter's default rules provide a uniform strategy for the completion of partial knowledge. Nevertheless, the effect of a normal default with no prerequisite : $\alpha(\overline{x})/\alpha(\overline{x})$ is to "maximize" α, i.e., to "minimize" $\neg\alpha$.

Circumscription has initiated the development of a variety of semantical approaches to nonmonotonic reasoning, which are usually referred to as *preferential models approaches.* These issues are discussed in the next section.

4.3 Nonmonotonic consequence relations

One important way to study formal behavior of the above systems for default reasoning is to investigate the corresponding inference procedures. The following general pattern emerges: Starting with a consequence operator Cn and an operation $\mathcal{E}$ that associates with each set S of premises a set $\mathcal{E}(S)$ of additional "explicit presumptions", one can define a new consequence operator $\mathrm{C}(A)=\mathrm{Cn}(S\cup\mathcal{E}(S))$. Thus, $\alpha\in\mathrm{C}(S)$, also denoted as $S\vdash_{\mathrm{C}}\alpha$, informally means that "α is believed in the context S".

In the 1930's, Tarski developed a very general approach to the concept of monotone consequence relation. He considered a logical system as an ordered pair (For, C) where For consists of formulas of a formal language and C is a consequence operator on For. We remind the reader that there exists a bijective correspondence between all consequence operators over For and all consequence relations over For. Each consequence operator $\mathrm{C}:\mathcal{P}(\mathrm{For})\rightarrow\mathcal{P}(\mathrm{For})$ defines the relation $\vdash_{\mathrm{C}}$ between subsets of For and elements of For, $S\vdash_{\mathrm{C}}\alpha$ iff $\alpha\in\mathrm{C}(S)$. Each consequence relation $\vdash$ determine the operator $\mathrm{C}_{\vdash}(S)=\{\alpha\mid S\vdash\alpha\}$, $S\subseteq\mathrm{For}$.

There are three properties that are the essential properties for being a notion of monotone consequence.

• Inclusion (alias Reflexivity):	
(1) If $\alpha \in A$ then $A \vdash \alpha$.	(i) $A \subseteq C(A)$
• Cumulative transitivity (alias Cut, or Idempotence):	
(2) If $A, \{\beta_i \mid i \in I\} \vdash \alpha, A \vdash \beta_i, i \in I,$ then $A \vdash \alpha$.	(ii) $A \subseteq B \subseteq C(A)$ implies $C(B) \subseteq C(A)$.
[(2′) If $\{\beta \mid A \vdash \beta\} \vdash \alpha$, then $A \vdash \alpha$.]	[(ii′) $C(C(A)) \subseteq C(A)$]
• Monotonicity:	
(3) If $A \subseteq B$ and $A \vdash \alpha$, then $B \vdash \alpha$.	(iii) $A \subseteq B$ implies $C(A) \subseteq C(B)$.

A general agreement among many researchers is that a good candidate for nonmonotonic consequence relation should have the following properties, see [19, 25, 31]:

- Inclusion: $A \subseteq C(A)$;
- Cumulative Transitivity: $A \subseteq B \subseteq C(A)$ implies $C(B) \subseteq C(A)$
- Cumulative Monotony: $A \subseteq B \subseteq C(A)$ implies $C(A) \subseteq C(B)$.

The last two properties can be combined into a single principle of cumulation: $A \subseteq B \subseteq C(A)$ implies $C(A) = C(B)$.

[31] showed that Reiter default inference does not need to satisfy neither Cumulative Monotony nor even Cumulative Transitivity, even when rules are normal. The same failures arise for many variants of autoepistemic logics. On the contrary, circumscriptive formalisms yield nonmonotonic consequence relations satisfying Inclusion, Cumulative Transitivity, Cumulative Monotony, and many other nice properties. These results somewhat justify the existence of two "orthogonal" approaches to modeling default reasoning.

In the next subsection, we focus on consequence relations between two formulas, rather than between a set of formulas and a single formula. As usual, we write $\mid\!\sim$ (called snake) for a nonmonotonic consequence relation.

4.3.1 *System* P

A consequence relation $\mid\!\sim$ is said to be *cumulative* if it contains all instances of the Reflexivity axiom:

REF $\alpha \mid\!\sim \alpha$ [reflexivity];

and it is closed under the so-called Gabbay-Makinson rules in Gentzen-style:

LLE $\dfrac{\models \alpha \leftrightarrow \beta \quad \alpha \mid\!\sim \gamma}{\beta \mid\!\sim \gamma}$ [left logical equivalence];

RW $\dfrac{\models \alpha \rightarrow \beta \quad \gamma \mid\!\sim \alpha}{\gamma \mid\!\sim \alpha}$ [right weakening];

CUT $\dfrac{\alpha \wedge \beta \mid\!\sim \gamma \quad \alpha \mid\!\sim \beta}{\alpha \mid\!\sim \gamma}$;

CM $\dfrac{\alpha \mid\!\sim \beta \quad \alpha \mid\!\sim \gamma}{\alpha \wedge \beta \mid\!\sim \gamma}$ [caution monotonicity].

These properties make an inference system known as System **C**. The rules Cut and Cautious monotony are particularly important, since they says that if β can be derived from α, then the set of conclusions derivable from $\alpha \wedge \beta$ is the same as that derivable from α alone.

A stronger system, denoted **P** (for preferential) has a central role in [25]. System **P** is an extension of **C** with the following rule:

OR $\dfrac{\alpha \mid\!\sim \gamma \quad \beta \mid\!\sim \gamma}{\alpha \vee \beta \mid\!\sim \gamma}$.

[26] suggested two rules that should be adding in the predicate calculus case:

EI $\dfrac{\alpha \mid\!\sim \beta}{\exists x\alpha \mid\!\sim \exists x\beta}$ [$\exists$-itroduction]

EE $\dfrac{\exists x\alpha \mid\!\sim \beta \quad x \text{ is not free in } \beta}{\alpha \mid\!\sim \beta}$ [$\exists$-elimination]

The authors also argued that the following rules should be rejected:

$$\frac{\alpha \mid\!\sim \beta}{\forall x\alpha \mid\!\sim \forall x\beta} \text{ and } \frac{\forall x\alpha \mid\!\sim \forall x\beta}{\alpha(x/t) \mid\!\sim \beta(x/t)}.$$

A preferential consequence relation that satisfies Rational Monotony:

RM $\dfrac{\alpha \mid\!\sim \gamma \quad \alpha \not\mid\!\sim \neg\beta}{\alpha \wedge \beta \mid\!\sim \gamma}$,

is called a rational consequence relation. The rule (RM) roughly says that $\alpha \mid\!\sim \gamma$ implies $\alpha \wedge \beta \mid\!\sim \gamma$ only in those cases when β is "irrelevant" to α, i.e., where "irrelevance" means that $\alpha \not\mid\!\sim \neg\beta$. Notice that the rule (RM) is structurally different from the rules of System **P** because it is non-Horn.

System **P** is often referred to as the core properties of nonmonotonic reasoning system or as the KLM properties. Such status of System **P** is underlined by the fact that numerous (quite different) semantics are adequate to characterize preferential consequence relations. Moreover, System **P** could be regarded as a fragment of conditional logic studied by [13].

4.3.2 *Conditional logic*

A natural framework for studying consequence relations could be the so-called flat (i.e. unnested) fragment of a conditional logic consisting of Boolean combinations of formulas $\alpha \mid\!\sim \beta$, where α, β do not contain $\mid\!\sim$. In this context, the operator $\mid\!\sim$ is seen as a binary connective, and it is usually denoted $\rightsquigarrow$ ($\rightarrowtail$, $\supset$, or something similar). Of course, we focus on conditional first-order logic, i.e., classical first-order logic augmented with a binary operator $\rightsquigarrow$. Although nesting of the conditional operator could be permitted, usually this is not important in the analysis of consequence relations. [14, 15] emphasized two benefits to studying conditional logics: 1) Conditional logics have an advantage over approaches such as Default Logic and Circumscription, in that one can reason about default conditionals, and it makes sense to talk of a set of defaults implying another, or of being unsatisfiable; 2) Given a conditional logic of defaults, one may subsequently determine what nonmonotonic inferences ought to obtain. The most basic form of default inference is given by:

β follows by default from α in theory T just when $T \models_C \alpha \rightsquigarrow \beta$,

where $\models_C$ denotes the corresponding entailment relation. Thus, conditional logics and their semantics played an important role in attempts to investigate nonmonotonic consequence relations.

There is a variety of approaches to nonmonotonic reasoning based on conditional logic. One of the most useful tools for the study of non-monotonic consequence relations is their semantical representations.

Preferential models give a best known model-theoretical representation of preferential consequence relation, see [25]. This semantics is based on a partial ordering on possible worlds (interpretations) over the same universe. A preferential model is a triple $\mathbf{W} = (W, <, M, \mathfrak{I})$ where W is a set, the elements of which will be called worlds, $<$ is a strict partial order on W, M is a set which is the universe for interpretations, $\mathfrak{I}$ assigns an interpretation over M to each world and satisfying the following two condition:

1) [smoothness condition] for each formula α, the $[\alpha]_{\mathbf{W}} = \{w \in W \mid \mathfrak{I}(w) \models \alpha\}$ is smooth, that is for all $w \in [\alpha]_{\mathbf{W}}$, either there is w' minimal in $[\alpha]_{\mathbf{W}}$ such that $w' < w$ or w is itself minimal in $[\alpha]_{\mathbf{W}}$;
2) [$\exists$-conditions]
   a. if a world w such that $\mathfrak{I}(w) = (\mathbf{M}, v)$ is minimal in $[\exists x\alpha]_{\mathbf{W}}$, then there exists a state w' minimal in $[\alpha]_{\mathbf{W}}$ such that $\mathfrak{I}(w') = (M, v[x := a])$ for some $a \in M$;
   b. if a world w such that $\mathfrak{I}(w) = (\mathbf{M}, v)$ is minimal in $[\alpha]_{\mathbf{W}}$, then there exists a state w' minimal in $[\exists x\alpha]_{\mathbf{W}}$ such that $\mathfrak{I}(w') = (M, v[x := a])$ for some $a \in M$.

The smoothness condition is a technical condition needed to deal with infinite sets of formulas; it is always satisfied in any finite preferential model, and in any model in which $<$ is well-founded (i.e., no infinite descending chains). Any preferential model $\mathbf{W}$ naturally defines a consequence relation:

$\alpha \mid\!\sim_{\mathbf{W}} \beta$ iff for any w minimal in $[\alpha]_{\mathbf{W}}$, $\mathfrak{I}(w) \models \beta$.

It turns out that each preferential relation is defined by a preferential model, see [25]. In [16], preferential models are further generalized in at least two respects:

- the domain of individuals may vary across possible worlds;
- the conditional operator may bind variables.

Delgrande studied semantics for classical first-order logic enriched with the binary operator $\rightsquigarrow$, a new quantifier symbol $\forall_C$ (besides the usual $\forall$), and a designated unary predicate symbol A whose interpretation at a world is the domain for $\forall_C$. Formulas are interpreted using models of the form $\mathbf{W} = (W, R, M, \mathfrak{I})$, where:

- W is a set of worlds;
- R is a ternary accessibility relation on the set of worlds, $R \subseteq W \times W \times W$; $R(w, w_1, w_2)$ has the informal reading "according to world w, w_2 is *more normal* than w_1"; instead of $R(w, w_1, w_2)$, it is convenient to use the more suggestive notation $w_1 <_w w_2$;
- M is a nonempty set that is the universe for interpretations;
- $\mathfrak{I}$ maps any world to an L-model over a subset M_w of M, such that interpretations of constant symbols are the same at any world, and the predicate symbol A is interpreted at a world w as M_w.

The intent is that there be a "universal" domain M, consisting of the set of all individuals. At each world w we have a nonempty set M_w of individuals "existing" or "actual" at w.

Given a model $\mathbf{W} = (W, R, M, \mathfrak{I})$ and a valuation v, truth at a world w is defined as usual with the new clause: the quantifier $\forall_C$ ranges over the set of individuals D_w at world w, that is

$\mathbf{M}, v, w \models \forall_C x \alpha$ iff for every $a \in A_w$ we have $\mathbf{M}, v(a/x), w \models \alpha$.

A sound and complete axiomatization with respect to the given semantics is given in [16]. The expressive power of this logic has been particularly approved by "variable binding" operators. The "variable binding" operator $\rightsquigarrow_{\overline{x}}$ is introduced by:

$$\alpha \rightsquigarrow_{\overline{x}} \beta \text{ is } \exists_C \overline{x} \alpha \rightsquigarrow \forall_C \overline{x}(\alpha \to \beta),$$

where $\exists_C$ is defined as $\neg\forall_C\neg$. In fact, the main role of $\exists_C$, $\forall_C$, and A is to explicate the meaning of $\rightsquigarrow_{\overline{x}}$, and there would be no need for these symbols to appear explicitly in the statement of a conditional (default) theory.

Delgrande's logic was partially inspired by Reiter's rules of the form $\frac{:\alpha(x)}{\alpha(x)}$, and their informal interpretation "normally, $\alpha(x)$ holds". According to the above semantics, $\alpha(x)$ holds normally (in a model) if there is a "large" subset od the universe of the model in which $\alpha(x)$ "really" holds. In other words, "normally" is read as a generalized quantifier. In the next subsection, we present some extensions of first-order logic with generalized quantifiers, and discuss possible applications in formalizing default reasoning.

4.3.3 Logics with generalized quantifiers

[46] formalizes the notion of "normally $\alpha(x)$" using restricted quantification: $\nabla x \alpha(x)$ is true if α holds on an "important" subset of the domain. A first order language L

extended with the quantifier ∇ is interpreted in a structure $(\mathbf{M}, \mathcal{N})$, where $\mathbf{M}$ is an L-model, and $\mathcal{N}$ is a family of *important* subsets of M that satisfies (slightly weaker axioms than the corresponding ones for filters):

- $M \in \mathcal{N}$;
- if $A \in \mathcal{N}$, $A \subseteq B \subseteq M$, then $B \in \mathcal{N}$;
- if $A, B \in \mathcal{N}$, then $A \cap B \neq \emptyset$ (thus $\emptyset \notin \mathcal{N}$, if $M \neq \emptyset$).

$\nabla x \alpha(x)$ is true in a structure $(\mathbf{M}, \mathcal{N})$ under a valuation v iff there exists $D \in \mathcal{N}$ for which α is true for every element of D. Such semantics covers two important extremes:

- fix one element a of the universe M, then $\mathcal{N}_a = \{A \subseteq M \mid a \in A\}$ will be a family of important subsets of M;
- let some probability measure μ be given on M, then $\mathcal{N}_\mu = \{A \subseteq M \mid \mu(A) > 0.5\}$ will be be a family of important subsets of M.

This approach was generalized in order to include conditionals of the form $\alpha \rightsquigarrow_x \beta$, but we omit details here.

[2] has introduced a binary quantifier to treat statements like "α's are typically β's". These statements were interpreted using models of the form $(\mathbf{M}, \mathcal{T})$, where $\mathbf{M}$ is a first-order model, and $\mathcal{T} : \mathcal{P}(M) \to \mathcal{P}(M)$ is a *choice* function which maps a subset S of M to the set $\mathcal{T}(S)$ of "typical S's". In the simplest case, $\mathcal{T}$ has to satisfy only one constraint: $\mathcal{T}(S) \subseteq S$. The satisfiability relation for the binary quantifier is defined as follows:

$$(\mathbf{M}, \mathcal{T}), v \models \alpha \rightsquigarrow_x \beta \text{ iff } \mathcal{T}\{a \in M \mid \mathbf{M}, v(a/x) \models \alpha\} \subseteq \{a \in M \mid \mathbf{M}, v(a/x) \models \beta\}.$$

The most influential semantics for $\rightsquigarrow_{\overline{x}}$ is based on ranking models, see [10], that is the class of models consisting of first-order structures whose elements are ranked. R is a ranking function on D if $R : D \to L$, where L is a totally ordered set. A ranked first-order structure is a pair $(\mathbf{M}, \mathcal{R})$, where $\mathbf{M}$ is a standard first-order structure and $\mathcal{R} = \{R_n \mid n \geq 1\}$, where for each n, R_n is a ranking function on M^n. The notion of satisfiability of α under v in a ranked first-order structure $(\mathbf{M}, \mathcal{R})$ is defined as follows:

$$(\mathbf{M}, \mathcal{R}), v \models \alpha \rightsquigarrow_{\overline{x}} \beta \text{ if for each } \overline{c} \in \min_{\overline{x}}(\alpha), (\mathbf{M}, \mathcal{R}), v(\overline{c}/\overline{x}) \models \beta,$$

where $\min_{\overline{x}}(\alpha)$ is the set of minimal elements of $\{\overline{a} \in M^{|\overline{x}|} \mid (\mathbf{M}, \mathcal{R}), v(\overline{a}/\overline{x}) \models \alpha\}$. The set $\min_{\overline{x}} \alpha$ can be empty even when $\exists \overline{x} \alpha$ holds, but the corresponding set defined above contains an infinite descending set of tuples. In order to exclude this case, only smooth rankings are usually considered.

A sound and complete axiomatization with respect to the given semantics is given in [10]. It is convenient to use the following shorthand notation:

$$\alpha \leq_{\overline{x}} \beta =_{\text{def}} \neg(\alpha \vee \beta \rightsquigarrow_{\overline{x}} \neg\alpha) \text{ and } \alpha <_{\overline{x}} \beta =_{\text{def}} \alpha \vee \beta \rightsquigarrow_{\overline{x}} \neg\beta.$$

Notice that $\alpha \leq_{\overline{x}} \beta$ is read "some normal tuples w.r.t. the property $\alpha \vee \beta$ satisfy α", and $\alpha <_{\overline{x}} \beta$ is read as "all normal tuples w.r.t. the property $\alpha \vee \beta$ satisfy $\neg\beta$".

FO All instance of usual first-order axioms;
Ref $\alpha \rightsquigarrow_{\overline{x}} \alpha$ [Reflexivity]
LE $\forall \overline{x}(\alpha \leftrightarrow \beta) \to ((\alpha \rightsquigarrow_{\overline{x}} \gamma) \leftrightarrow (\beta \rightsquigarrow_{\overline{x}} \gamma))$ [Left Equivalence]
RW $\forall \overline{x}(\alpha(\overline{x}) \to \beta(\overline{x})) \to ((\gamma \rightsquigarrow_{\overline{x}} \alpha) \leftrightarrow (\gamma \rightsquigarrow_{\overline{x}} \beta))$ [Right Weakening]
CM $(\alpha \rightsquigarrow_{\overline{x}} \beta) \wedge (\alpha \rightsquigarrow_{\overline{x}} \gamma) \to (\alpha \wedge \beta \rightsquigarrow_{\overline{x}} \gamma)$ [Cautious Monotony]
And $(\alpha \rightsquigarrow_{\overline{x}} \beta) \wedge (\alpha \rightsquigarrow_{\overline{x}} \gamma) \to (\alpha \rightsquigarrow_{\overline{x}} \beta \wedge \gamma)$ [And]
Or $(\alpha \rightsquigarrow_{\overline{x}} \gamma) \wedge (\beta \rightsquigarrow_{\overline{x}} \gamma) \to (\alpha \vee \beta \rightsquigarrow_{\overline{x}} \gamma)$ [Or]
RM $(\alpha \rightsquigarrow_{\overline{x}} \gamma) \wedge \neg(\alpha \wedge \beta \rightsquigarrow_{\overline{x}} \gamma) \to (\alpha \rightsquigarrow_{\overline{x}} \neg\beta)$ [Rational Monotony]
UW $\forall \overline{x}\beta \to (\alpha \rightsquigarrow_{\overline{x}} \beta)$ [Universal Weakening]
Ins $(\alpha \rightsquigarrow_{\overline{x}} \beta) \to (\exists \overline{x}\alpha \to \exists \overline{x}(\alpha \wedge \beta))$ [Instantiation]
Ren $(\alpha \rightsquigarrow_{\overline{x}} \beta) \to (\alpha(\overline{y}/\overline{x}) \rightsquigarrow_{\overline{y}} \beta(\overline{y}/\overline{x}))$,
where $\overline{y}$ does not occur in α and β [Renaming]
UI $\forall \overline{y}(\alpha \rightsquigarrow_{\overline{x}} \beta) \leftrightarrow (\alpha \rightsquigarrow_{\overline{x}} \forall \overline{y}\beta)$,
whenever $\overline{y} \cap (\mathrm{fv}(\alpha) \cup \overline{x}) = \emptyset$ [Universal Interchange]
Per $(\alpha \rightsquigarrow_{\overline{x},\overline{y}} \beta) \leftrightarrow (\alpha \rightsquigarrow_{\overline{y},\overline{x}} \beta)$ [Permutation]
WC $(\alpha \leq_{\overline{x}} \beta) \wedge (\alpha' \leq_{\overline{y}} \beta') \to ((\alpha \wedge \beta) \leq_{\overline{x},\overline{y}} (\alpha' \wedge \beta')$,
where $\overline{x} \cap \overline{y} = \emptyset, \overline{x} \cap \mathrm{fv}(\alpha' \wedge \beta') = \emptyset, \overline{y} \cap \mathrm{fv}(\alpha \wedge \beta) = \emptyset$, [Weak Concatenation]
SC $(\alpha <_{\overline{x}} \beta) \wedge (\alpha' \leq_{\overline{y}} \beta') \to ((\alpha \wedge \beta) <_{\overline{x},\overline{y}} (\alpha' \wedge \beta')$,
where $\overline{x} \cap \overline{y} = \emptyset, \overline{x} \cap \mathrm{fv}(\alpha' \wedge \beta') = \emptyset, \overline{y} \cap \mathrm{fv}(\alpha \wedge \beta) = \emptyset$, [Strong Concatenation]
Dis $(\alpha \wedge \alpha' \rightsquigarrow_{\overline{x},\overline{y}} \gamma) \to (\alpha \rightsquigarrow_{\overline{x}} (\alpha' \rightsquigarrow_{\overline{y}} \gamma))$,
where $\overline{x} \cap \overline{y} = \emptyset, \overline{x} \notin \mathrm{fv}(\alpha'), \overline{y} \notin \mathrm{fv}(\alpha)$, [Distribution]
Pro $\alpha \rightsquigarrow_{\overline{x},\overline{y}} ((\alpha \rightsquigarrow_{\overline{x}} \beta) \to \beta)$, where $\overline{x} \cap \overline{y} = \emptyset$ [Projection].

Modus ponens is the only inference rules.

There are a number of variants of the previous logic. One of the most prominent is certainly probabilistic semantics.

Probabilistic semantics for conditional logics has been pointed out by a number of authors: [1], [37], [27], etc.). The intuition behind this semantics is as follows: a default $\alpha \rightsquigarrow_x \beta$ denotes that "the probability of $\beta(x)$ given $\alpha(x)$ is *very close* to 1", that is "with probability almost one, any element that has the property $\alpha(x)$, has the property $\beta(x)$". In order to formalize this intuition we have to say 1) what we mean by "high" ("almost all"), and 2) how we would treat tuples of domain elements. The latter problem can be solved by defining separate probability distributions for each tuple size. A somewhat nicer solution would be to use the probability distribution over the domain to induce a probability distribution over tuples, e.g. $\Pr(d_1,\dots,d_k) = \Pr(d_1)\cdots\Pr(d_k)$. The former problem is more interesting. There are several ways to resolve it.

[20] considered semantics based on a sequence of probability assignments in order to capture the notion of "high probability". The statement "$\mathrm{P}(\beta(x) \mid \alpha(x))$ is almost 1" is interpreted with $\lim_{n\to\infty} \mathrm{P}_n(\beta(x) \mid \alpha(x)) = 1$. We find that an alternative approach is more natural. Semantics for defaults based on infinitesimal probabilities are discussed at length and shown capable of serving as universal core for a variety of nonmonotonic logics. The meaning of "very close to 1" involves probability measures whose range is a non-Archimedean field, i.e., i an ordered field containing

infinitesimals[1]. As is shown in [27] (following the ideas from [1]), rational consequence relations are naturally generated by non-standard probability spaces developed within Robinson's infinitesimal analysis, see [24, 43]. It is enough to use much simpler non-Archimedean field.

4.4 Probabilistic first-order logic $L_{\omega\omega}^{\mathrm{P},\mathbb{I}}$

In this section, we describe a probabilistic extension of ordinary first-order logic, introduced in [23], which enables us to generalize considerations from the previous section. The language of the logic is rich enough so we can represent in it simultaneously statistical knowledge, imprecise probabilities, beliefs and defaults.

The semantics of our logic is based on probability measures whose range is the field $\mathbb{Q}(\varepsilon)$, i.e., the smallest field generated by adding to the rationals a single infinitesimal ε. Elements of $\mathbb{Q}(\varepsilon)$ are rational expressions of the form $\frac{p(\varepsilon)}{q(\varepsilon)}$, where $p(\varepsilon)$ and $q(\varepsilon)$ are polynomials in ε over $\mathbb{Q}$, and $q(\varepsilon)$ is not identically equal to zero. Two rational expressions $\frac{p(\varepsilon)}{q(\varepsilon)}$ and $\frac{p_1(\varepsilon)}{q_1(\varepsilon)}$ are equal if polynomials $p(\varepsilon)q_1(\varepsilon)$ and $p_1(\varepsilon)q(\varepsilon)$ have the same non-zero coefficients. Addition and multiplication are defined in the usual way. Any element η of $\mathbb{Q}(\varepsilon)$ can be transformed into the normalized form:

$$(*) \qquad \eta = \frac{a\varepsilon^k + \sum_{i=k+1}^{n} a_i\varepsilon^i}{1+\sum_{j=1}^{m} b_j\varepsilon^j}, k<n, 0<m$$

for some unique integer k and some unique leading coefficient a such that $a \neq 0$ unless $\eta = 0$. The ordering $<$ on $\mathbb{Q}(\varepsilon)$ is defined so that $\eta > 0$ iff $a > 0$. This makes $\mathbb{Q}(\varepsilon)$ a non-Archimedean ordered field since ε is an infinitesimal. Given the normalized form $(*)$ of η, the unique integer k is the *order* of η, written $\mathrm{ord}(\eta)$, with $\mathrm{ord}(0) = \infty$. If $k = 0$, then η is infinitesimaly different from a non-zero rational number called the *standard part* of η and denoted $\mathrm{st}(\eta) = a$; if $k > 0$, then $\mathrm{st}(\eta) = 0$. For $k < 0$, η is an infinite number. The monad (halo) of a rational $q \in [0,1]$, defined by $\mathrm{monad}(x) = \{y \mid y \approx x\}$, can be characterized by:

$$\mathrm{monad}(q) = \bigcap_{n\in\mathbb{N}^+} \left[\max\left\{0, q-\frac{1}{n}\right\}, \min\left\{1, q+\frac{1}{n}\right\}\right].$$

Notice that $\mathbb{Q}(\varepsilon)$ is countable and recursive, i.e., its operations are computable and its ordering is decidable. Thus, $\mathbb{Q}(\varepsilon)$ can be directly included in our Syntax. In the sequel, we focus on $\mathbb{Q}(\varepsilon)$-probabilistic spaces of the form $(W, \mathcal{F}, \mu)$ with a finitely additive probability measure $\mu : \mathcal{F} \to \mathbb{I}$, where $\mathbb{I}$ is the unit interval of $\mathbb{Q}(\varepsilon)$.

[1] An element ε is infinitesimal if $|\varepsilon| < 1/n$, for every positive integer n.

4.4.1 Syntax and semantics

Let L be a first-order language. The logic $L^{\mathrm{P},\mathbb{I}}_{\omega\omega}$ is an extension of first-order logic containing probabilistic quantifiers $(\mathrm{CP}\overline{x} \geqslant r)$ and $(\mathrm{CP}\overline{x} \leqslant r)$, for every $r \in \mathbb{I}$, and $(\mathrm{CP}\overline{x} \approx q)$, for every $q \in \mathbb{I} \cap \mathbb{Q}$, where $\overline{x} = (x_1, \dots, x_n)$ is a tuple of pairwise distinct variables.

The set of terms and the set of atomic formulas of $L^{\mathrm{P},\mathbb{I}}_{\omega\omega}$ are the same as in first-order logic.

Definition 4.1. The set of formulas of $L^{\mathrm{P},\mathbb{I}}_{\omega\omega}$ is the least set such that:

- each atomic formula of first-order logic is a formula of $L^{\mathrm{P},\mathbb{I}}_{\omega\omega}$;
- if α is a formula of $L^{\mathrm{P},\mathbb{I}}_{\omega\omega}$, then $\neg\alpha$ is a formula of $L^{\mathrm{P},\mathbb{I}}_{\omega\omega}$;
- if α and β are formulas of $L^{\mathrm{P},\mathbb{I}}_{\omega\omega}$, then so is $\alpha \wedge \beta$;
- if α is a formula of $L^{\mathrm{P},\mathbb{I}}_{\omega\omega}$, and x is a variable, then $\forall x\alpha$ is a formula of $L^{\mathrm{P},\mathbb{I}}_{\omega\omega}$;
- if α and β are formulas of $L^{\mathrm{P},\mathbb{I}}_{\omega\omega}$, and $(\mathrm{CP}\overline{x} \diamond r)$ is a quantifier of $L^{\mathrm{P},\mathbb{I}}_{\omega\omega}$ ($\diamond$ is a placeholder for $\leqslant, \geqslant, \approx$), then $(\mathrm{CP}\overline{x} \diamond r)(\alpha \mid \beta)$ is a formula of $L^{\mathrm{P},\mathbb{I}}_{\omega\omega}$. ■

The other classical connectives and the existential quantifier $\exists$ are defined as usual, while $\bot$ and $\top$ denote $\alpha \wedge \neg\alpha$ and $\neg\bot$, respectively. We also use the following abbreviations:

- $(\mathrm{CP}\overline{x} < r)(\alpha \mid \beta)$ for $\neg(\mathrm{CP}\overline{x} \geqslant r)(\alpha \mid \beta)$;
- $(\mathrm{CP}\overline{x} > r)(\alpha \mid \beta)$ for $\neg(\mathrm{CP}\overline{x} \leqslant r)(\alpha \mid \beta)$;
- $(\mathrm{CP}\overline{x} = r)(\alpha \mid \beta)$ for $(\mathrm{CP}\overline{x} \geqslant r)(\alpha \mid \beta) \wedge (\mathrm{CP}\overline{x} \leqslant r)(\alpha \mid \beta)$;
- $(\mathrm{P}\overline{x} \diamond r)\alpha$ for $(\mathrm{CP}\overline{x} \diamond r)(\alpha \mid \top)$, $\diamond \in \{\leqslant, =, \geqslant, \approx, <, >\}$.

The notions of free and bound variables are defined as usual, with the quantifier $(\mathrm{CP}\overline{x} \diamond r)$ binding all the variables in the tuple $\overline{x}$, $\diamond \in \{\leqslant, \geqslant, \approx\}$.

Example 4.4. Let U be a unary and B a binary relation symbol. The following are $L^{\mathrm{P},\mathbb{I}}_{\omega\omega}$-formulas:

- $(\mathrm{CP}x \approx 0.5)(U(x) \to B(x,x) \mid \neg B(x,y))$ (y occurs free in this formula);
- $(\mathrm{CP}xy \geq \frac{2-2\varepsilon}{3})(\forall z(B(x,z) \to B(z,y)) \mid x = y)$ (this formula is a sentence);
- $(\mathrm{P}x > \frac{2}{3})(\mathrm{CP}y \approx 0)(B(y,x) \wedge \neg B(x,x) \mid (\mathrm{CP}z \geq 0.99)(U(z) \mid B(x,y)))$, etc. ■

Definition 4.2. A model for $L^{\mathrm{P},\mathbb{I}}_{\omega\omega}$ is a structure $\overline{\mathbf{M}} = \langle \mathbf{M}, \mathcal{F}_n, \mu_n \rangle_{n\in\mathbb{N}}$ such that

- $\mathbf{M} = (M, S^{\mathbf{M}})_{S\in L}$ is a classical L-model;
- for all $n \geqslant 1$, $(M^n, \mathcal{F}_n, \mu_n)$ is a finitely additive probability space, where $\mathcal{F}_n$ is a field of subsets of M^n and μ_n is a finitely additive probability measure whose domain is $\mathcal{F}_n$, and whose range is $\mathbb{I}$ (i.e., $\mu_n(X) \geq 0$, $\mu_n(X \cup Y) = \mu_n(X) + \mu_n(Y)$ if $X \cap Y = \emptyset$, and $\mu_n(M^n) = 1$); moreover,
  - for each n-ary function symbol f of L, the graph of $f^{\mathbf{M}}$,
  $$\{(\overline{x}, y) \in M^{n+1} \mid f^{\mathbf{M}}(\overline{x}) = y\}$$
  is in $\mathcal{F}_{n+1}$;

- for each n-ary relation symbol R of L, $R^{\mathbf{M}} \in \mathcal{F}_n$;
- for all $i, j \leqslant n$, $\{(x_1, \ldots, x_n) \in M^n \mid x_i = x_j\} \in \mathcal{F}_n$;
- if $X \in \mathcal{F}_n$, then $M \times X \in \mathcal{F}_{n+1}$;
- if $X \in \mathcal{F}_{n+1}$ and $\Pi : M^{n+1} \to A^n$ is the projection map

$$\Pi(x_1, \ldots, x_n, x_{n+1}) = (x_1, \ldots, x_n),$$

  then $\Pi(X) \in \mathcal{F}_n$;
- if $X \in \mathcal{F}_{n+m}$ and $\overline{c} \in M^m$, then $\{\overline{a} \in M^n \mid (\overline{a}, \overline{c}) \in X\} \in \mathcal{F}_n$;
- if $X \in \mathcal{F}_{n+m}$, then $\{\overline{a} \in M^n \mid \mu_m\{\overline{c} \in M^m \mid (\overline{a}, \overline{c}) \in X\} \diamond r\} \in \mathcal{F}_n$, where $r \in \mathbb{I}$ and $\diamond \in \{\leqslant, \geqslant\}$, or $r \in \mathbb{I} \cap \mathbb{Q}$ and $\diamond = \approx$;
- each μ_n is invariant under permutation: for every permutation π of $\{1, 2, \ldots, n\}$ and $X \in \mathcal{F}_n$, if

$$X^\pi = \{(a_{\pi(1)}, \ldots, a_{\pi(n)}) \in M^n \mid (a_1, \ldots, a_n) \in X\},$$

  then $X^\pi \in \mathcal{F}_n$ and $\mu_n(X^\pi) = \mu_n(X)$;
- if $X \in \mathcal{F}_n$, then $\mu_{n+1}(M \times X) = \mu_n(X)$. ■

The conditions for $\mathcal{F}_n$, $n \geqslant 1$, are of essential importance for the definition of the satisfiability relation (Definition 4.3). These conditions ensures that all definable sets are measurable, that is $\mathcal{F}_n$s contain all sets definable by $L^{\mathrm{P},\mathbb{I}}_{\omega\omega}$-formulas (with or without parameters).

The sequence of probabilities $(\mu_n : n = 1, 2, \ldots)$ is not assumed to be a sequence of product measures. Instead of the conditions:

(i) $\mathcal{F}_{m+n}$ is generated by the set $\{X \times Y \mid X \in \mathcal{F}_m, Y \in \mathcal{F}_n\}$, and
(ii) the measure μ_{m+n} is the product probability measure of μ_m and μ_n,

$$\mu_{m+n}(X \times Y) = \mu_m(X) \cdot \mu_n(Y).$$

we take their weakening:

(1) if $X \in \mathcal{F}_n$, then $A \times X \in \mathcal{F}_{n+1}$,
(2) if $X \in \mathcal{F}_n$, then $\mu_{n+1}(A \times X) = \mu_n(X)$.

Of course, we could restrict the class of $L^{\mathrm{P},\mathbb{I}}_{\omega\omega}$-models to those structures whose sequence of probabilities is a sequence of product measures and give a complete axiomatization for them (Theorem 4.4).

Example 4.5. The weak requirements (1) and (2) are fulfilled in the trivial spaces $(M^n, \{\emptyset, M^n\}, \mu_n)$, where $\mu_n(M^n) = 1$, $\mu_n(\emptyset) = 0$. Note also that for any n, μ_n is invariant under permutations: for any permutation $\pi : \{1, 2, \ldots, n\} \to \{1, 2, \ldots, n\}$, $(M^n)^\pi = \{(a_{\pi(1)}, \ldots, a_{\pi(n)}) \mid (a_1, \ldots, a_n) \in M^n\} = M^n$ and $\emptyset^\pi = \emptyset$.

Let us define a more illustrative example. Let $(M, \mathcal{P}(M), \mu_1)$ be a probability space, where $M = \{1, 2, 3, 4\}$ and $\mu_1 : \mathcal{P}(M) \to [0, 1]$ is a probability measure defined by a nonuniform distribution on the singletons:

$$\mu_1 = \begin{pmatrix} 1 & 2 & 3 & 4 \\ \frac{\varepsilon}{3} & \frac{1+\varepsilon}{3} & \frac{1}{3} & \frac{1-2\varepsilon}{3} \end{pmatrix}, \mu_1(X) = \sum_{x \in X} \mu_1(\{x\}), X \subseteq A.$$

There are many possibilities to define a probability measure on $\mathcal{P}(M^2)$ such that the conditions (1) and (2) are satisfied. We define two probability measures by distributions on singletons:

μ_2	1	2	3	4
1	$\frac{\varepsilon}{3}$	0	0	0
2	0	$\frac{1+\varepsilon}{3}$	0	0
3	0	0	$\frac{1}{3}$	0
4	0	0	0	$\frac{1-2\varepsilon}{3}$

and

μ_2'	1	2	3	4
1	$\frac{\varepsilon^2}{9}$	$\frac{\varepsilon+\varepsilon^2}{9}$	$\frac{\varepsilon}{9}$	$\frac{\varepsilon-2\varepsilon^2}{9}$
2	$\frac{\varepsilon+\varepsilon^2}{9}$	$\frac{(1+\varepsilon)^2}{9}$	$\frac{1+\varepsilon}{9}$	$\frac{1-\varepsilon-2\varepsilon^2}{9}$
3	$\frac{\varepsilon}{9}$	$\frac{1+\varepsilon}{9}$	$\frac{1}{9}$	$\frac{1-2\varepsilon}{9}$
4	$\frac{\varepsilon-2\varepsilon^2}{2}$	$\frac{1-\varepsilon-2\varepsilon^2}{9}$	$\frac{1-2\varepsilon}{9}$	$\frac{(1-2\varepsilon)^2}{9}$

Notice that μ_2' is the product probability measure: $\mu_2'(X \times Y) = \mu_1(X) \cdot \mu_1(Y)$, $X, Y \subseteq M$. Analogously, we can define $(M^n, \mathcal{P}(M^n), \mu_n)$ and $(M^n, \mathcal{P}(M^n), \mu_n')$, for $n \geq 3$. ■

Definition 4.3. Let $\overline{\mathbf{M}}$ be an $L_{\omega\omega}^{\mathrm{P,I}}$-model and v a valuation. The satisfiability relation $\overline{\mathbf{M}}, v \models \varphi$ is defined as for first-order logic with the new clauses:

- $\overline{\mathbf{M}}, v \models (\mathrm{CP}\overline{x} \leqslant r)(\alpha \mid \beta)$ iff

  $\mu_n\{\overline{a} \in M^n \mid \overline{\mathbf{M}}, v(\overline{a}/\overline{x}) \models \beta\} = 0$ and $r = 1$
  or
  $$\mu_n\{\overline{a} \in M^n \mid \overline{\mathbf{M}}, v(\overline{a}/\overline{x}) \models \beta\} > 0 \text{ and } \frac{\mu_n\{\overline{a} \in M^n \mid \overline{\mathbf{M}}, v(\overline{a}/\overline{x}) \models \alpha \wedge \beta\}}{\mu_n\{\overline{a} \in M^n \mid \overline{\mathbf{M}}, v(\overline{a}/\overline{x}) \models \beta\}} \leqslant r;$$

- $\overline{\mathbf{M}}, v \models (\mathrm{CP}\overline{x} \geqslant r)(\alpha \mid \beta)$ iff

  $\mu_n\{\overline{a} \in M^n \mid \overline{\mathbf{M}}, v(\overline{a}/\overline{x}) \models \beta\} = 0$
  or
  $$\mu_n\{\overline{a} \in M^n \mid \overline{\mathbf{M}}, v(\overline{a}/\overline{x}) \models \beta\} > 0 \text{ and } \frac{\mu_n\{\overline{a} \in M^n \mid \overline{\mathbf{M}}, v(\overline{a}/\overline{x}) \models \alpha \wedge \beta\}}{\mu_n\{\overline{a} \in M^n \mid \overline{\mathbf{M}}, v(\overline{a}/\overline{x}) \models \beta\}} \geqslant r;$$

- $\overline{\mathbf{M}}, v \models (\mathrm{CP}\overline{x} \approx q)(\alpha \mid \beta)$ iff

  $\mu_n\{\overline{a} \in M^n \mid \overline{\mathbf{M}}, v(\overline{a}/\overline{x}) \models \beta\} = 0$ and $q = 1$
  or
  $$\mu_n\{\overline{a} \in M^n \mid \overline{\mathbf{M}}, v(\overline{a}/\overline{x}) \models \beta\} > 0 \text{ and } \frac{\mu_n\{\overline{a} \in M^n \mid \overline{\mathbf{M}}, v(\overline{a}/\overline{x}) \models \alpha \wedge \beta\}}{\mu_n\{\overline{a} \in M^n \mid \overline{\mathbf{M}}, v(\overline{a}/\overline{x}) \models \beta\}} \approx q.$$

If Γ is a set of formulas, we write $\overline{\mathbf{M}}, v \models \Gamma$ to mean that $\overline{\mathbf{M}}, v \models \gamma$ for every formula γ in Γ. ■

All clauses in the previous definition are formulated on the assumption that the conditional probability is 1, whenever the condition has the probability 0. That approach is fully in line with probabilistic interpretation of defaults, see [27, 38, 39].

Moreover, if we had kept the standard definition according to which conditional probability is undefined when the condition has probability 0, then many formulas would not have truth-values, i.e., for some formulas the question whether they are satisfied in a given model would make no sense.

Example 4.6. The language $L = \{U, B\}$, from Example (4.4), is interpreted on $M = \{1,2,3,4\}$: $U^{\mathbf{M}} = \{2,3\}$ and $B^{\mathbf{M}} = \{(1,1),(1,3),(2,4)\}$. We check whether some sentences are satisfiable in the $L^{\mathrm{P},\mathbb{I}}_{\omega\omega}$-models

$$\overline{\mathbf{M}} = (M, U^{\mathbf{M}}, B^{\mathbf{M}}, \mathcal{P}(M^n), \mu_n)_{n\geq 1} \text{ and } \overline{\mathbf{M}}' = (M, U^{\mathbf{M}}, B^{\mathbf{M}}, \mathcal{P}(M^n), \mu'_n)_{n\geq 1},$$

where μ_n and μ'_n are measures defined in Example 4.5. It is not hard to verify that:

- $\overline{\mathbf{M}}, v(1/y) \not\models (\mathrm{CP}x \approx 0.5)(U(x) \to B(x,x) \mid \neg B(x,y))$, and $\overline{\mathbf{M}}, v(2/y) \models (\mathrm{CP}x \approx 0.5)(U(x) \to B(x,x) \mid \neg B(x,y))$, for any valuation v;
- $\overline{\mathbf{M}} \models (\mathrm{CP}xy \geq \frac{2-2\varepsilon}{3})(\forall z(B(x,z) \to B(z,y)) \mid x = y)$, and $\overline{\mathbf{M}}' \not\models (\mathrm{CP}xy \geq \frac{2-2\varepsilon}{3})(\forall z(B(x,z) \to B(z,y)) \mid x = y)$;
- $\overline{\mathbf{M}} \models (\mathrm{P}x > \frac{2}{3})(\mathrm{CP}y \approx 0)(B(y,x) \wedge \neg B(x,x) \mid (\mathrm{CP}z \geq 0.99)(U(z) \mid B(x,y)))$, and $\overline{\mathbf{M}}' \models (\mathrm{P}x > \frac{2}{3})(\mathrm{CP}y \approx 0)(B(y,x) \wedge \neg B(x,x) \mid (\mathrm{CP}z \geq 0.99)(U(z) \mid B(x,y)))$. ■

Example 4.7. Now, we present some examples of what can be represented by $L^{\mathrm{P},\mathbb{I}}_{\omega\omega}$-formulas choosing the suitable language.

- "Most birds fly": $(\mathrm{CP}x > 0.5)(\texttt{fly}(x) \mid \texttt{bird}(x))$, where "> 0.5" corresponds to "Most",
- "90% of birds can fly": $(\mathrm{CP}x = 0.9)(\texttt{fly}(x) \mid \texttt{bird}(x))$,
- "Approximately 90% of birds fly": $(\mathrm{CP}x \approx 0.9)(\texttt{fly}(x) \mid \texttt{bird}(x))$,
- "More than 90% of birds can fly": $(\mathrm{CP}x > 0.9)(\texttt{fly}(x) \mid \texttt{bird}(x))$,
- "Almost all birds fly": $(\mathrm{CP}x \approx 1)(\texttt{fly}(x) \mid \texttt{bird}(x))$, etc. ■

Example 4.8. In medicine, two commonly used statistics, when considering some diagnostic test as an indicator that a patient has some disease, are so called sensitivity and specificity. Sensitivity is defined as the percentage of true positive cases relative to the sum of true positives and false negatives (i.e., the total number of tested patients having the disease). Specificity is defined as the percentage of true negative cases relative to the sum of true negatives and false positives (i.e., the total number of tasted patients who do not have the disease). Let the formulas $\texttt{tested}(x)$, $\texttt{positive}(x)$, $\texttt{disease}(x)$ have the following meaning:

$\texttt{tested}(x)$: the patient x has been tested,
$\texttt{positive}(x)$: the patient x was positive on the test,
$\texttt{disease}(x)$: the patient x has the disease.

We can express the requirement that the sensitivity of the test is at least 95% by

$$(\mathrm{CP}x \geqslant 0.95)(\texttt{positive}(x) \mid \texttt{tested}(x) \wedge \texttt{disease}(x)).$$

If we know that the specificity of the test is higher than 90%, this can be stated as $(\mathrm{CP}x > 0.9)(\neg\texttt{positive}(x) \mid \texttt{tested}(x) \wedge \neg\texttt{disease}(x))$. ■

Definition 4.4. A formula α is true in an $L^{\mathrm{P},\mathbb{I}}_{\omega\omega}$-model $\overline{\mathbf{M}}$, denoted by $\overline{\mathbf{A}} \models \alpha$, if $\overline{\mathbf{M}}, v \models \alpha$ for every valuation $v : \mathrm{Var} \to M$.

A formula is valid if it is true in every $L^{\mathrm{P},\mathbb{I}}_{\omega\omega}$-model. ■

In the case that some free variables occur in a valid formula, we treat them as universally quantified.

Example 4.9. The formula $(\mathrm{P}x > 0)\alpha \wedge (\mathrm{CP}x \geqslant r)(\beta \mid \alpha) \to (\mathrm{CP}x \leqslant 1-r)(\neg\beta \mid \alpha)$ is $L^{\mathrm{P},\mathbb{I}}_{\omega\omega}$-valid, but $(\mathrm{CP}x \geqslant r)(\beta \mid \alpha) \to (\mathrm{CP}x \leqslant 1-r)(\neg\beta \mid \alpha)$ is not. ■

4.4.2 Axiomatization

The set of all valid formulas can be characterized by the following set of axiom schemata:

(FO) all $L^{\mathrm{P},\mathbb{I}}_{\omega\omega}$-instances of the axioms for $L_{\omega\omega}$;
(CP1) $(\mathrm{CP}\overline{x} \geqslant 0)(\alpha \mid \beta)$;
(CP2) $(\mathrm{CP}\overline{x} \leqslant r_1)(\alpha \mid \beta) \to (\mathrm{CP}\overline{x} < r_2)(\alpha \mid \beta)$, $r_1 < r_2$;
(CP3) $(\mathrm{CP}\overline{x} < r)(\alpha \mid \beta) \to (\mathrm{CP}\overline{x} \leqslant r)(\alpha \mid \beta)$;
(CP4) $(\mathrm{CP}\overline{x} \approx q)(\alpha \mid \beta) \to (\mathrm{CP}\overline{x} \geqslant q - 1/n)(\alpha \mid \beta)$, for every positive integer n such that $0 \leqslant q - \frac{1}{n}$;
(CP5) $(\mathrm{CP}\overline{x} \approx q)(\alpha \mid \beta) \to (\mathrm{CP}\overline{x} \leqslant q + 1/n)(\alpha \mid \beta)$, for every positive integer n such that $q + \frac{1}{n} \leqslant 1$;
(CP6) $(\mathrm{P}\overline{x} = 0)\beta \to (\mathrm{CP}\overline{x} = 1)(\alpha \mid \beta)$;
(CP7) $(\mathrm{P}\overline{x} = r_1)\beta \wedge (\mathrm{P}\overline{x} = r_2)(\alpha \wedge \beta) \to (\mathrm{CP}\overline{x} = r_2/r_1)(\alpha \mid \beta)$, $r_1 \neq 0$;
(P1) $(\mathrm{P}\overline{x} \geqslant 1)(\alpha \leftrightarrow \beta) \to ((\mathrm{P}\overline{x} = r)\alpha \to (\mathrm{P}\overline{x} = r)\beta)$;
(P2) $(\mathrm{P}\overline{x} \leqslant r)\alpha \leftrightarrow (\mathrm{P}\overline{x} \geqslant 1-r)\neg\alpha$;
(P3) $((\mathrm{P}\overline{x} = r_1)\alpha \wedge (\mathrm{P}\overline{x} = r_2)\beta \wedge (\mathrm{P}\overline{x} = 0)(\alpha \wedge \beta)) \to (\mathrm{P}\overline{x} = \min\{1, r_1 + r_2\})(\alpha \vee \beta)$;
(P4) $(\mathrm{P}x_1 \cdots x_i \cdots x_n \diamond r)\alpha \leftrightarrow (\mathrm{P}x_1 \cdots y \cdots x_n \diamond r)\alpha(y/x_i)$, where y is a variable that does not occur in α, and $\diamond \in \{\leqslant, \geqslant, \approx\}$;
(P5) $(\mathrm{P}x_1 \cdots x_n \diamond r)\alpha \leftrightarrow (\mathrm{P}x_{\pi(1)} \cdots x_{\pi(n)} \diamond r)\alpha$, where $\diamond \in \{\leqslant, \geqslant, \approx\}$ and π is a permutation of $\{1, \ldots, n\}$;
(P6) $(\mathrm{P}\overline{x} \diamond r)\alpha(\overline{x}) \leftrightarrow (\mathrm{P}\overline{xy} \diamond r)\alpha(\overline{x})$, where $\diamond \in \{\leqslant, \geqslant, \approx\}$, and variables $\overline{y}$ are not free in α

and the inference rules **(MP)** (modus ponens) and **(Gen)** (generalization) plus the following:

(Nec) $\dfrac{(\forall x)\alpha}{(\mathrm{P}x = 1)\alpha}$;

(Ran) $\dfrac{\alpha \to (\mathrm{P}\overline{x} \neq r)\beta, r \in \mathbb{I}}{\alpha \to \bot}$;

(Approx) For every $q \in \mathbb{I}_{\mathbb{Q}} \setminus \{0, 1\}$

$$\frac{\gamma \to (\mathrm{CP}\overline{x} \geqslant q - \frac{1}{n})(\alpha \mid \beta), n \geqslant \frac{1}{q} \quad \gamma \to (\mathrm{CP}\overline{x} \leqslant q + \frac{1}{n})(\alpha \mid \beta), n \geqslant \frac{1}{1-q}}{\gamma \to (\mathrm{CP}\overline{x} \approx q)(\alpha \mid \beta)},$$

$$\frac{\gamma \to (\mathrm{CP}\overline{x} \geqslant 1 - \frac{1}{n})(\alpha \mid \beta), n \geqslant 1}{\gamma \to (\mathrm{CP}\overline{x} \approx 1)(\alpha \mid \beta)} \text{ and } \frac{\gamma \to (\mathrm{CP}\overline{x} \leqslant \frac{1}{n})(\alpha \mid \beta), n \geqslant 1}{\gamma \to (\mathrm{CP}\overline{x} \approx 0)(\alpha \mid \beta)},$$

where the range of the parameter n is the set of naturals $\mathbb{N}$.

The axioms (**CP4**), (**CP5**), and the rule (**Approx**) describe the relationship between the standard conditional probability and the conditional probability infinitesimally close to some rational q. Rule (**Ran**) defines, at the syntax level, the range of probability functions to be the set $\mathbb{I}$. The rules (**Ran**) and (**Approx**) are infinitary in the sense that they have an infinite number of premises.

The logic $L_{\omega\omega}^{\mathrm{P},\mathbb{I}}$ is not compact. For example, the set $T = \{(\mathrm{P}x \neq r)U(x) \mid r \in \mathbb{I}\}$, where U is a unary relation symbol, is not satisfiable, although every finite subset of T has an $L_{\omega\omega}^{\mathrm{P},\mathbb{I}}$-model. The lack of compactness forces us to consider infinitary inference rules in order to solve one of the main proof-theoretical problem: providing an axiomatic system which would be strongly complete.

Definition 4.5. A proof of a formula α from a set Γ of formulas is a countable sequence of formulas α_κ indexed by countable ordinal numbers such that the last formula is α, and each formula in the sequence is an axiom, or a formula in Γ or it is derived from the preceding formulas by a rule of inference with no application of (**Gen**) to a formula when the variable is free in formulas of Γ. A formula α is deducible from Γ ($\Gamma \vdash \alpha$) if there is a proof of α from Γ. A formula α is a theorem ($\vdash \alpha$) if it is deducible from the empty set.

A set Γ of formulas is consistent if there is at least one $L_{\omega\omega}^{\mathrm{P},\mathbb{I}}$ formula that is not deducible from Γ, otherwise Γ is inconsistent.

A set Γ is maximal consistent iff Γ is consistent and for every formula α, either $\alpha \in \Gamma$ or $\neg\alpha \in \Gamma$. Note that if Γ is maximal consistent and $\Gamma \vdash \alpha$ then $\alpha \in \Gamma$. ■

Many general meta-theorems about deductions can be proven as in first-order logic including the so called "generalization on constants": if c is a constant symbol, Γ is a set of formulas in which c does not occur, and $\alpha(c/x)$ is a formula such that $\Gamma \vdash \alpha(c/x)$, then there is a deduction of $(\forall x)\alpha(x)$ from Γ in which c does not occur.

Theorem 4.1 (Deduction theorem). *If Γ is a set of formulas and $\Gamma, \varphi \vdash \psi$, then $\Gamma \vdash \varphi \to \psi$.*

The proof of the Deduction theorem is a straightforward application of transfinite induction on the length of the proof of ψ from Γ, φ, see [23].

Lemma 4.1 gives some auxiliary statements which are needed for the proof of the completeness theorem.

Lemma 4.1. *For all formulas α and β:*

1. $\vdash (\mathrm{CP}\overline{x} \geqslant r_1)(\alpha \mid \beta) \to (\mathrm{CP}\overline{x} \geqslant r_2)(\alpha \mid \beta)$, $r_1 > r_2$
2. $\vdash (\mathrm{CP}\overline{x} \leqslant r_1)(\alpha \mid \beta) \to (\mathrm{CP}\overline{x} \leqslant r_2)(\alpha \mid \beta)$, $r_1 < r_2$

3. $\vdash (\mathrm{CP}\overline{x} = r_1)(\alpha \mid \beta) \to \neg(\mathrm{CP}\overline{x} = r_2)(\alpha \mid \beta),\ r_1 \neq r_2$
4. $\vdash (\mathrm{CP}\overline{x} = r_1)(\alpha \mid \beta) \to \neg(\mathrm{CP}\overline{x} \geqslant r_2)(\alpha \mid \beta),\ r_1 < r_2$
5. $\vdash (\mathrm{CP}\overline{x} = r_1)(\alpha \mid \beta) \to \neg(\mathrm{CP}\overline{x} \leqslant r_2)(\alpha \mid \beta),\ r_1 > r_2$
6. $\vdash (\mathrm{CP}\overline{x} = q)(\alpha \mid \beta) \to (\mathrm{CP}\overline{x} \approx q)(\alpha \mid \beta),\ q \in \mathbb{I}_{\mathbb{Q}}$
7. $\vdash (\mathrm{CP}\overline{x} \approx q_1)(\alpha \mid \beta) \to \neg(\mathrm{CP}\overline{x} \approx q_2)(\alpha \mid \beta),\ q_1, q_2 \in \mathbb{I}_{\mathbb{Q}},\ q_1 \neq q_2$
8. $\vdash (\mathrm{P}\overline{x} = 0)\beta \to \neg(\mathrm{CP}\overline{x} \leqslant r)(\alpha \mid \beta),\ r < 1$
9. $\vdash (\mathrm{P}\overline{x} \leqslant 1)\alpha$

Proof. As an illustration, we prove the statement (6). For shortness, we omit details related to the obvious arguments.

1. $(\mathrm{CP}\overline{x} = q)(\alpha \mid \beta) \leftrightarrow (\mathrm{CP}\overline{x} \geqslant q)(\alpha \mid \beta) \wedge (\mathrm{CP}\overline{x} \leqslant q)(\alpha \mid \beta)$

2_n. $(\mathrm{CP}\overline{x} \geqslant q)(\alpha \mid \beta) \to (\mathrm{CP}\overline{x} \geqslant q - \frac{1}{n})(\alpha \mid \beta),\ n \geqslant \frac{1}{q}$ [(1) of this lemma]

3_n. $(\mathrm{CP}\overline{x} \leqslant q)(\alpha \mid \beta) \to (\mathrm{CP}\overline{x} \leqslant q + \frac{1}{n})(\alpha \mid \beta),\ n \geqslant \frac{1}{1-q}$ [(2) of this lemma]

4_n. $(\mathrm{CP}\overline{x} = q)(\alpha \mid \beta) \to (\mathrm{CP}\overline{x} \geqslant q - \frac{1}{n})(\alpha \mid \beta),\ n \geqslant \frac{1}{q}$

5_n. $(\mathrm{CP}\overline{x} = q)(\alpha \mid \beta) \to (\mathrm{CP}\overline{x} \leqslant q + \frac{1}{n})(\alpha \mid \beta),\ n \geqslant \frac{1}{1-q}$

6. $(\mathrm{CP}\overline{x} = q)(\alpha \mid \beta) \to (\mathrm{CP}\overline{x} \approx q)(\alpha \mid \beta)$
[from 4_n, $n \geqslant \frac{1}{q}$ and 5_n, $n \geqslant \frac{1}{1-q}$ by (**Approx**)]

Note that, by restricting β to $\top$, we obtain the analogous statements for unconditional probabilities (except the statements 8 and 9, of course). ■

Example 4.10. From "Most birds fly" $(\mathrm{CP}x > 0.5)(\mathtt{fly}(x) \mid \mathtt{bird}(x))$, and "Penguins do not fly" $\forall x(\mathtt{penguin}(x) \to \neg\mathtt{fly}(x))$, we can deduce "Most birds are not penguins" $(\mathrm{CP}x > 0.5)(\neg\mathtt{penguin}(x) \mid \mathtt{bird}(x))$. ■

4.4.3 Soundness and completeness

Soundness of our system follows from the soundness of $L_{\omega\omega}$, and from the properties of probabilistic measures. More precisely, every instance of an axiom schema is true in every $L^{\mathrm{P},\mathbb{I}}_{\omega\omega}$-model, while the inference rules preserve truth in a model (if the premise(s) of a rule are true in a model, then the conclusion of that rule is true in the model).

Theorem 4.2 (Soundness theorem). *The axiomatic system for $L^{\mathrm{P},\mathbb{I}}_{\omega\omega}$ is sound with respect to the class of $L^{\mathrm{P},\mathbb{I}}_{\omega\omega}$-models, i.e., each theorem is valid.*

The proof of Soundness theorem is straightforward, and we omit it. The completeness theorem for $L^{\mathrm{P},\mathbb{I}}_{\omega\omega}$ can be proved by using the Henkin style procedure and the methodology already described in Chapter 1.

Theorem 4.3 (Completeness theorem). *If T is a consistent set of formulas, then T has a model.*

Proof. We sketch out the proof. All details can be found in [23].

First, we extend L to a new language $L \cup C$ by adding a denumerable set of new constant symbols $C = \{c_n \mid n = 0, 1, \ldots\}$. Let $(\alpha_n : n = 0, 1, 2, \ldots)$ be an enumeration of all $(L \cup C)^{\mathrm{P},\mathbb{I}}_{\omega\omega}$-formulas.

Next, we extend T to a maximal consistent set of formulas T^* which satisfies two conditions:

- if $\exists x \alpha \in T^*$ then for some constant symbol $c \in C$, $\alpha(c/x) \in T^*$,
- for each formula $\alpha(\overline{x})$ there exists r in $\mathbb{I}$ such that $(\mathrm{P}\overline{x} = r)\alpha(\overline{x}) \in T^*$.

Using T^*, we define a model for T. The construction of the classical model $\mathbf{M}$ from the constants $c \in C$ by taking the equivalence classes $[c]$ is standard. If $\overline{c} = (c_1, \ldots, c_n)$ is a tuple of constant symbols, we write $\overline{[c]}$ for the tuple $([c_1], \ldots, [c_n])$. For every formula $\varphi(x_1, \ldots, x_n)$, let

$$\langle \varphi(\overline{x}) \rangle = \left\{ \overline{[c]} \in M^n \mid \varphi(\overline{c}/\overline{x}) \in T^* \right\}.$$

Let $\mathfrak{F}_n$ be the collection of subsets of M^n of the form $\langle \varphi(\overline{x}) \rangle$. Each $\mathfrak{F}_n$ is a field of subsets of M^n and that all the clauses of Definition 4.2 are fulfilled. Now, we define the probabilities $\mu_n : \mathfrak{F}_n \to \mathbb{I}$:

$$\mu_n(\langle \varphi(\overline{x}) \rangle) = r \ \text{ iff } \ (\mathrm{P}\overline{x} = r)\varphi(\overline{x}) \in T^*.$$

For each n, μ_n is a finitely additive probability measure.

$\overline{\mathbf{M}} = \langle \mathbf{M}, \mathfrak{F}_n, \mu_n \rangle_{n \in \omega}$ is an $L^{\mathrm{P},\mathbb{I}}_{\omega\omega}$-model that satisfies T. ■

If we extend the list of axioms and rules of $L^{\mathrm{P},\mathbb{I}}_{\omega\omega}$ with the following axiom

(Prod) $$(\mathrm{P}\overline{x} = r)\alpha(\overline{x}) \wedge (\mathrm{P}\overline{y} = s)\beta(\overline{y}) \to (\mathrm{P}\overline{x}\overline{y} = rs)(\alpha(\overline{x}) \wedge \beta(\overline{y})),$$

where all variables in $\overline{x}, \overline{y}$ are distinct, we are able to prove the completeness theorem for the class of product $L^{\mathrm{P},\mathbb{I}}_{\omega\omega}$-models.

Definition 4.6. A product model is an $L^{\mathrm{P},\mathbb{I}}_{\omega\omega}$-model $\overline{\mathbf{M}} = \langle \mathbf{M}, \mathfrak{F}_n, \mu_n \rangle_{n \in \mathbb{N}}$ such that the sequence of probabilities $(\mu_n : n = 1, 2, \ldots)$ is a sequence of product measures: for any two sets $X \subseteq M^m$ and $Y \subseteq M^n$, and their Cartesian product $X \times Y \subseteq M^{m+n}$, if $X \in \mathfrak{F}_m$ and $Y \in \mathfrak{F}_n$, then $X \times Y \in \mathfrak{F}_{m+n}$ and $\mu_{m+n}(X \times Y) = \mu_m(X) \cdot \mu_n(Y)$. ■

Theorem 4.4. *If $T \cup \{$***Prod***$\}$ is consistent, then T has a product $L^{\mathrm{P},\mathbb{I}}_{\omega\omega}$-model.*

A somewhat different approach to developing first-order probability logic was considered in [5, 6, 21]. [6] introduced numerical terms of the form $[\alpha \mid \beta]_{\overline{x}}$ which should denote the conditional probability of α given β, for arbitrary first-order formulas α and β. The formula $[\alpha \mid \beta]_{\overline{x}} \leqslant r$ corresponds to our formula $(\mathrm{CP}\overline{x} \leqslant r)(\alpha \mid \beta)$. Also, instead of infinitesimals, [6] introduces an infinite sequence of "approximate" relations $\leqslant_i$ and $=_i$ with associated sequence τ of real numbers as "tolerance factor". So, $t_1 =_i t_2$ is interpreted as $|t_1 - t_2| \leqslant \tau_i$. The system in [6] is two-sorted first-order logic.

4.4.4 Decidable fragments of $L_{\omega\omega}^{P,\mathbb{I}}$

First-order classical logic is undecidable, and thus the same holds for $L_{\omega\omega}^{P,\mathbb{I}}$. However, there are numerous decidable fragments of first-order logic (cf. [8]). The most traditional ones are determined by restrictions on the quantifiers prefix and language. In the search for decidable fragments of $L_{\omega\omega}^{P,\mathbb{I}}$, our idea is to forbid the nesting of probability quantifiers and try to find a decidable fragment of classical first-order logic to which a technique for proving decidability, developed in [38], could be applied. [23] presents two decidable fragments of $L_{\omega\omega}^{P,\mathbb{I}}$. One of them is based on a subfragment of the fragment defined by Gödel, Kalmár, and Schütte, for which decidability can be proved. Gödel, Kalmár, and Schütte discovered a decision procedure for the satisfiability of sentences from the class $[\exists^*\forall^2\exists^*, \text{all}]$ which consists of the sentences of the form $\exists x_1 \dots \exists x_k \forall y_1 \forall y_2 \exists z_1 \dots \exists z_\ell \varphi$ where φ is quantifier free and it contains any number of relation symbols of any arity, but it does not have function symbols of arity $\geqslant 1$ and the equality sign.

Since the equality may be important for some applications, we also define a subfragment of the fragment introduced by Shelah, which allows less quantifiers, but it admits one unary function symbol and equality sign, which might make it more suitable for some applications. The proof of decidability is practically the same for both fragments. Note that both fragments may include individual constants.

Theorem 4.5 (Gödel 1932, Kalmár 1933, Schütte 1934). *The satisfiability and the finite satisfiability problems are decidable for the class* $[\exists^*\forall^2\exists^*, \text{all}]$.

Theorem 4.6 (Shelah 1977). *The satisfiability and the finite satisfiability problems are decidable for the classes* $[\exists^*\forall\exists^*, \text{all}, (1)]_=$, *i.e., for the class of formulas whose prenex form has the quantifier prefix of the form* $\exists^*\forall\exists^*$ *and whose language contains equality sign, arbitrary relation and constant symbols, and at most one unary function symbol.*

Theorem 4.7. *[23] Let L be a first-order language without function symbols and equality sign. Let $\mathcal{G}_L$ be the class of all first-order formulas whose prenex form has the quantifier prefix from the set*

$$\{\exists, \forall, \exists\exists, \exists\forall, \forall\exists, \forall\forall, \exists\exists\forall, \exists\forall\forall, \forall\exists\exists, \forall\forall\exists, \exists\exists\forall\forall, \forall\forall\exists\exists\},$$

and let $\mathcal{G}_L^{\text{prob}}$ denote the class of all $L_{\omega\omega}^{P,\mathbb{I}}$-sentences which are Boolean combinations of probabilistic formulas of the form $(\mathrm{CP}\overline{x} \diamond r)(\varphi(\overline{x}) \mid \psi(\overline{x}))$, $\varphi(\overline{x}), \psi(\overline{x}) \in \mathcal{G}_L$. The satisfiability problem for $\mathcal{G}_L^{\text{prob}}$ is decidable.

Theorem 4.8. *[23] Let L be a language with at most one unary function symbol and no function symbols of arity ≥ 2 (there are no restrictions on relation and constant symbols). Let $\mathcal{S}_L$ denote the class of all $L_{\omega\omega}$-formulas possibly with equality whose prenex form has the quantifier prefix from the set $\{\exists, \forall, \exists\forall, \forall\exists\}$, and let $\mathcal{S}_L^{\text{prob}}$ denote the class of all $L_{\omega\omega}^{P,\mathbb{I}}$-sentences which are Boolean combinations of probabilistic for-*

mulas of the form $(\mathrm{CP}\overline{x}\diamond r)(\varphi(\overline{x}) \mid \psi(\overline{x}))$, *where* $\varphi(\overline{x}), \psi(\overline{x}) \in \mathcal{S}_L$. *The satisfiability problem for* $\mathcal{S}_L^{\mathrm{prob}}$ *is decidable.*

In the next example we outline the main steps of the corresponding decision procedures.

Example 4.11. Let $L = \{A, B, C\}$ consists of three relation symbols, $\mathrm{ar}(A) = \mathrm{ar}(B) = 2$ and $\mathrm{ar}(C) = 1$.

For shortness, let $\varphi(x) = \forall y(A(x,y) \to \neg B(x,y))$, $\psi(x) = \exists y A(x,y)$, $\theta(x) = C(x)$. Notice that $\neg\varphi(x) \leftrightarrow (\exists y)(A(x,y) \wedge B(x,y))$ and $\neg\psi(x) \to \varphi(x)$ are valid first order formulas. Let α be the conjunction of the following three formulas:

$$(\mathrm{CP}x \approx 1)(\varphi(x) \mid \psi(x)), (\mathrm{CP}x \approx 1)(\psi(x) \mid \theta(x)), (\mathrm{CP}x \approx 1)(\neg\varphi(x) \mid \theta(x)).$$

In the search for a model of α we start with six "atoms" that should be considered:

$$\begin{aligned}
\alpha_1(x) &= \neg\theta(x) \wedge \neg\psi(x) \wedge \varphi(x),\\
\alpha_2(x) &= \neg\theta(x) \wedge \psi(x) \wedge \neg\varphi(x),\\
\alpha_3(x) &= \neg\theta(x) \wedge \psi(x) \wedge \varphi(x),\\
\alpha_4(x) &= \theta(x) \wedge \neg\psi(x) \wedge \varphi(x),\\
\alpha_5(x) &= \theta(x) \wedge \psi(x) \wedge \neg\varphi(x),\\
\alpha_6(x) &= \theta(x) \wedge \psi(x) \wedge \varphi(x).
\end{aligned}$$

It is easy to find an L-model for $\exists x_1 \cdots \exists x_6 \bigwedge_{i=1}^{6} \alpha_i(x_i)$ over $M = \{1,2,3,4,5,6\}$:

$$A^{\mathbf{M}} = \{(2,2),(3,3),(5,5),(6,6)\}, B^{\mathbf{M}} = \{(2,2),(5,5)\}, C^{\mathbf{M}} = \{4,5,6\},$$

The next step is to solve the system in unknowns y_i, $i = 1,\ldots,6$ which represent the measure of the atoms $\alpha_i(x)$, $i = 1,\ldots,6$, respectively:

$$\begin{aligned}
&y_i \geqslant 0, i = 1,\ldots,6,\ y_7 = y_8 = 0,\\
&\sum_{i=1}^{8} y_i = 1,\ y_2+y_3+y_5+y_6 > 0,\ y_4+y_5+y_6 > 0,\\
&\frac{y_3+y_6}{y_2+y_3+y_5+y_6} \approx 1,\ \frac{y_5+y_6}{y_4+y_5+y_6} \approx 1,\ \frac{y_5}{y_4+y_5+y_6} \approx 1.
\end{aligned}$$

Note that $y_2+y_3+y_5+y_6 > 0$ says that $\psi(x)$, which is equivalent to $\alpha_2(x) \wedge \alpha_3(x) \wedge \alpha_5(x) \wedge \alpha_6(x)$, has positive measure, while $\frac{y_3+y_6}{y_2+y_3+y_5+y_6} \approx 1$ corresponds to the formula $(\mathrm{CP}x \approx 1)(\varphi(x) \mid \psi(x))$, and similarly for the other (in)equalities. After eliminating the sign $\approx$, and performing the Fourier-Motzkin procedure, see [23, 38], we finish with the true condition. Thus, α is satisfiable. Using the results obtained during the solving of the system, we can define a measure on M, e.g. $\mu(1) = \frac{3}{4} - \varepsilon - 2\varepsilon^2$, $\mu(2) = \varepsilon$, $\mu(3) = \frac{1}{4} - \varepsilon$, $\mu(4) = \varepsilon^2$, $\mu(5) = \varepsilon$, $\mu(6) = \varepsilon^2$. ■

4.5 Some extensions of the logic $L_{\omega\omega}^{\mathrm{P},\mathbb{I}}$

Adding qualitative quantifiers could be very important for AI applications. The list of logical symbols of $L_{\omega\omega}^{\mathrm{P},\mathbb{I}}$ can be extended with a range of new binary (qualitative) quantifiers, e.g., $\preceq_{\overline{x}}$, and $\trianglelefteq_{\overline{x}}$, with the intended meanings:

- $\mathbf{M}, v \models \alpha \preceq_{\overline{x}} \beta$ iff $\mu_n([\alpha]_{\mathbf{M},v}^{\overline{x}}) \leqslant \mu_n([\beta]_{\mathbf{M},v}^{\overline{x}})$;
- $\mathbf{M}, v \models \alpha \trianglelefteq_{\overline{x}} \beta$ iff $\mu_n([\alpha \wedge \neg\beta]_{\mathbf{M},v}^{\overline{x}}) \leqslant \mu_n([\alpha \wedge \beta]_{\mathbf{M},v}^{\overline{x}})$, etc.

It is obvious that these new quantifiers can be described by infinitarly $L_{\omega\omega}^{\mathrm{P},\mathbb{I}}$-formulas. For example,

$$\alpha \preceq_{\overline{x}} \beta \Leftrightarrow \bigwedge_{q \in \mathbb{I}} (\mathrm{P}\overline{x} = q)\alpha \to (\mathrm{P}\overline{x} \geqslant q)\beta.$$

Having in mind this meta-equivalence, we add the following axiom schema:

$$\alpha \preceq_{\overline{x}} \beta \to ((\mathrm{P}\overline{x} = q)\alpha \to (\mathrm{P}\overline{x} \geqslant q)\beta), q \in \mathbb{I}$$

and the rule:

$$\frac{\varphi \to ((\mathrm{P}\overline{x} = q)\alpha \to (\mathrm{P}\overline{x} \geqslant q)\beta),\ q \in \mathbb{I}}{\varphi \to \alpha \preceq_{\overline{x}} \beta}$$

The axiomatic systems obtained in this way are sound with respect to the class of $L_{\omega\omega}^{\mathrm{P},\mathbb{I}}$-models, and the proofs of the completeness theorems are analogous to the proof of Theorem 4.3.

We can extend the logic $L_{\omega\omega}^{\mathrm{P},\mathbb{I}}$ with a list of new quantifiers $(\mathrm{CP} \cdot \approx_k 0)$, $k \in \mathbb{N}$, whose meaning is given by the following clause:

$$\mathbf{M}, v \models (\mathrm{CP}\overline{x} \approx_k 0)(\alpha \mid \beta) \text{ iff } \mu_n([\beta]_{\mathbf{M},v}^{\overline{x}}) > 0 \text{ and } \mathrm{ord}\left(\frac{\mu_n([\alpha]_{\mathbf{M},v}^{\overline{x}} \cap [\beta]_{\mathbf{M},v}^{\overline{x}})}{\mu_n([\beta]_{\mathbf{M},v}^{\overline{x}})}\right) \geqslant k.$$

These quantifiers express orders of an infinitesimal and increase the expressive power of the language. Note that $(\mathrm{CP} \cdot \approx_0 0)$ and $(\mathrm{CP} \cdot \approx_1 0)$ are the quantifiers $(\mathrm{CP} \cdot \not\approx 0)$ and $(\mathrm{CP} \cdot \approx 0)$, respectively. It is easy to see that each $(\mathrm{CP} \cdot \approx_k 0)$, $k \geqslant 2$, can be introduced by an infinitary formula:

$$(\mathrm{CP}x \approx_k 0)(\alpha \mid \beta) \Leftrightarrow \bigvee_{r \in \mathbb{I}} (\mathrm{CP}x = r\varepsilon^k)(\alpha \mid \beta),$$

i.e.,

$$\neg(\mathrm{CP}x \approx_k 0)(\alpha \mid \beta) \Leftrightarrow \bigwedge_{r \in \mathbb{I}} (\mathrm{CP}x \neq r\varepsilon^k)(\alpha \mid \beta),$$

since the set of infinitesimals whose order is at least k can be defined by:

$$\{x \in \mathbb{Q}(\varepsilon) \mid (\exists r \in \mathbb{Q}(\varepsilon))\, x = r\varepsilon^k\}.$$

Similarly as above, using this meta-equivalence, we add the following axiom schema:

$$(\mathrm{CP}\overline{x}=r\varepsilon^k)(\alpha \mid \beta) \to (\mathrm{CP} \approx_k \overline{x})(\alpha \mid \beta),\, r \in \mathbb{I},$$

and the rule:

$$\frac{\gamma \to (\mathrm{CP}\overline{x} \neq r\varepsilon^k)(\alpha \mid \beta), r \in \mathbb{I}}{\gamma \to \neg(\mathrm{CP} \approx_k \overline{x})(\alpha \mid \beta)}.$$

It is straightforward to introduce a binary quantifier $\prec_{\overline{x}}$ that may be interesting in formal treatment of κ-calculus from [37]:

$$\mathbf{M}, v \models \alpha \prec_{\overline{x}} \beta \text{ iff } \mathrm{ord}(\mu_n([\alpha]^{\overline{x}}_{\mathbf{M},v})) < \mathrm{ord}(\mu_n([\beta]^{\overline{x}}_{\mathbf{M},v})).$$

To facilitate the reading in what follows, we use the abbreviation $(\mathrm{O}\overline{x} = k)(\alpha \mid \beta)$ for $(\mathrm{CP}x \approx_k 0)(\alpha \mid \beta) \wedge \neg(\mathrm{CP}x \approx_{k+1} 0)(\alpha \mid \beta)$, and $(\mathrm{O}\overline{x} = k)\alpha$ for $(\mathrm{O}\overline{x} = k)(\alpha \mid \top)$. The meta-equivalence

$$\alpha \prec_{\overline{x}} \beta \Leftrightarrow \bigwedge_{k \in \mathbb{N}} ((\mathrm{O}\overline{x} = k)\alpha \to (\mathrm{P}\overline{x} \approx_{k+1} 0)\beta)$$

gives the following axiom schema:

$$\alpha \prec_{\overline{x}} \beta \to ((\mathrm{O}\overline{x} = k)\alpha \to (\mathrm{P}\overline{x} \approx_{k+1} 0)\beta)\,,\, k \in \mathbb{N}$$

and the rule:

[(κ)] from $\gamma \to ((\mathrm{O}\overline{x} = k)\alpha \to (\mathrm{P}\overline{x} \approx_{k+1} 0)\beta)$, $k \in \mathbb{N}$, infer $\gamma \to \alpha \prec_{\overline{x}} \beta$.

It is not hard to prove that the axiomatic system extended in this way is sound and complete with respect to the class of $L^{\mathrm{P},\mathbb{I}}_{\omega\omega}$-models. The proofs are straightforward modifications of the corresponding proofs for the logic $L^{\mathrm{P},\mathbb{I}}_{\omega\omega}$.

4.6 Applications

As is suggested in [22, 23], the logic $L^{\mathrm{P},\mathbb{I}}_{\omega\omega}$ can capture many ideas behind defaults. Even the decidable fragments $\mathcal{G}^{\mathrm{prob}}_L$ and $\mathcal{S}^{\mathrm{prob}}_L$ are rich enough for most practical purposes. First, iterated probabilities ("the probability of probability ...") rarely occur in practical considerations, so one probabilistic quantifier should suffice in most situations. Second, many real statements does not involve more than three (blocks of same) quantifiers.

As is mentioned in the previous sections, the quantifier $(\mathrm{CP}\cdot \approx 1)$ plays a central role in modeling nonmonotonic reasoning. A conditional assertion $\alpha \rightsquigarrow_{\overline{x}} \beta$, i.e., a sentence of the form "almost all $\overline{x}$'s that satisfy α, also satisfy β" can be translated into the logic $L^{\mathrm{P},\mathbb{I}}_{\omega\omega}$ to the following formula $(\mathrm{CP}\overline{x} \approx 1)(\beta \mid \alpha)$.

Example 4.12. If $\overline{\mathbf{M}}$ is the $L^{\mathrm{P},\mathbb{I}}_{\omega\omega}$-structure from Example 4.5, we have:

$$\overline{\mathbf{M}} \models B(x,y) \rightsquigarrow_{x,y} U(x) \vee U(y),$$

where $B(x,y) \rightsquigarrow_{x,y} U(x) \vee U(y)$ can be read "almost all pairs (x,y) satisfying $B(x,y)$, also satisfy $U(x) \vee U(y)$". It is not difficult to verify that:

$$\overline{\mathbf{M}} \models B(x,x) \rightsquigarrow_x U(x) \qquad \overline{\mathbf{M}} \not\models \exists x B(x,y) \rightsquigarrow_y U(y)$$
$$\overline{\mathbf{M}} \not\models \forall x (B(x,y) \rightsquigarrow_y U(y)) \qquad \overline{\mathbf{M}} \models (\mathrm{CP}x \approx \tfrac{2}{3})(B(x,y) \rightsquigarrow_y U(y))$$ ■

All the rules of System **P** are satisfied under the above translation.

Theorem 4.9. *(We assume that all the free variables of the formulas α, β, γ are contained in the fixed tuple $\overline{x}$.)*

1. $\alpha \to \gamma, \beta \rightsquigarrow_{\overline{x}} \alpha \vdash \beta \rightsquigarrow_{\overline{x}} \gamma$ *[Right Weakening]*
2. $\vdash \beta \rightsquigarrow_{\overline{x}} \beta$ *[Reflexivity]*
3. $\beta \leftrightarrow \beta', \beta \rightsquigarrow_{\overline{x}} \alpha \vdash \beta' \rightsquigarrow_{\overline{x}} \alpha$ *[Left Logical Equivalence]*
4. $\beta \rightsquigarrow_{\overline{x}} \alpha, \alpha \wedge \beta \rightsquigarrow_{\overline{x}} \gamma \vdash \beta \rightsquigarrow_{\overline{x}} \gamma$ *[Cut]*
5. $\beta \rightsquigarrow_{\overline{x}} \alpha, \beta \rightsquigarrow_{\overline{x}} \gamma \vdash \beta \wedge \alpha \rightsquigarrow_{\overline{x}} \gamma$ *[Cautions Monotonicity]*

Proof. The statements (1-3) are obvious.

We use the completeness theorem to get properties (4) and (5). Let $\overline{\mathbf{M}}$ be an $L_{\omega\omega}^{\mathrm{P,I}}$-model, and v a valuation. Set $A = [\alpha]^{\overline{x}}_{\overline{\mathbf{M}},v}$, $B = [\beta]^{\overline{x}}_{\overline{\mathbf{M}},v}$, $C = [\gamma]^{\overline{x}}_{\overline{\mathbf{M}},v}$. For shortness we omit the subscript in the notation of the measure. The claims (4) and (5) are obviously true if $\mu(B) = 0$. So, suppose that $\mu(B) = b > 0$.

(4) Assume that $\overline{\mathbf{M}} \models (\mathrm{CP}\overline{x} \approx 1)(\alpha(\overline{x}) \mid \beta(\overline{x})), (\mathrm{CP}\overline{x} \approx 1)(\gamma(\overline{x}) \mid \alpha(\overline{x}) \wedge \beta(\overline{x}))$. From $\mu(A \mid B) \approx 1$, we have $\mu(A \mid B) = 1 - \varepsilon'$, where ε' is an infinitesimal. Thus, $\mu(A \cap B) = b - \varepsilon' b$. Reasoning in the same manner, we obtain that $\mu(A \cap B \cap C) = b - \varepsilon' b - \varepsilon''(b - \varepsilon' b)$, for some infinitesimal ε''. Now, by the monotonicity of the probability, we have

$$b - (\varepsilon' - \varepsilon'' + \varepsilon'\varepsilon'')b = \mu(A \cap B \cap C) \leqslant \mu(B \cap C) \leqslant \mu(B) = b,$$

Thus, $\mu(B \cap C) \approx b$, and hence $\mu(C \mid B) \approx 1$.

(5) Assume that $\overline{\mathbf{M}} \models (\mathrm{CP}\overline{x} \approx 1)(\alpha(\overline{x}) \mid \beta(\overline{x})), (\mathrm{CP}\overline{x} \approx 1)(\gamma(\overline{x}) \mid \beta(\overline{x}))$. Note that $\mu(A \cap B) \neq 0$ since $\mu(A \mid B) \approx 1$ and $\mu(B) > 0$. From $\mu(A \mid B) \approx 1$, as in the proof of the previous statement, we have $\mu(A \cap B) = b - \varepsilon' b$, where ε' is an infinitesimal, and consequently $\mu(A^c \cap B) = \varepsilon' b$. Also, there is an infinitesimal ε'' such that $\mu(B \cap C) = b - \varepsilon'' b$. It is easy to see that $b - \varepsilon' b - \varepsilon'' b \leqslant \mu(A \cap B \cap C) \leqslant b - \varepsilon' b$. Therefore, $1 - \frac{\varepsilon'' b}{b - \varepsilon' b} \leqslant \frac{\mu(A \cap B \cap C)}{\mu(A \cap B)} \leqslant 1$. ■

Example 4.13. One of the problems with System **P** mentioned above is the so called "inheritance blocking". If the elements of some set A have a certain property B, we expect that the elements of a subset $A \cap C$ will "inherit" the property B. However, this would be in conflict with the essence of nonmonotonicity. Namely, if we were allowed to infer $\alpha \wedge \gamma \mid\sim \beta$ from $\alpha \mid\sim \beta$, for arbitrary γ, the system would be monotonous. Therefore, as expected, System **P** does not allow such derivations. The logic $L_{\omega\omega}^{\mathrm{P,I}}$ offers an opportunity to formulate some possible solutions.

Let us consider the set Δ consisting of the following two defaults:

- the Swedes are blond, i.e., $\mathtt{Swede}(x) \rightsquigarrow_x \mathtt{blond}(x)$, and

- the Swedes are tall, i.e., $\texttt{Swede}(x) \rightsquigarrow_x \texttt{tall}(x)$.

Because of the inheritance blocking problem, in P it is not possible to conclude that Swedes who are not tall are blond ($\texttt{Swede}(x) \wedge \neg\texttt{tall}(x) \rightsquigarrow_x \texttt{blond}(x)$). In this particular case, it may turn out that the short Swedes are exactly the ones which are not blond. A solution might be to add a clause which excludes such possibility, for example: $(\mathrm{CP}x \approx 0)(\neg\texttt{blond}(x) \mid \texttt{Swede}(x) \wedge \neg\texttt{tall}(x))$. ■

In the suitable extensions of $L^{\mathbf{P},\mathbb{I}}_{\omega\omega}$, described in Section 4.5, we could be able to introduce new default relations between classical first-order formulas. Namely, for each $k \geqslant 1$ we have $\beta \rightsquigarrow^{\geqslant k}_{\overline{x}} \alpha$ and $\beta \rightsquigarrow^{=k}_{\overline{x}} \alpha$ denoting respectively

$$(\mathrm{CP}\overline{x} \approx_k 0)(\neg\alpha \mid \beta) \text{ and } (\mathrm{CP}\overline{x} = k)(\neg\alpha \mid \beta).$$

It is easy to prove that the relations $\rightsquigarrow^{\geqslant k}_{\overline{x}}$ and $\rightsquigarrow^{=k}_{\overline{x}}$ for each $n \geqslant 1$, satisfy all the rules of System **P**. Studying properties of these two relations shows that they may be useful in characterizations of various types of weakened monotonicity that are significant in nonmonotonic reasoning. We believe that more detailed research in this direction will be very fruitful.

Example 4.14. Let us consider again the set Δ introduced in Example 4.13, consisting of the following two defaults: $\texttt{Swede}(x) \rightsquigarrow_x \texttt{blond}(x)$ and $\texttt{Swede}(x) \rightsquigarrow_x \texttt{tall}(x)$. If we assume that the starting defaults are of different strength, we can avoid the inheritance blocking and obtain the desired conclusion. For example, from $\texttt{Swede}(x) \rightsquigarrow^{=1}_{\overline{x}} \texttt{tall}(x)$ and $\texttt{Swede}(x) \rightsquigarrow^{=2}_{\overline{x}} \texttt{blond}(x)$, it follows that $\texttt{Swede}(x) \wedge \neg\texttt{tall}(x) \rightsquigarrow_x \texttt{blond}(x)$. ■

Example 4.15. When we say that birds fly, we assume that typical birds do fly, but there may exist some atypical specimens, e.g., penguins, which are not able to fly. Expectation would be that we may apply the usual reasoning to typical specimens while with atypical ones we have to be careful. So, if we know that "Tweety is a bird" we assume that "Tweety flies" but this conclusion becomes unacceptable if we discover that "Tweety is a penguin". The problem is that no available formalism for defaults offers a way of discussing whether Tweety is a typical bird. Our rich language makes it possible to try to do this.

Let $!\alpha(c/x)$ denote $(\mathrm{CP}x \approx 0)(x = c \mid \alpha(x))$. We find that $\neg!\alpha(c/x)$, i.e., $(\mathrm{CP}x \not\approx 0)(x = c \mid \alpha(x))$ is closely related to the statement "the element c is a typical element of (witness for) the nonempty set $\{x : \alpha(x)\}$", which may suggest the following rule of inference

$$\frac{\beta(x) \rightsquigarrow \alpha(x) \quad (\mathrm{P}x > 0)\beta(x) \quad \neg!\beta(x := c)}{\alpha(x := c)}.$$

This rule may be interpreted as saying that a default rule can be applied to a particular individual only in the case when that individual is typical. Using this rule we might be able to elucidate some of problems listed in [7]. ■

4.7 Conclusions

The logic $L^{\mathrm{P},\mathbb{I}}_{\omega\omega}$ and some of its extensions occupy the central position in this chapter. Our general hypothesis is that probabilistic first-order logics are expressive enough to capture many aspects of formal systems for default reasoning, roughly described in Sections 4.2 and 4.3. There are a number of close parallels between probabilistic logics and formalisms for default reasoning, and detailed comparison of the corresponding logical systems would be very interesting and useful, but it is left for future work.

References

[1] E. W. Adams. *The logic of Conditional.* Reidel, Dordrecht, 1975.

[2] N. Alechina. For All Typical. In *Proceedings ECSQARU'95: Symbolic and Quantitative Approach to Reasoning and Uncertainty*, volume 946, pages 1–8, 1995.

[3] G. Antoniou and K. Wang. Default logic. In D. Gabbay and J. Woods, editors, *The Many Valued and Nonmonotonic Turn in Logic*, pages 517–555. Elsevier, North-Holland, 2007.

[4] F. Baader and B. Hollunder. Embedding defaults into terminological knowledge representation formalisms. *Journal of Automated Reasoning*, 14:149–180, 1995.

[5] F. Bacchus. *Lp*, A Logic for Representing and Reasoning with Statistical Knowledge. *Computational Intelligence*, 6 (4):209–231, 1990.

[6] F. Bacchus, A. J. Grove, J. Y. Halpern, and D. Koller. From Statistical Knowledge Bases to Degrees of Belief. *Artificial Intelligence*, 87 (1-2):75–143, 1996.

[7] S. Benferhat, A. Saffiotti, and P. Smets. Belief functions and default reasoning. *Artificial Intelligence*, 122:1–69, 2000.

[8] E. Börger, E. Grädel, and Y. Gurevich. *The Classical Decision Problem.* Springer, Berlin, Heidelberg, 1997.

[9] R. Brachman and H. Levesque. *Knowledge Representation and Reasoning.* Morgan Kaufmann, San Francisco, CA, USA, 2004.

[10] R. I. Brafman. A first-order conditional logic with qualitative statistical semantics. *Journal of Logic and Computation*, 7 (6):777–803, 1997.

[11] G. Brewka. *Nonmonotonic Reasoning: Logical Foundations of Commonsense.* Cambridge, United Kingdom, 2010.

[12] G. Brewka, I. Niemelä, and M. Truszczyński. Nonmonotonic reasoning. In V. Lifschitz, B. Porter, and H. van Harmelen, editors, *Handbook of Knowledge Representation*, pages 239–284. Elsevier, San Diego, USA, 2007.

[13] J.P. Burgess. Quick completeness proofs for some logics of conditionals. *Notre Dame Journal of Formal Logic*, 22:76–84, 1981.

[14] J. Delgrande. A First-Order Conditional Logic for Prototypical Properties. *Artificial Intelligence*, 33, 1987.

[15] J. Delgrande. An Approach to Default Reasoning Based on a First-Order Conditional Logic: Revised Report. *Artificial Intelligence*, 36, 1988.

[16] J. Delgrande. On first-order conditional logics. *Artificial Intelligence*, 105, 1998.

[17] H. D. Ebbinghaus, J. Flum, and W. Thomas. *Mathematical Logic*. Springer, New York, USA, 1994. second edition.

[18] H. Etherington. *Reasoning with Incomplete Information: Investigations of Non-Monotoninc Reasoning*. PhD thesis, Department of Computer Science, University of British Columbia, Vancouver, BC, 1986.

[19] D. Gabbay. Theoretical foundations for non-monotonic reasoning in expert systems. In K. R. Apt, editor, *Logics and models of concurrent systems*, pages 439–457. Springer, Berlin, Heidelberg, 1985.

[20] M. Goldszmidt, P. Morris, and J. Pearl. A maximum entropy approach to nonmonotonic reasoning. *IEEE Transactions on Pattern Analysis and Machine Intelligence*, 15 (3):220–232, 1993.

[21] J.Y. Halpern. An analysis of first-order logics of probability. *Artificial Intelligence*, 46:311–350, 1990.

[22] N. Ikodinović, Z. Ognjanović, M. Rašković, and Z. Marković. First-order probabilistic logics and their applications. In *Zbornik radova, subseries Logic in computer science, Vol. 18 (26)*, pages 37–78, Belgrade, Serbia, 2015. Matematički institut SANU.

[23] N. Ikodinović, M. Rašković, Z. Marković, and Z. Ognjanović. A first-order probabilistic logic with approximate conditional probabilities. *Logic Journal of the IGPL*, 22(4):539–564, 2014.

[24] H.J. Keisler. *Elementary calculus. An infinitesimal approach*. Prindle, Weber & Schmidt, Boston, Massachusetts, 1986. second edition.

[25] S. Kraus, D. Lehmann, and M. Magidor. Nonmonotonic reasoning, preferential models and cumulative logics. *Artificial Intelligence*, 44 (1-2):167–207, 1990.

[26] D. Lehmann and M. Magidor. Preferential logics: the predicate calculus case. In *Theoretical Aspects of Reasoning about Knowledge: Proc. 3rd conf.*, pages 57–72, 1990.

[27] D. Lehmann and M. Magidor. What does a conditional knowledge base entail? *Artificial Intelligence*, 55:1–60, 1992.

[28] V. Lifschitz. On open defaults. In J. Lloyd, editor, *Computational Logic: Symposium Proceedings*, pages 80–95. Springer, 1990.

[29] V. Lifschitz. Circumscription. In D. Gabbay, C. J. Hogger, and J. A. Robinson, editors, *Handbook of Logic in Artificial Intelligence and Logic Programming, Volume 3: Nonmonotonic Reasoning and Uncertain Reasoning*, pages 298–352. Oxford University Press, New York, USA, 1994.

[30] W. Lukasiewicz. Two results on default logic. In *IJCAI'85 Proc. of the 9th international joint conference on Artificial Intelligence*, pages 459–461, San Francisco, CA, USA, 1985. Morgan Kaufmann Publishers Inc.

[31] D. Makinson. General theory of cumulative inference. In M. Reinfrank, J. de Kleer, M.L. Ginsberg, and E. Sandewall, editors, *Proceedings of the Second*

International Workshop on Non-Monotonic Reasoning, volume 346 of *Lecture Notes in Artificial Intelligence*, pages 1–18, Berlin, Heidelberg, 1989. Springer.

[32] J. McCarthy. Circumscription: A Form of Non-Monotonic Reasoning. *Artificial Intelligence*, 13, 1980.

[33] J. McCarthy. Applications of circumscription to formalizing common sense knowledge. *Artificial Intelligence*, 26, 1986.

[34] D. McDermott and J. Doyle. Nonmonotonic logic I. *Artificial Intelligence*, 13, 1980.

[35] R.C. Moore. Possible-world semantics for autoepistemic logic. In *Proceedings of AAAI Workshop on Nomonotonic Reasoning*, pages 369–401, 1984.

[36] R.C. Moore. Semantical considerations on nonmonotonic logic. *Artificial Intelligence*, 25 (1):75–94, 1985.

[37] J. Pearl. System Z: A natural ordering of defaults with tractable application to default reasoning. In *Proc. of Theoretical Aspects of Reasoning about Knowledge*, pages 121–135, 1990.

[38] M. Rašković, Z. Marković, and Z. Ognjanović. A logic with approximate conditional probabilities that can model default reasoning. *International Journal of Approximate Reasoning*, 49(1):52–66, 2008.

[39] M. Rašković, Z. Ognjanović, and Z. Marković. A Probabilistic Approach to Default Reasoning. In *Proceeding of the NMR '04*, pages 335–341, 2004.

[40] R. Reiter. On closed world data bases. In *Logics and Databases*, pages 55–76. Plenum Press, New York, USA, 1978.

[41] R. Reiter. A logic for default reasoning. *Artificial Intelligence*, 13(1-2):81–132, 1980.

[42] R. Reiter. Nonmonotonic reasoning. *Annual Review of Computer Science*, 2:147–187, 1987.

[43] A. Robinson. *Non-standard analysis*. North-Holland, Amsterdam, 1966.

[44] E. Sandewall. An approach to the frame problem and its implementation. In B. Meltzer and D. Michie, editors, *Machine Intelligence 7*, pages 195–204. Edinburgh University Press, Edinburgh, 1972.

[45] T. Schaub. *The automation of reasoning with incomplete information: from semantic foundations to efficient computation*. Springer, Berlin, Heidelberg, 1997.

[46] K. Schlechta. Defaults as Generalized Quantifiers. *Journal of Logic and Computation*, 5 (4):473–494, 1995.

Chapter 5
Some New Probability Operators

Zoran Marković, Miodrag Rašković

Abstract This chapter is dedicated to the formalization of the reasoning about probability involving qualitative statements (e.g. the probability of p is greater than the probability of q), probabilistic quantification (e.g. the probability of p is equal to $\frac{3n+1}{4n+7}$ for some positive integer n), conditional statements (e.g., the conditional probability of p given q is at least r), confirmation (e.g., p confirms q), and independence. The common restriction is that iterations and nesting of probabilistic operators are not allowed. In other words, admissible statements are Boolean combinations of the atomic probabilistic assessments. Beside the presentation of complete axiomatizations, we shall also discuss the hierarchical structure of the introduced logics in terms of their expressiveness.

5.1 Introduction

The present chapter is focused on reasoning about probability involving the following atomic statements:

- The probability of α is lesser or equal than probability of β, denoted by $\alpha \leq \beta$.
- The conditional probability of α given β is at least r, denoted by $CP_{\geq r}(\alpha, \beta)$.
- The probability of α is in the given set F, denoted by $Q_F \alpha$.
- α weakly confirms β, denoted by $\alpha \nearrow \beta$.
- α weakly disconfirms β, denoted by $\alpha \searrow \beta$.

Zoran Marković
Mathematical Institute of the Serbian Academy of Sciences and Arts, Kneza Mihaila 36, 11000 Belgrade, Serbia, e-mail: zoranm@mi.sanu.ac.rs

Miodrag Rašković
Mathematical Institute of the Serbian Academy of Sciences and Arts, Kneza Mihaila 36, 11000 Belgrade, Serbia, e-mail: miodragr@mi.sanu.ac.rs

Z. Ognjanović (ed.), *Probabilistic Extensions of Various Logical Systems*,
https://doi.org/10.1007/978-3-030-52954-3_5

The intended meaning of the weak confirmation proposition $\alpha \nearrow \beta$ is that the conditional probability of β given α is greater or equal to the probability of β ($\mu(\beta|\alpha) \geq \mu(\beta)$). Similarly, $\alpha \searrow \beta$ reads $\mu(\beta|\alpha) \leq \mu(\beta)$.

The main characteristic of the probability operators $\leq$, $\neg Q_F$, $\nearrow$ and $\searrow$ is the fact that they are Π_1–definable with respect to the basic probabilistic operators $P_{\geq r}$, i.e. they are equivalent to countable conjunctions of Boolean combinations of the basic probabilistic formulas of the form $P_{\geq r}$. Indeed:

- $\alpha \leq \beta \iff \bigwedge_{r \in [0,1]_\mathbb{Q}} (P_{\geq r}\alpha \to P_{\geq r}\beta)$;

- $\neg Q_F \alpha \iff \bigwedge_{r \in F} P_{\neq r}\alpha$. Recall that, by definition, $P_{\neq r}\alpha$ is equivalent to

$$(P_{\geq r}\alpha \wedge \neg P_{\geq 1-r}\neg\alpha) \vee (P_{\geq r}\neg\alpha \wedge \neg P_{\geq 1-r}\alpha);$$

- $\alpha \nearrow \beta \iff \bigwedge_{r,s \in [0,1]_\mathbb{Q}} ((P_{\geq r}\alpha \wedge P_{\geq s}\beta) \to P_{\geq rs}(\alpha \wedge \beta))$;

- $\alpha \searrow \beta \iff \bigwedge_{r,s \in [0,1]_\mathbb{Q}} ((P_{\leq r}\alpha \wedge P_{\leq s}\beta) \to P_{\leq rs}(\alpha \wedge \beta))$.

The common restriction on the application of the above operators is that neither nesting, nor iteration is allowed. In other words, admissible formulas are (finite) Boolean combinations of the basic probabilistic formulas. Hence, the base for formalizations presented here is the LPP_2–logic.

5.1.1 Qualitative probability

In [5] Bruno de Finetti introduced a notion of the comparative (later renamed qualitative) probability as a binary relation $\leq$ on a subalgebra H of the power set algebra $\mathscr{P}(W)$ ($W \neq \emptyset$ is at most countable) satisfying the following conditions:

1. Non-triviality: $\emptyset < W$;
2. Reflexivity: $x \leq x$;
3. Transitivity: $(x \leq y \wedge y \leq z) \implies x \leq z$;
4. Linearity: $x \leq y \vee y \leq x$;
5. Lower bound: $\emptyset \leq x$;
6. Additivity: if $(x \cup y) \cap z = \emptyset$, then

$$x \leq y \iff x \cup z \leq y \cup z.$$

Note that the monotonicity property $x \subseteq y \implies x \leq y$ and the upper bound property $x \leq W$ can be easily derived from the above conditions. The rationale for this approach is that we may consider some events more probable than some other events, without knowing their exact probabilities.

A function $d : H \longrightarrow [0,1]$, $d(\emptyset) = 0$ and $d(W) = 1$, is a distribution of the qualitative probability relation $\leq$ iff, for all $x, y \in H$,

$$x \leq y \iff d(x) \leq d(y).$$

De Finetti conjectured that d must be a probability (hence the term qualitative probability). In [27] a counterexample was constructed on a finite algebra and conditions on qualitative probabilities were given that force only probabilistic distributions. Further improvements, generalizations and formalizations were given in [2, 3, 6, 8, 13, 15–17, 21, 29, 31, 44, 45]. Some work in this line of research is presented in [7, 35, 38, 39].

5.1.2 Conditional probabilities

Conditional probability can be founded in two different ways. The first one, proposed by Kolmogorov, starts with the probability space (W, H, μ) and an event α of positive measure. Then, the conditional probability $\mu_{/\alpha}$ is a probability defined on the quotient algebra $H_{/\alpha} = \{[\alpha \wedge \beta] : \alpha \in For_C\}$ by

$$\mu_{/\alpha}([\alpha \wedge \beta]) = \frac{\mu([\alpha \wedge \beta])}{\mu([\alpha])}.$$

Instead of $\mu_{/\alpha}([\alpha \wedge \beta])$ we write the usual $\mu([\beta] \mid [\alpha])$.

An alternative approach introduced by de Finetti (so called coherent conditional probability) starts with conditional probability. The difference is that in de Finetti's axiomatization conditional probability $\mu(\mid)$ maps $H \times G$ into $[0,1]$, where H is a Boolean algebra, and $G = H \setminus \{\emptyset\}$. In addition, $\mu(,)$ satisfies the following conditions:

1. $\mu([\alpha] \mid [\alpha]) = 1$ for all $[\alpha] \in G$,
2. $\mu(\mid [\alpha]) : H \longrightarrow [0,1]$ is a probability measure for all $[\alpha] \in G$,
3. If $[\alpha] \in H$ and $[\beta], [\gamma], [\beta \wedge \gamma] \in G$, then

$$\mu([\alpha \wedge \beta] \mid [\gamma]) = \mu([\beta] \mid [\gamma]) \cdot \mu([\alpha] \mid [\beta \wedge \gamma]).$$

It is known that de Finetti's approach imposes serious difficulties for iterations of conditional probabilities. Here we shall present formalization of Kolmogorov's concept of conditional probability.

Several formal systems for reasoning about Kolmogorov's and de Finetti's conditional probabilities are given in [13, 16, 25, 32–34, 40, 41].

5.1.3 Independence and confirmation

The notion of independence is one of the fundamental notions of probability theory. In terms of conditional probability it is expressed by

$$\mu(x|y) = \mu(x) \text{ and } \mu(y|x) = \mu(y),$$

which in terms of pure probability is rewritten as

$$\mu(x \cap y) = \mu(x)\mu(y).$$

Closely related notion is Carnap's notion of confirmation, see [4]:

> "The explication of certain concepts which are connected with the scientific procedure of confirming or disconfirming hypotheses with the help of observations and which we therefore briefly call concepts of confirmation."

Carnap proposes three different semantical concepts of confirmation:

1. The classificatory concept: a hypothesis is confirmed by some evidence;
2. The comparative concept: a hypothesis A is confirmed by some evidence B at least as strongly as C is confirmed by D;
3. The quantitative concept: for a hypothesis A and an evidence B, a real number $c(A,B) \in [0,1]$ represents the corresponding degree of confirmation.

There is no general agreement on the basic properties of the confirmation measure c. However, all different approaches agree that $c(A,B) > 0$ iff the posterior probability of A on the evidence B is greater that the prior probability of A.

In probabilistic terms, A confirms B iff $\mu(A \cap B) > \mu(A)\mu(B)$. Some of formal systems that can represent either independence, or confirmation, or both of them are given in [12–14, 16, 18, 19]. Complete axiomatizations are presented in [9].

5.1.4 Probabilistic operators Q_F

Our work on probabilistic operators Q_F introduced in [36] is inspired by Keisler's line of research on generalized quantifiers presented in [26]. A significant technical influence is Hoover's paper [22]. Hierarchy of probabilistic logics with operators of the form Q_F was given in [24, 37].

The purpose of the Q_F operators is to restrict probability functions to specific predetermined values. The prime example of this kind is the $Q_{[0,1]_{\mathbb{Q}}}$ operator that restricts probability values to the set of rational numbers from the real unit interval. Some connections between this operator and the qualitative probability operator will be discussed in the final section.

5.1.5 Organization of the chapter

In Section 5.2 we provide an integral version of syntax, semantic and axiom systems for the logics containing studied operators and give their precise definitions. In Section 5.3 we discuss the completeness and decidability status of the introduced logics. We outline the proof of the strong completeness theorem. Decidability status is discussed informally through an illustrative example showing different places in the hierarchy (NP, PSPACE, recursively denumerable). In section 5.4 we present logics $LPP_{1,\nearrow,\searrow}$ and $LPP_{1,CP}$ that allow iterations and nesting of the probabilistic operators. In Section 5.5 a hierarchy of the $LPP_{2,O}$–logics is discussed. Concluding remarks are in the final section 5.6.

5.2 Formal languages

Here we shall present an integral formal language that contains all particular languages studied in this chapter. Then we shall specify the corresponding fragment for each of them. We shall end this section with the common semantics and an integral list of axioms and inference rules appearing in them. Each logic will be specified by the corresponding axioms and rules. In order to have Kolmogorov's conditional probabilities defined for all pairs of events, the convention $\mu([\alpha],\emptyset) = 1.$ is used. To justify such a choice, note that conditional probabilities are also used as the truth functions for conditional events (conditionals) $\beta \mid\!\sim \alpha$. In particular, valid conditional $\bot \mid\!\sim \alpha$ is evaluated by 1.

As before, *Var* is a countably infinite set of atomic propositions (variables, letters). Variables for propositional letters are p and q, indexed if necessary. Classical propositional formulas are finite Boolean combination of atomic propositions and are denoted by α, β and γ, indexed if necessary. By For_C we denote the set of all classical propositional formulas.

The set For_P of probabilistic formulas is inductively defined as follows:

$$P_{\geq r}\alpha \mid CP_{\geq r}(\alpha,\beta) \mid CP_{\leq r}(\alpha,\beta) \mid Q_F\alpha \mid \alpha \leq \beta \mid \alpha \nearrow \beta \mid \alpha \searrow \beta \mid \neg\phi \mid \phi \wedge \psi.$$

Here r can be an arbitrary rational number from $[0,1]_{\mathbb{Q}}$ and F can be any recursive subset of $[0,1]_{\mathbb{Q}}$. Note that the operators $CP_{\geq r}$ and $CP_{\leq r}$ are not mutually definable since $CP_{\leq r}(\alpha,\bot)$ is not equivalent to $CP_{\geq 1-r}(\neg\alpha,\bot)$.

Variables for probabilistic formulas are ϕ, ψ and θ, indexed if necessary. The corresponding fragment for the given logic L is denoted by For_P^L.

Derived operators $P_{\leq r}, P_{>r}, P_{<r}, P_{=r}, P_{\neq r}, Q_{F\cap G}, Q_{F\cup G}, Q_{F\setminus G}, \geq, <, >, =$ and $\neq$ are introduced in the natural way suggested by their designation. On the other hand, operators Q_{1-F}, $\perp$, $\uparrow$ and $\downarrow$ are defined as follows:

- $Q_{1-F}\alpha \Longleftrightarrow Q_F\neg\alpha$;
- $\alpha\perp\beta \Longleftrightarrow \alpha \nearrow \beta \wedge \alpha \searrow \beta$ (α and β are independent);

- $\alpha \uparrow \beta \Longleftrightarrow \neg(\alpha \searrow \beta)$ (α confirms β);
- $\alpha \downarrow \beta \Longleftrightarrow \neg(\alpha \nearrow \beta)$ (α disconfirms β).

As is usual in logic, sets of formulas are called theories. The class of models coincides with the class of models for the LPP_2–logic. For the given model $M = (W,H,v,\mu)$, the satisfiability relation $\models$ is extended to non-LPP_2–formulas inductively as follows:

- $M \models CP_{\geq r}(\alpha,\beta) \Longleftrightarrow \mu([\beta]) = 0$, or $\mu([\beta]) > 0$ and $\frac{\mu([\alpha\wedge\beta])}{\mu([\beta])} \geq r$,
- $M \models CP_{\leq r}(\alpha,\beta) \Longleftrightarrow \mu([\beta]) > 0$ and $\frac{\mu([\alpha\wedge\beta])}{\mu([\beta])} \leq r$,
- $M \models Q_F\alpha \Longleftrightarrow \mu([\alpha]) \in F$,
- $M \models \alpha \leq \beta \Longleftrightarrow \mu([\alpha] \leq \mu([\beta]))$,
- $M \models \alpha \nearrow \beta \Longleftrightarrow \mu([\alpha\wedge\beta]) \geq \mu([\alpha])\mu([\beta])$,
- $M \models \alpha \searrow \beta \Longleftrightarrow \mu([\alpha\wedge\beta]) \leq \mu([\alpha])\mu([\beta])$,
- $M \models \neg\phi \Longleftrightarrow M \not\models \phi$,
- $M \models \phi \wedge \psi \Longleftrightarrow M \models \phi \ \wedge M \models \psi$.

The notions of valid formulas and semantical consequences are introduced as usual. The integral list of axioms and inference rules appearing in the studied logics is given below:

LPP_2 axioms:

A0: Tautology instances
A1: $P_{\geq 0}\alpha$
A2: $P_{\geq r}\alpha \to P_{>s}\alpha$, $s < r$
A3: $P_{>r}\alpha \to P_{\geq r}\alpha$
A4: $(P_{\geq r}\alpha \wedge P_{\geq s}\beta \wedge P_{=0}(\alpha\wedge\beta)) \to P_{\geq r+s}(\alpha\vee\beta)$, $r+s \leq 1$ and
A5: $(P_{\leq r}\alpha \wedge P_{<s}\beta) \to P_{<r+s}(\alpha\vee\beta)$, $r+s \leq 1$.

Q_F axiom:

A6: $P_{=r}\alpha \to Q_F\alpha$, $r \in F$.

$\leq$ axiom:

A7: $\alpha \leq \beta \to (P_{\geq r}\alpha \to P_{\geq r}\beta)$.

$\nearrow$ and $\searrow$ axioms:

A8: $\alpha \nearrow \beta \to (P_{\geq r}\alpha \wedge P_{\geq s}\beta \to P_{\geq rs}(\alpha\wedge\beta))$ and
A9: $\alpha \searrow \beta \to (P_{\leq r}\alpha \wedge P_{\leq s}\beta \to P_{\leq rs}(\alpha\wedge\beta))$.

Conditional probability axioms:

A10: $P_{=0}\beta \to CP_{\geq 1}(\alpha,\beta)$
A11: $CP_{\leq 0}(\alpha,\beta) \leftrightarrow (P_{>0}\beta \wedge P_{=0}(\alpha\wedge\beta))$
A12: $(CP_{\geq r}(\alpha,\beta) \wedge P_{\geq s}\beta) \to P_{\geq rs}(\alpha\wedge\beta)$, for $s > 0$
A13: $P_{>0}\beta \to (CP_{\geq r}(\alpha,\beta) \leftrightarrow CP_{\leq 1-r}(\neg\alpha,\beta))$.

LPP_2 rules:

R1 (Modus Ponens for For_C): from α and $\alpha \rightarrow \beta$ infer β.
R2 (Modus Ponens for For_P): from ϕ and $\phi \rightarrow \psi$ infer ψ.
R3 (Necessitation): from α infer $P_{\geq 1}\alpha$.
R4 (Archimedean rule): from the set of premises

$$\{\phi \rightarrow P_{\geq s}\alpha \,:\, s \in [0,r)_{\mathbb{Q}}\}$$

infer $\phi \rightarrow P_{\geq r}\alpha$.

Q_F rule:

R5: from the set of premises

$$\{\phi \rightarrow P_{\neq r}\alpha \,:\, r \in F\}$$

infer $\phi \rightarrow \neg Q_F\alpha$.

$\leq$ rule:

R6: from the set of premises

$$\{\phi \rightarrow (P_{\geq r}\alpha \rightarrow P_{\geq r}\beta) \,: r \in [0,1]_{\mathbb{Q}}\}$$

infer $\phi \rightarrow \alpha \leq \beta$.

$\nearrow$ and $\searrow$ rules:

R7: from the set of premises

$$\{\phi \rightarrow (P_{\geq r}\alpha \wedge P_{\geq s}\beta \rightarrow P_{\geq rs}(\alpha \wedge \beta)) \,:\, r \in [0,1]_{\mathbb{Q}}\}$$

infer $\phi \rightarrow \alpha \nearrow \beta$.
R8: from the set of premises

$$\{\phi \rightarrow (P_{\leq r}\alpha \wedge P_{\leq s}\beta \rightarrow P_{\leq rs}(\alpha \wedge \beta)) \,:\, r \in [0,1]_{\mathbb{Q}}\}$$

infer $\phi \rightarrow \alpha \searrow \beta$.

Conditional probability (CP) rule:

R9: from the set of premises

$$\{\gamma \rightarrow (P_{\geq r}\beta \rightarrow P_{rs}(\alpha \wedge \beta)) \,:\, r \in [0,1]_{\mathbb{Q}} \setminus \{0\}\}$$

infer $\gamma \rightarrow CP_{\geq s}(\alpha,\beta)$.

Note that the rules R4 – R9 are infinitary, i.e., with countable sets of premises. The notions of inference, theorem and maximal consistent theory are introduced in the same way as for the LPP_2–logic. Particular logics are defined as follows:

- Logic LPP_2: $\mathrm{For}_P^{LPP_2}$ admits only the basic probability operators $P_{\geq r}$. The axiom system is formed by axioms A0–A5 and rules R1–R4;
- Logic $LPP_{2,O}$: $\mathrm{For}_P^{LPP_{2,O}}$ admits the basic probability operators $P_{\geq r}$ and Q_F, where F ranges over the recursive family O of recursive subsets of $[0,1]_{\mathbb{Q}}$. The axiom system is obtained by adding the axiom A6 and the rule R5 to the axiom system of the LPP_2–logic;
- Logic $LPP_{2,\leq}$: $\mathrm{For}_P^{LPP_{2,\leq}}$ admits the basic probability operators $P_{\geq r}$ and $\leq$. The axiom system is obtained by addition of the axiom A7 and the rule R6 to the LPP_2–axiomatization;
- Logic $LPP_{2,\nearrow,\searrow}$: $\mathrm{For}_P^{LPP_{2,\nearrow,\searrow}}$ admits the basic probability operators $P_{\geq r}$, $\nearrow$ and $\searrow$. The axiom system is obtained by adding axioms A8 and A9 and rules R7 and R8 to the LPP_2–axiomatization;
- Logic $LPP_{2,CP}$: $\mathrm{For}_P^{LPP_{2,CP}}$ admits the basic probability operators $P_{\geq r}$, $CP_{\geq r}$ and $CP_{\leq r}$. The axiom system is obtained by addition of axioms A10–A13 and the rule R9 to the LPP_2–axiomatization.

The strong completeness theorems for $LPP_{2,\leq}$ and $LPP_{2,\nearrow,\searrow}$, decidability and complexity estimation can be obtained in a way similar as in the case of the LPP_2–logic, see [9, 35]. The case of the $LPP_{2,O}$–logic is slightly different: decidability can fail, see [36]. We shall give some general discussion in the next section.

5.3 Completeness and decidability

We start from the fact that all infinitary inference rules R4–R7 have the same form: from the set of premises

$$\{\phi \to \theta_n \ : \ n \in \omega\}$$

infer $\phi \to \theta$. The following lemma is true for all studied logics:

Lemma 5.1. Suppose that T is a consistent theory, $T \cup \{\phi \to \theta\}$ is inconsistent and that $\phi \to \theta$ is derived from the set of premises $\{\phi \to \theta_n \ : \ n \in \omega\}$ by a direct application of some inference rule R. Then there is $n \in \omega$ such that $T \cup \{\phi, \neg\theta_n\}$ is consistent.

Proof. Suppose that $T \cup \{\phi, \neg\theta_n\}$ is inconsistent for all n. Then, for all n,

$$T \vdash \phi \to \theta_n.$$

By the rule R, we obtain that $T \vdash \phi \to \theta$, which contradicts the assumed consistency of T. Hence, there is a positive integer n such that $T \cup \{\phi, \neg\theta_n\}$ is inconsistent. ■

The next step is to prove the Lindenbaum's theorem. Here we shall give its extended version, i.e., for the logic that contains all of the above axioms and inference rules.

Theorem 5.1 (Lindenbaum's theorem). Each consistent theory can be extended to a maximal consistent theory.

Proof. Let T bi a consistent theory and let ϕ_n, $n \in \omega$ be an enumeration of For_P. We define the sequence of theories T_n, $n \in \omega$ inductively as follows:

1. $T_0 = T \cup \{\alpha \,:\, T \vdash \alpha\} \cup \{P_{=1}\alpha \,:\, T \vdash \alpha\}$;
2. If ϕ_n is consistent with T_n, then $T_{n+1} = T_n \cup \{\phi_n\}$;
3. ϕ_n is inconsistent with T_n and it is not a conclusion of any of infinitary inference rules. Then $T_{n+1} = T_n \cup \{\neg\phi_n\}$;
4. ϕ_n is inconsistent with T_n and it is a conclusion $\psi \to \theta$ of the infinitary rule R applied to the set of premises $\{\psi \to \theta_k \,:\, k \in \omega\}$. Let m be the least nonnegative integer such that $T_n, \neg(\psi \to \theta_m)$ is consistent (the existence of such m is provided by the lemma 5.1). Then $T_{n+1} = T_n \cup \{\neg\phi_n, \psi, \neg\theta_m\}$.

The theory T_ω is defined as the union of all theories T_n. By the standard argument it can be shown that it is deductively closed, i.e. that $T_\omega \vdash \phi \Longrightarrow \phi \in T_\omega$. Since each T_n is consistent, it follows that T_ω must be consistent as well. The maximality is the direct consequence of the construction. ■

Now the construction of the canonical model follows the same steps as in the case of the LPP_2–logic. The set of worlds W is the set of all classical evaluations. The valuation v is defined by $v(w, p) = 1$ iff $w \models p$, i.e. $w(p) = 1$. The algebra H is the algebra of all definable sets

$$H = \{[\alpha] \,:\, \alpha \in \mathrm{For}_C\},$$

where $[\alpha] = \{w \in W \,:\, w \models \alpha\}$. The measure $\mu : H \longrightarrow [0,1]$ is defined by

$$\mu([\alpha]) = \sup\{r \in [0,1]_{\mathbb{Q}} \,:\, T_\omega \vdash P_{\geq r}\alpha\}.$$

By the argument similar to the one presented in [37] we can show correctness of the above construction and the fact that the constructed canonical model is indeed the model of theory T_ω.

Example 5.1. In this example we shall illustrate the expressiveness of the introduced probabilistic languages by coding the game of tossing the pair of fair dices.

Let $p_1, \dots, p_6, q_1, \dots, q_6$ be pairwise distinct propositional variables. The intended meaning of p_i is that i landed on the first dice, while the intended meaning of the q_j is that j landed on the second dice. Let α be the conjunction of the following formulas:

- $p_i \to \neg p_j$, $i \neq j$. The intended meaning is that only one number can land at each toss of the first dice;
- $q_i \to \neg q_j$, $i \neq j$;
- $P_{=\frac{1}{36}}(p_i \wedge q_j)$. The intention is to enforce discrete uniform distribution on joint tosses.

By completeness theorem, $\alpha \models P_{=\frac{1}{6}} p_i$ and $\alpha \models P_{=\frac{1}{6}} q_j$, which together with $\alpha \models P_{=\frac{1}{36}}(p_i \wedge q_j)$ gives that

$$\alpha \models p_i \perp q_j$$

for all $i, j \in \{1, \ldots, 6\}$.   ∎

The status of decidability we shall discus through the following example: let ϕ be the conjunction of the formulas:

1. $P_{\geq 0.5}(p \wedge q)$;
2. $CP_{\geq 0.7}(p, q)$;
3. $p \leq q$;
4. $p \nearrow q$;
5. $Q_{\{\frac{n}{2n+3} \,:\, n \in \omega\}} p$.

If by x_1, x_2, x_3 and x_4 we denote the unknown measures of the sets $[p \wedge q]$, $[p \wedge \neg q]$, $[\neg p \wedge q]$ and $[\neg p \wedge \neg q]$, then the satisfiability of the above formula is equivalent to the satisfiability of the system

$$x_1 + x_2 + x_3 + x_4 = 1$$
$$x_1 \geq 0$$
$$x_2 \geq 0$$
$$x_3 \geq 0$$
$$x_4 \geq 0$$
$$x_1 \geq 0.5$$
$$10x_1 \geq 7x_2 + 7x_3$$
$$x_1 \geq x_1^2 + x_1 x_3 + x_2 x_3$$
$$(\exists n \in \omega)(2n+3)(x_1 + x_2) = n.$$

Note that satisfiability of the conjunctions of the first two formulas is reducible to satisfiability of systems of linear inequalities. Following the argumentation for the decidability of logic for linear weight formulas presented in [13], we obtain both decidability of $LPP_{2,\leq}$ and the complexity estimation (NP). Adding the third formula preserves decidability, but increases complexity (PSPACE, see decidability of logic with polynomial weight formulas in [13]).

However, the last formula shows us that we can code Diophantine equations, so by the Matiyasevich theorem [30], we get undecidability of $LPP_{2,O}$.

5.4 Logics with iterations and nesting of probability operators

In this section we shall present logics $LPP_{1,CP}$ and $LPP_{1,\nearrow,\searrow}$ that allow iterations and nesting of the corresponding probability operators, see [9, 32]. As in the previous section, we start with the definition of the joint probabilistic language, then specify the concrete logics.

The set of formulas For is defined inductively as follows:

$$p \mid P_{\geq r}\alpha \mid CP_{\geq r}(\alpha,\beta) \mid CP_{\leq r}(\alpha,\beta) \mid \alpha \nearrow \beta \mid \alpha \searrow \beta \mid \neg\alpha \mid \alpha \wedge \beta.$$

The class of models is the same as for the LPP_1–logic, see section 1.2.5. More precisely, a model is a structure $M = (W, Prob, v)$, where W is a nonempty set whose elements are called worlds, $v : W \times var \longrightarrow \{0,1\}$ is a truth evaluation, and $Prob$ to each world $w \in W$ assigns a probability space $(W(w), H(w), \mu(w)$ with the following properties:

- $W(w)$ is a nonempty subset of W;
- $H(w)$ is a subalgebra of the powerset algebra $\mathscr{P}(W(w))$;
- $\mu(w) : H(w) \longrightarrow [0,1]$ is a finitely additive probability measure.

The satisfiability relation is defined as follows:

- $M,w \models p$ iff $v(w,p) = 1$;
- $M,w \models P_{\geq r}\alpha$ iff $\mu(w)([\alpha]_w) \geq r$, where $[\alpha]_w = \{x \in W(w) \,:\, M,x \models \alpha\}$;
- $M,w \models CP_{\geq r}(\alpha,\beta)$ iff $\mu(w)([\beta]_w) = 0$, or $\mu(w)([\alpha \wedge \beta]_w) \geq r \cdot \mu(w)([\beta]_w)$ and $\mu(w)([\beta]_w) > 0$;
- $M,w \models CP_{\leq r}(\alpha,\beta)$ iff $\mu(w)([\beta]_w) > 0$ and $\mu(w)([\alpha \wedge \beta]_w) \leq r \cdot \mu(w)([\beta]_w)$
- $M,w \models \alpha \nearrow \beta$ iff $\mu(w)([\alpha \wedge \beta]_w) \geq \mu(w)([\alpha]_w) \cdot \mu(w)([\beta]_w)$;
- $M,w \models \alpha \searrow \beta$ iff $\mu(w)([\alpha \wedge \beta]_w) \leq \mu(w)([\alpha]_w) \cdot \mu(w)([\beta]_w)$;
- $M,w \models \neg\alpha$ iff $M,w \not\models \alpha$;
- $M,w \models \alpha \wedge \beta$ iff $M,w \models \alpha$ and $M,w \models \beta$.

The above mentioned two logic are defined in the following way:

- Logic $LPP_{1,CP}$: $For^{LPP_{1,CP}}$ admits basic probability operators $P_{\geq r}$ and conditional probability operators $CP_{\geq r}$ and $CP_{\leq r}$. The axiom system is the same as for the $LPP_{2,CP}$–logic;
- Logic $LPP_{1,\nearrow,\searrow}$: $For^{LPP_{1,\nearrow,\searrow}}$ admits basic probability operators $P_{\geq r}$ and confirmation operators $\nearrow$ and $\searrow$. The axiom system is the same as for the $LPP_{2,\nearrow,\searrow}$–logic.

The proof of the corresponding strong completeness theorems is slightly different than the one presented in the previous section. The worlds in the canonical model are maximal consistent sets of formulas (complete theories for short). For the given complete theory w, $W(w) = W$ is the set of all complete theories, $H(w) = H = \{[\alpha] \,:\, \alpha \in For\}$, where $[\alpha] = \{x \in W \,:\, \alpha \in x\}$ and $\mu(w)(\alpha) = \sup\{r \in [0,1]_{\mathbb{Q}} \,:\, P_{\geq r}\alpha \in w\}$.

Using techniques presented in earlier chapters of this book we can show the correctness of the above construction and that any consistent theory T is satisfied in its each complete extension w.

The method of filtration can be adopted for the proof of the decidability of the satisfiability problem PSAT for $LPP_{1,\nearrow,\searrow}$, see [9].

5.5 Hierarchies of probability logics

The starting point in the classification of $LPP_{2,O}$ logics is the characterization of their mutual expressiveness. We say that the logic L_2 is more expressive than the logic L_1 iff for each L_1-formula ϕ exists an L_2-formula ψ such that $\mathcal{M}(\phi) = \mathcal{M}(\psi)$. Here $\mathcal{M}(\phi)$ denotes the class of all models of ϕ.

Operators $Q_{[r,s]}$, $Q_{(r,s]}$, $Q_{[r,s)}$, $Q_{(r,s)}$ and $Q_{[0,1]\backslash F}$ are introduced in the natural way:

- $Q_{[r,s]}\alpha \Longleftrightarrow_{\text{def}} P_{\geq r}\alpha \wedge P_{\leq s}\alpha$;
- $Q_{(r,s]}\alpha \Longleftrightarrow_{\text{def}} P_{>r}\alpha \wedge P_{\leq s}\alpha$;
- $Q_{[r,s)}\alpha \Longleftrightarrow_{\text{def}} P_{\geq r}\alpha \wedge P_{<s}\alpha$;
- $Q_{(r,s)}\alpha \Longleftrightarrow_{\text{def}} P_{>r}\alpha \wedge P_{<s}\alpha$;
- $Q_{[0,1]\backslash F}\alpha \Longleftrightarrow_{\text{def}} \neg Q_F\alpha$.

The closure $\overline{O}$ of the recursive family O is the minimal superset of the set

$$O \cup \{[r,s] \;:\; r,s \in [0,1]_{\mathbb{Q}} \text{ and } r < s\}$$

that is closed under finite unions, intersections, set difference and the quasi complement $1-$. Moreover, we say that family O_1 is representable in O_2 iff $O_1 \subseteq \overline{O_2}$.

Example 5.2. Fix a positive integer $k > 0$, sets

$$F_1 = \{2^{-i} \;:\; i = k, k+1, \ldots\} \cup \{\tfrac{3^i-1}{3^i} : i = k, k+1, \ldots\},$$
$$F_2 = \{2^{-i} \;:\; i = 1, 2, \ldots\},$$
$$F_3 = \{3^{-i} \;:\; i = 1, 2, \ldots\},$$

and the family $O_2 = \{F_2, F_3\}$. Note that F_1 is representable in O_2 since

$$F_1 = (F_2 \cap [0, 2^{-k}]) \cup ((1 - F_3) \cap [1 - 3^{-k}, 1]).$$

On the other hand, the set $F_4 = \{2^{-2i} \;:\; i = 1, 2, \ldots\}$ is not representable in O_2. ■

We say that L_2 is more expressive than L_1 and write $L_1 \leqslant L_2$ iff for every L_1–formula ϕ there is an L_2–formula ψ such that

$$\mathcal{M}(\phi) = \mathcal{M}(\psi).$$

The next two theorems provide the characterization of the expressiveness relation $\leq$ between $LPP_{2,O}$–logics. The corresponding proofs can be found in [37].

Theorem 5.2. $L_1 \leqslant L_2 \Longrightarrow \overline{O}_1 \subseteq \overline{O}_2$. ■

Theorem 5.3. Let O_1 and O_2 be recursive families of recursive rational subsets of $[0,1]$, and L_1 and L_2 be the corresponding logics. The family O_1 is representable in the family O_2 iff $L_1 \leqslant L_2$. ■

Example 5.3. Let $O_1 = \{F\}$ and $O_2 = \{G\}$, where

$$F = \left\{ \frac{k}{2n} : k, n \in \omega, n > 0, 0 \le k \le 2n \right\}$$

and

$$G = \left\{ \frac{k}{2n-1} : k, n \in \omega, n > 0, 0 \le k \le 2n \right\}.$$

Since

$$1 - F = \left\{ \frac{2n-k}{2n} : k, n \in \omega, n > 0, 0 \le k \le 2n \right\} = F,$$

we have that $X \in \overline{O_1}$ iff either X or its complement (with respect to $[0,1]$) are finite/cofinite subsets of F. Hence, $G \notin \overline{O_1}$. Similarly, $F \notin \overline{O_2}$, so by Theorem 5.3 the corresponding logics L_1 and L_2 are incomparable. ■

5.5.1 Upper hierarchy

Theorem 5.3 gives the essential connection between the notions of expressive power of logics and representability of the corresponding families of Q_F operators.

Definition 5.1. Let O_1 and O_2 be recursive families of recursive subsets of $[0,1]_{\mathbb{Q}}$. The binary relation $\sim$ is defined by $O_1 \sim O_2$ iff $\overline{O_1} = \overline{O_2}$. ■

It is easy to see that the introduced relation $\sim$ is an equivalence relation on the set $\mathscr{O}$ of all recursive families of subsets of $[0,1]_{\mathbb{Q}}$. By $\mathscr{O}_{/\sim}$ the corresponding quotient set is denoted. Each equivalence class $o \in \mathscr{O}_{/\sim}$ contains a unique maximal family O_o such that $O_o = \overline{O_o}$. For such an equivalence class o and the corresponding family O_o we say that O_o represents o. Let the set $\{\overline{O_o} : o \in \mathscr{O}_{/\sim}\}$ be denoted by $\mathscr{O}^*$. Note that $\mathscr{O}$ and $\mathscr{O}^*$ are countable.

Definition 5.2. Let O_1 and O_2 be different families from $\mathscr{O}^*$. Then $O_1 \leqslant O_2$ iff O_1 is representable in O_2. ■

The following theorems present the key properties of the structure $(\mathscr{O}^*, \leqslant)$. The corresponding proofs can be found in [24, 37].

Theorem 5.4. Let O_1 and O_2 be different families from $\mathscr{O}^*$. Then $O_1 \leqslant O_2$ iff $\overline{O_1} \subseteq \overline{O_2}$. ■

Theorem 5.5. The structure $(\mathscr{O}^*, \leqslant)$ is a lattice. ■

The meet $(\cdot)$ and join $(+)$ operations are defined in the usual way:

- $O_1 \cdot O_2 = \overline{O_1 \cap O_2}$;
- $O_1 + O_2 = \overline{O_1 \cup O_2}$.

Since every set that is representable both in O_1 and in O_2 is also representable in $O_1 \cap O_2$, we have $\overline{O_1 \cap O_2} = O_1 \cap O_2$, and $O_1 \cdot O_2 = O_1 \cap O_2$. Note that the join operation and the set union do not coincide, since there are examples of $O_1, O_2 \in \mathscr{O}^*$ that $O_1 \cup O_2 \neq \overline{O_1 \cup O_2}$.

Theorem 5.6. The lattice $(\mathscr{O}^*, \leqslant)$ is not modular. ■

Theorem 5.7. $\overline{\emptyset}$ is the smallest element of $(\mathscr{O}^*, \leqslant)$. ■

Theorem 5.8. The following propositions are satisfied:

1. A necessary and sufficient condition that an $O \in \mathscr{O}^*$ be an atom is that $O = \overline{\{F\}}$, where F is a recursive set with only one accumulation point. The lattice $(\mathscr{O}^*, \leqslant)$ is non-atomic.
2. There is no greatest element in $(\mathscr{O}^*, \leqslant)$. Consequently, the lattice $\mathscr{O}^*$ is σ-incomplete, i.e. there exists a countable increasing chain $L_0 < L_1 < L_2 < \cdots$ that has no upper bound in $\mathscr{O}^*$. ■

We can define a hierarchy of the $LPP_{2,O}$-logics as follows: L_1 is less expressive than L_2 ($L_1 \leqslant L_2$) iff the corresponding families O_1 and O_2 of subsets of $[0,1]_{\mathbb{Q}}$ satisfy a similar requirement ($O_1 \leqslant O_2$). The hierarchy of the probability logics is isomorphic to $\langle \mathscr{O}^*, \leqslant \rangle$. For instance, the probability logic LPP_2 is the minimum of the hierarchy of the LPP_{2O}-logics and corresponds to the 0-element of $\langle \mathscr{O}^*, \leqslant \rangle$.

The natural maximum of $\langle \mathscr{O}^*, \leqslant \rangle$ would be the minimal extension of all $LPP_{2,O}$ logics, i.e. the logic $LPP_{2,\mathscr{O}}$.

5.5.2 Lower hierarchy

Here we shall study the hierarchy of $LPP_2^{\mathrm{Fr}(n)}$ logics. The syntax is the same as for the LPP_2–logic. The class of models is restricted to probabilities with ranges in the set

$$Fr(n) = \left\{0, \frac{1}{n}, \frac{2}{n}, \ldots, \frac{n-1}{n}, 1\right\}.$$

The axiomatization is given below:

Axioms:

A1: Tautology instances;
A2: $P_{\geqslant s}\alpha \to P_{>r}\alpha$, $r < s$;
A3: $P_{>s}\alpha \to P_{\geqslant s}\alpha$;
A4: $P_{\geqslant 0}\alpha$;
A5: $(P_{\geqslant s}\alpha \wedge P_{\geqslant r}\beta \wedge P_{\geqslant 1}(\neg\alpha \vee \neg\beta)) \to P_{\geqslant \min(1,s+r)}(\alpha \vee \beta)$;
A6: $(P_{\leqslant s}\alpha \wedge P_{<r}\beta) \to P_{\leqslant \min(1,s+r)}(\alpha \vee \beta)$;
A7: $\bigvee_{k=0}^{n} P_{=\frac{k}{n}}\alpha$;

Inference rules:

R1 (Modus Ponens for For_C): from α and $\alpha \to \beta$ infer β;
R2 (Modus Ponens for For_P): from ϕ and $\phi \to \psi$ infer ψ;
R3 (Necessitation): from α infer $P_{\geq 1}\alpha$.

The purpose of the axiom, A7 is to ensure that ranges of all probability functions appearing in models are subsets of the set $Fr(n) = \{0, \frac{1}{n}, \frac{2}{n}, \ldots, \frac{n-1}{n}, 1\}$. Consequently, all operators of the form Q_F become redundant in this context, since they are equivalent to finite disjunctions of the operators $P_{=r}$.

We shall define the lower hierarchy in the same manner as the upper hierarchy.

Definition 5.3. Let L_1 and L_2 be arbitrary $LPP_2^{\mathrm{Fr}(n)}$-logics. We say that the logic L_2 is more expressible than L_1 and write $L_1 \leqslant L_2$ iff for each L_1-formula ϕ exists an L_2 formula ψ such that $\mathcal{M}(\phi) = \mathcal{M}(\psi)$ (i.e. ϕ and ψ have the same models). ■

Note that the introduced relation is reflexive and transitive. Moreover, for any $LPP_2^{Fr(n)}$-formula ϕ, an LPP_2-formula ψ defined by

$$\psi =_{\mathrm{def}} \phi \wedge \bigwedge_{\alpha \in \mathrm{For}_C(\phi)} \bigvee_{k=0}^{n} P_{=\frac{k}{n}} \alpha$$

have the same models as ϕ (here $\mathrm{For}_C(\phi)$ is the set of all classical propositional formulas appearing in ϕ), so we can naturally consider the upper hierarchy as an end-extension of the lower hierarchy.

We shall show that the characterization theorem for the upper hierarchy (Theorem 5.4) has the natural counterpart in the lower hierarchy.

Theorem 5.9. Suppose that L_1 and L_2 are arbitrary $LPP_2^{\mathrm{Fr}(n)}$-logics. Then, $L_1 \leqslant L_2$ if and only if $\mathrm{Fr}(n_1) \subseteq \mathrm{Fr}(n_2)$.

Proof. Suppose that $\mathrm{Fr}(n_1) \subseteq \mathrm{Fr}(n_2)$ and let ϕ be an arbitrary L_1-formula. As above, we define an L_2-formula ψ by

$$\psi =_{\mathrm{def}} \phi \wedge \bigwedge_{\alpha \in \mathrm{For}_C(\phi)} \bigvee_{k=0}^{n_1} P_{=\frac{k}{n_1}} \alpha.$$

Clearly, ϕ and ψ have the same models, so $L_1 \leqslant L_2$.

Conversely, let $\mathrm{Fr}(n_1) \not\subseteq \mathrm{Fr}(n_2)$. Then, we can chose $s \in \mathrm{Fr}(n_1) \setminus \mathrm{Fr}(n_2)$. Let p be an arbitrary propositional letter. Then, $P_{=s}p$ is satisfiable as L_1-formula, while by A8, $\vdash_{L_2} \neg P_{=s}p$. Hence, $L_1 \nleqslant L_2$. ■

Since $Fr(1) \subseteq Fr(n)$ for all positive integers n, the $LPP_2^{\mathrm{Fr}(1)}$ logic is the minimum of the lower hierarchy. Moreover, $Fr(n)$ is a proper subset of $Fr(2n)$ for all positive integers n, so the lower hierarchy has no maximal elements.

Note that logics L_1 and L_2 are incomparable if and only if the symmetric difference of $\mathrm{Fr}(n_1)$ and $\mathrm{Fr}(n_2)$ is nonempty. Thus, the hierarchy contains incomparable elements (for instance, $\mathrm{Fr}(2) = \{0, \frac{1}{2}, 1\}$ and $\mathrm{Fr}(3) = \{0, \frac{1}{3}, \frac{2}{3}, 1\}$).

Another immediate consequence of Theorem 5.9 is the fact that the lower hierarchy is a lattice. Namely, the greatest lower bound of L_1 and L_2 is determined by $\mathrm{Fr}(n_1) \cap \mathrm{Fr}(n_2) = \mathrm{Fr}(\mathrm{GCD}(n_1, n_2))$, while the least upper bound of L_1 and L_2 is determined by $\mathrm{Fr}(n_1) \cup \mathrm{Fr}(n_2) = \mathrm{Fr}(\mathrm{LCM}(n_1, n_2))$. Note that $L_1 \leqslant L_2$ iff $n_1 | n_2$ (n_1 divides n_2).

Theorem 5.10. The lower hierarchy is atomic and non-modular.

Proof. Let The N_5 lattice can be embedded into the lower hierarchy as follows:

- $LPP_2^{Fr(1)} < LPP_2^{Fr(2)} < LPP_2^{Fr(4)} < LPP_2^{Fr(12)}$;
- $LPP_2^{Fr(1)} < LPP_2^{Fr(3)} < LPP_2^{Fr(12)}$;
- Logics $LPP_2^{Fr(12)}$ and $LPP_2^{Fr(3)}$ are incomparable;
- Logics $LPP_2^{Fr(3)}$ and $LPP_2^{Fr(4)}$ are incomparable.

Thus, the lower hierarchy is not modular. ■

It is quite natural to merge the upper and the lower hierarchy into the single hierarchy of probability logics due to the same definition of $\leqslant$. Since each $LPP_2^{fr(n)}$-logic can be embedded into any $LPP_{2,O}$-logic in the same manner as we have demonstrated for the LPP_2 logic, the upper hierarchy is an end-extension of the lower hierarchy.

5.6 Discussion and concluding remarks

The first topic of our discussion is the relationship between the qualitative probability operator $\preceq$ and the basic probability operators $P_{\geq r}$. As we stated in the introduction,

$$\alpha \preceq \beta \Longleftrightarrow \bigwedge_{r \in [0,1]_{\mathbb{Q}}} (P_{\geq r}\alpha \to P_{\geq r}\beta),$$

so the "if" direction can be formally stated by axioms, while the "only if" direction by the adequate infinitary inference rule (Section 5.2).

But what about the strict inequalities, can we formally represent Σ_1–predicates? More precisely, can we formally express the equivalence

$$\alpha \prec \beta \Longleftrightarrow \bigvee_{r \in (0,1)_{\mathbb{Q}}} (P_{<r}\alpha \wedge P_{>r}\beta)$$

using LPP_2–theories? It seems reasonable to conjecture that the answer is negative, i.e. that the class of models of $\alpha \prec \beta$ differs from the class of models of any LPP_2–theory T.

Scott gave a set of necessary and sufficient first order conditions in [43] for a finite Boolean algebra with ordering relation to have a probability representation, i.e., a probability function d such that:

1. if $\alpha < \beta$, then $d(\alpha) < d(\beta)$, and
2. if $\alpha \leq \beta$ and $\beta \leq \alpha$, then $d(\alpha) = d(\beta)$.

In [31] Narens showed that this cannot be extended to infinite Boolean algebras and that there is no first order axiomatization of Boolean algebras with ordering which have probability representation. He proposed instead to use "weak probability representations" where condition (1) is replaces with

1'. if $\alpha < \beta$, then $d(\alpha) \leq d(\beta)$.

He proved that Boolean algebras with ordering and weak probability representation are first order axiomatizable, but the problem is that weak representations need not to be unique. Furthermore, Narens showed that while the class of Boolean algebras with the unique weak probability representation is first order axiomatizable (with an infinite set of of axioms), the class of those with non-unique representations is not first order axiomatizable.

Closely related to this is the question whether $\alpha < \beta$ can be expressed by an LPP_2–formula. It seems that the answer is negative due to topological properties of the corresponding set of models. More precisely, a space of all models $\mathcal{M}$ can be seen as a compact subspace of the Hilbert cube $[0,1]^{At}$, where At is the set of all atoms. Note that $\mu \in \mathcal{M}$ iff:

1. $\sum_{\alpha \in At(X)} \mu(\alpha) = 1$ for all finite nonempty $X \subseteq Var$;
2. $\mu(\alpha) = \mu(\alpha \wedge p) + \mu(\alpha \wedge \neg p)$ for all $\alpha \in At$ and all $p \notin Var(\alpha)$.

It turns out that $\mathcal{M}(p < q)$ is a proper Σ_1–set in the corresponding Borel hierarchy, i.e. that it cannot be expressed as a finite Boolean combination of the basic open sets.

The situation changes in the presence of Q_F operators. For example, in the subclass of all rational-valued models, qualitative formula $p < q$ have the same models as the theory

$$T = \{Q_{[0,1]_{\mathbb{Q}}}\alpha \ : \ \alpha \in \mathrm{For}_C\} \cup \{P_{\geq r}p \to P_{>r}q \ : \ r \in [0,1)_{\mathbb{Q}}\}.$$

In the corresponding axiomatizations of logics with the fixed countable range of probability functions (say $[0,1]_{\mathbb{Q}}$), we replace the Archimedean rule with the rule

$$\frac{\{\phi \to P_{\neq r}\alpha \ : \ r \in [0,1]_Q\}}{\neg\phi}$$

that formally fixes the range of the probability functions to the set of rational numbers.

Finally, note that all formulas from this chapter are evaluated in the models of the LPP_2–logic (or LPP_1–logic, if iterations of the operators are allowed), using a fixed measure from a model. On the other hand, there are situations of interest where sharp probabilities are too simple, and uncertainty is modeled using sets of probability measures instead of one measure. Instead of referring to individual measures form the set, the uncertainty is sometimes represented by two simple boundaries, called lower and upper probability [23, 28]. Given a set P of probability measures, the former one assigns to an event X the infimum of the probabilities assigned to X by the measures in P, while the latter one returns their supremum.

The first logical formalization of those notions is given in [20]. Later, it is also shown that the techniques presented here can be modified for obtaining strong completeness [10, 11, 42]. Obviously, the logical formalization of lower and upper probabilities requires a generalization of LPP_2−models by replacing a single measure μ with a set of probability measures P. Syntactically, the operators of the form $P_{\geq r}$ are replaced with two types of operators, $L_{\geq r}$ and $U_{\geq r}$, such that, for a given model $M = (W,H,v,P)$,

- $M \models L_{\geq r}\alpha$ iff $\inf\{\mu([\alpha]_M) \mid \mu \in P\} \geq r$, and
- $M \models U_{\geq r}\alpha$ iff $\sup\{\mu([\alpha]_M) \mid \mu \in P\} \geq r$.

The axiom system for the logic is presented in [42], and the multi-agent variant with iterations of the operators is given in [10, 11]. Although the axiomatization techniques are similar to those for LPP_2 and LPP_1, there are still significant differences. Namely, since lower and upper probabilities are not finitely additive, the axioms A4 and A5 are not applicable. Instead, apart form the straightforward modification of LPP_2 inference rules and the axioms A0–A3, the complete axiomatization from [42] contains two additional axioms that formalize a characterization result for upper probabilities from [1]:

- $(U_{\leq r_1}\alpha_1 \wedge \cdots \wedge U_{\leq r_m}\alpha_m) \rightarrow U_{\leq r}\alpha$, if the formulas $\alpha \rightarrow \bigvee_{J\subseteq\{1,\dots,m\},|J|=k+n} \bigwedge_{j\in J} \alpha_j$ and $\bigvee_{J\subseteq\{1,\dots,m\},|J|=k} \bigwedge_{j\in J} \alpha_j$ are tautologies, where $r = \frac{\sum_{i=1}^{m} r_i - k}{n}$, $n \neq 0$
- $\neg(U_{\leq r_1}\alpha_1 \wedge \cdots \wedge U_{\leq r_m}\alpha_m)$, if $\bigvee_{J\subseteq\{1,\dots,m\},|J|=k} \bigwedge_{j\in J} \alpha_j$ is an instance of the classical propositional tautology and $\sum_{i=1}^{m} r_i < k$,

and the following approximation of K-axiom:

- $L_{=1}(\alpha \rightarrow \beta) \rightarrow (U_{\geq s}\alpha \rightarrow U_{\geq s}\beta)$.

References

[1] B. Anger and J. Lembcke. Infinitely subadditive capacities as upper envelopes of measures. *Zeitschrift für Wahrscheinlichkeitstheorie und verwandte Gebiete*, 68(3):403–414, Sep 1985.

[2] R. I. Brafman. A first-order conditional logic with qualitative statistical semantics. *Journal of Logic and Computation*, 7 (6):777–803, 1997.

[3] J.P. Burgess. Axiomatizing the Logic of Comparative Probability. *Notre Dame Journal of Formal Logic*, 51(1):119–126, 2010.

[4] R. Carnap. *Logical Foundations of Probability.* The University of Chicago Press, 1962. 2nd edition, 1st edition 1950.
[5] B. de Finetti. La Prévision: Ses Lois Logiques, Ses Sources Subjectives (Foresight. Its Logical Laws, Its Subjective Sources). *Annales de l'Institut Henri Poincaré*, 7:1–68, 1937. Also in: Also in: H. Kyburg and H. Smokler (editors) Studies in Subjective Probability, 53-118, 1980.
[6] J. Delgrande, B. Renne, and J. Sack. The logic of qualitative probability. *Artificial Intelligence*, 2019.
[7] D. Doder. A logic with big-stepped probabilities that can model nonmonotonic reasoning of system P. *Publications de l'Institut Mathématique*, Ns. 90(104):13–22, 2011.
[8] D. Doder, B. Marinković, P. Maksimović, and A. Perović. A Logic with Conditional Probability Operators. *Publications de l'Institut Mathématique*, Ns. 87(101):85–96, 2010.
[9] D. Doder and Z. Ognjanović. Probabilistic logics with independence and confirmation. *Studia Logica*, 105(5):943–969, 2017.
[10] D. Doder, N. Savić, and Z. Ognjanović. A Decidable Multi-agent Logic with Iterations of Upper and Lower Probability Operators. In F. Ferrarotti and S. Woltran, editors, *Proceedings of the 10th International Symposium Foundations of Information and Knowledge Systems, FoIKS 2018, Budapest, Hungary, May 1418, 2018*, volume 10833 of *Lecture Notes in Computer Science*, pages 170–185. Springer, 2018.
[11] D. Doder, N. Savić, and Z. Ognjanović. Multi-agent Logics for Reasoning About Higher-Order Upper and Lower Probabilities. *Journal of Logic Language and information*, 29:77–107, 2020.
[12] R. Fagin and J.Y. Halpern. Reasoning about knowledge and probability. *Journal of the ACM*, 41 (2):340–367, 1994.
[13] R. Fagin, J.Y. Halpern, and N. Megiddo. A logic for reasoning about probabilities. *Information and Computation*, 87:78–128, 1990.
[14] T. Flaminio. Strong non-standard completeness for fuzzy logics. *Soft Computing*, 12(4):321–333, 2008.
[15] P. Gärdenfors. Qualitative probability as an intensional logic. *Journal of Philosophical Logic*, 4(2):171–185, 1975.
[16] L. Godo and E. Marchioni. Coherent conditional probability in a fuzzy logic setting. *Logic Journal of IGPL*, 14 (3):457–481, 2006.
[17] D. Guelev. A propositional dynamic logic with qualitative probabilities. *Journal of Philosophical Logic*, 28:575–605, 1999.
[18] T. Hailperin. *Sentential Probability Logic. Origins, development, current status and technical applications.* Lehigh University Press, 1996.
[19] P. Hajek. *Metamathematics of fuzzy logic.* Kluwer Academic Publishers, 1998.
[20] J.Y. Halpern and R. Pucella. A Logic for Reasoning about Upper Probabilities. *Journal of Artificial Intelligence Research*, 17:57–81, 2002.
[21] J.Y. Halpern and R. Pucella. A logic for reasoning about evidence. *Journal of Artificial Intelligence Research*, 26:1–34, 2006.

[22] D. Hoover. Probability logic. *Annals of mathematical logic*, 14:287–313, 1978.

[23] P.J. Huber. *Robust Statistics*. Wiley Series in Probability and Statistics - Applied Probability and Statistics Section Series. Wiley, 2004.

[24] N. Ikodinović, Z. Ognjanović, M. Rašković, and A. Perović. Hierarchies of probabilistic logics. *International Journal of Approximate Reasoning*, 55(9):1830–1842, 2014.

[25] N. Ikodinović and Z. Ognjanović. A logic with coherent conditional probabilities. In L. Godo, editor, *Proceedings of the 8th European Conference Symbolic and Quantitative Approaches to Reasoning with Uncertainty, ECSQARU 2005; Barcelona; Spain; 6-8 July 2005*, volume 3571 of *Lecture Notes in Computer Science*, pages 726–736. Springer, 2005.

[26] H.J. Keisler. Logic with the quantifier "there exist uncountably many". *Annals of Mathematical Logic*, 1(1):1–93, 1970.

[27] C.H Kraft, J.W Pratt, and A. Seidenberg. Intuitive probability on finite sets. *The Annals of Mathematical Statistics*, 30(2):408–419, 1959.

[28] H.E. Kyburg Jr. *Probability and the Logic of Rational Belief.* Wesleyan University Press, 1961.

[29] D. Lehmann. Generalized qualitative probability: Savage revised. In *Proceedings of 12th Conference on Uncertainty in Artificial Intelligence (UAI-96)*, pages 381–388, 1996.

[30] Y. V. Matiyasevich. Enumerable sets are diophantine. In *Doklady Akademii Nauk SSSR*, volume 11, pages 354–358, 1970.

[31] L. Narens. On qualitative axiomatizations for probability theory. *Journal of Philosophical Logic*, 9(2):143–151, 1980.

[32] Z. Ognjanović and N. Ikodinović. A logic with higher order conditional probabilities. *Publications de l'Institut Mathématique*, Ns. 82(96):141–154, 2007.

[33] Z. Ognjanović, N. Ikodinović, and Z. Marković. A logic with Kolmogorov style conditional probabilities. In *Proceedings of the 5th Panhellenic logic symposium, Athens, Greece*, pages 111–116, 2005.

[34] Z. Ognjanović, Z. Marković, and M. Rašković. Completeness theorem for a Logic with imprecise and conditional probabilities. *Publications de l'Institut Mathématique*, Ns. 78(92):35–49, 2005.

[35] Z. Ognjanović, A. Perović, and M. Rašković. Logics with the Qualitative Probability Operator. *Logic Journal of IGPL*, 16(2):105–120, 2008.

[36] Z. Ognjanović and M. Rašković. Some probability logics with new types of probability operators. *Journal of Logic and Computation*, 9(2):181–195, 1999.

[37] Z. Ognjanović, M. Rašković, and Z. Marković. *Probability logics. Probability-Based Formalization of Uncertain Reasoning*. Springer, Cham, Switzerland, 2016.

[38] A. Perović, Z. Ognjanović, M. Rašković, and Z. Marković. A Probabilistic Logic with Polynomial Weight Formulas. In L.S. Hartmann and G. Kern-Isberner, editors, *Proceedings of the 5th International Symposium Foundations of Information and Knowledge Systems FoIKS 2008, Pisa, Italy, February 11-*

15, 2008, volume 4932 of *Lecture Notes in Computer Science*, pages 239–252, Berlin, Heidelberg, 2008. Springer.

[39] A. Perović, Z. Ognjanović, M. Rašković, and Z. Marković. How to Restore Compactness into Probabilistic Logics? In S. Hölldobler, C. Lutz, and H. Wansing, editors, *Proceedings of the 11th European Conference, JELIA 2008, Dresden, Germany, September 28-October 1, 2008*, volume 5293 of *Lecture Notes in Computer Science*, pages 338–348, Berlin, Heidelberg, 2008. Springer.

[40] M. Rašković, Z. Marković, and Z. Ognjanović. A logic with approximate conditional probabilities that can model default reasoning. *International Journal of Approximate Reasoning*, 49(1):52–66, 2008.

[41] M. Rašković, Z. Ognjanović, and Z. Marković. A Logic with Conditional Probabilities. In *Proceedings of the European Workshop on Logics in Artificial Intelligence, JELIA'04*, volume 3229 of *Lecture Notes in Computer Science*, pages 226–238. Springer, 2004.

[42] N. Savić, D. Doder, and Z. Ognjanović. Logics with lower and upper probability operators. *International Journal of Approximate Reasoning*, 88:148–168, 2017.

[43] D. Scott. Measuremant models and linear inequalities. *Journal of mathematical psychology*, 1:233 –247, 1964.

[44] D. Scott and P. Krauss. Assigning probabilities to logical formulas. In J. Hintikka and P. Suppes, editors, *Aspects of inductive logic*, pages 219–264. North-Holland, Amsterdam, 1966.

[45] K. Segerberg. Qualitative Probability in a Modal Setting. In E. Fenstad, editor, *Proceedings 2nd Scandinavian Logic Symposium*, volume 63 of *Studies in Logic and the Foundations of Mathematics*, pages 341–352, Amsterdam, 1971. North-Holland.

Chapter 6
Applications of Logics About Simple Probabilities

Dragan Doder and Aleksandar Perović

Abstract This chapter presents applications of infinitary inference rules in axiomatization of big-stepped probabilities representation of weighted propositional logics. It also presents a method for extension of the initial $[0,1]$-evaluation of propositional letters to probability measures utilizing product and Gödel's t-norms and linear convex combinations of obtained measures. Applications in classification problems are illustrated with examples. The expression "simple probabilities" refers to the convention that iteration or nesting of probability operators is not allowed.

6.1 Introduction

This chapter is mostly dedicated to extensions and applications of the infinitary inference rules introduced in previous chapters in modeling of different types of reasoning such as:

- Preferential reasoning on finite languages characterized by big-stepped probabilities, and
- Providing a general formal framework for interpretation of various weighted logics.

However, we start with the purely semantical topic of merging fuzzy and probabilistic semantics. More precisely, starting from the initial evaluation of the set of propositional letters, we propose constructive methods for probabilistic extensions based on the product t-norm, Gödel's t-norm and their linear convex combinations.

Dragan Doder
Utrecht University, Princetonplein 5, 3584 CC Utrecht, The Netherlands, e-mail: d.doder@uu.nl

Aleksandar Perović
Faculty of Transport and Traffic Engineering, University of Belgrade, Vojvode Stepe 305, 11000 Belgrade, Serbia, e-mail: pera@sf.bg.ac.rs

Z. Ognjanović (ed.), *Probabilistic Extensions of Various Logical Systems*,
https://doi.org/10.1007/978-3-030-52954-3_6

Applications are illustrated with several examples, including the well known Grabisch's example from [18].

In order to avoid the repetition of the material form the previous chapters, we have omitted proofs of completeness and decidability theorems. The only exception is the case of weighted logics (Section 6.3).

6.1.1 Related work

Regarding the work relevant to the research presented in this chapter, we have restricted ourselves only to the most relevant ones. All of them can be classified in the field of uncertain reasoning. For the sake of convenience, we have provided further classification according to the relevant subcategories.

6.1.1.1 Probability logic

The most influential and lasting impact on our work is done by Hoover in [22], where the Archimedean rule appears for the first time, though in a different semantical setting that is appropriate for probabilistic quantifiers. On the other hand, Keisler's work on hyperfinite model theory and probability quantifiers presented in [24, 25] has motivated us to axiomatize probabilistic operators of the form Q_F, with the intended meaning that the probability of a formula is in the set F, see [33]. Seminal work of Fagin, Halpern and Megiddo p resented in [14] has inspired us to extend that line of research in various directions, see [7, 8, 31, 36, 37].

6.1.1.2 Fuzzy logic

Hajek's book [20] is an excellent introductory textbook for both propositional and predicate fuzzy logics. The expressive power of the $L\Pi\frac{1}{2}$-logic (similar to the existential fragment of the theory of real closed fields) is given in [13, 29]. Strong completeness with respect to nonarchimedean semantics is given in [16]. Representation of simple probabilities and coherent (de Finetti) conditional probabilities within various fuzzy logics is presented in [17, 21, 28]. Some work on axiomatization of de Finetti and Kolmogorov conditional probabilities are given in [23, 30].

6.1.1.3 Default reasoning

The most influential work is provided by Kraus, Lehmann and Magidor in their seminal papers [26, 27], where they established the essential connection between the rational relations (preference relations that satisfy rational monotonicity condition) and the nonstandard probabilities. A hierarchy of preference relations together with the characterization theorems was given in [3]. Probabilistic representation of preference relations for finite propositional languages was given in [1, 2]. Some of our work in this field is given in [6, 9, 41]. In Chapter 4 a first order Keisler and Hoover-style probability logic for modeling first order defaults is presented.

6.1.1.4 Possibility and necessity theory

Possibility and necessity theory is one of the prominent research tracks in the field of uncertainty reasoning. The main representatives of this particular branch of research are Didier Dubois and Henri Prade. References that are closely related are [1, 2, 5, 10–12, 15]. Our contribution to the field inspired by the mentioned papers is given in [38].

6.1.2 Organization of the chapter

The lengthiest section 6.2 is dedicated to combination of probability and fuzzy semantics in order to model classifications tasks characteristic for the automated control theory. The main idea is to provide a simple method for extending the initial evaluation of propositional letters to an evaluation of arbitrary formulas.

On the level of conjunctions of pairwise distinct propositional letters we use product and Gödel t-norm for the evaluation. In the case of more complex formulas, we use additivity of a measure to reduce the initial value to linear combination of values of conjunctions of pairwise distinct propositional letters.

The next section 6.3 is dedicated to the presentation of the general framework for modeling reasoning in so called weighted logics (e.g probability logics, fuzzy logics, possibility and necessity logics). Since the formalism presented here substantially differs from the ones presented elsewhere in this book, we give the proof of the strong completeness theorem.

In the section 6.4 we present a complete axiomatization of so called big-stepped probabilities, i.e., discrete probability distributions on finite Boolean algebras such that measures of all atoms form a big-stepped sequence. A big-stepped sequence is a sequence of the form $0 < x_1 < x_2 < \cdots < x_n$ such that, for all $1 < k \leq n$, $x_1 + \cdots + x_{k-1} < x_k$.

6.2 Charges generated by t-norms

In this section we shall present results published in [39]. Recall that a charge (see [4]) is by definition a finitely additive probability measure. A t-norm is a function $\otimes : [0,1]^2 \longrightarrow [0,1]$ such that the structure $([0,1], \otimes, 1, \leq)$ is a commutative ordered monoid, i.e., that $\otimes$ is commutative and associative, 1 is the neutral element for $\otimes$, and that $\otimes$ is compatible with $\leq$ (i.e., $x \leq x_1 \wedge y \leq y_1$, then $x \otimes y \leq x_1 \otimes y_1$).

The main application of t-norms in logic is as truth functions for conjunctions in various fuzzy logics. Our aim is to use them in extending of an evaluation $e : Var \longrightarrow [0,1]$ to a charge $\bar{e} : For_C \longrightarrow [0,1]$ following the truth functionality principle as much as possible. Let us illustrate this with the following example.

Example 6.1. Suppose that $e : Var \longrightarrow [0,1]$ is any evaluation and that $\bar{e}$ is a charge on For_C that extends e. Moreover, let p and q be distinct propositional letters and let α be the formula

$$p \wedge \neg q.$$

Since $\bar{e}$ is a charge, we have that

$$\begin{aligned}\bar{e}(\alpha) &= \bar{e}(p \wedge \neg q)\\ &= \bar{e}(p) - \bar{e}(p \wedge q)\\ &= e(p) - \bar{e}(p \wedge q).\end{aligned}$$

Hence, the only possible place where we can apply the truth functionality principle in the computation of $\bar{e}(\alpha)$ is in the computation of the value of $\bar{e}(p \wedge q)$. ■

The previous example indicates how to apply the truth functionality principle in construction of charges. Namely, if μ is a charge on For_C, using complete disjunctive normal form and finite additivity, it is easy to see that $\mu(\alpha)$ is a finite sum of measures of atoms

$$\mu(\pm p_1 \wedge \cdots \wedge \pm p_n).$$

Here $p_1, \ldots, p_n$ are all propositional letters appearing in α, and $+p_i$ and $-p_i$ denote p_i and $\neg p_i$, respectively. Since

$$\mu(\neg\beta) = 1 - \mu(\beta)$$

and

$$\mu(\beta \wedge \neg\gamma) = \mu(\beta) - \mu(\beta \wedge \gamma),$$

negation can be eliminated from each $\mu(\pm p_1 \wedge \cdots \wedge \pm p_n)$, so $\mu(\alpha)$ depends only on the logical structure (e.g., normal form) of α and μ-values of finite conjunctions of pairwise distinct propositional letters.

The natural truth functions for conjunctions are t-norms. In particular, if $\otimes$ is any t-norm and $e : Var \longrightarrow [0,1]$ is any evaluation, then there is a unique finitely additive real-valued function $e^{\otimes}$ on For_C which extends e and that

$$e^{\otimes}(p_1 \wedge \cdots \wedge p_n) = e(p_1) \otimes \cdots \otimes e(p_n),$$

for pairwise distinct propositional letters $p_1, \ldots, p_n$.

Unfortunately, $e^{\otimes}$ needs not to be a charge, as will be illustrated by the next example.

Example 6.2. Let $\otimes_L$ be the Łukasiewicz t-norm

$$x \otimes_L y = \max(0, x+y-1).$$

Suppose that $e(p) = 0.5$ for all $p \in Var$ and that p_1, p_2 and p_3 are pairwise distinct propositional letters. Let $e^L = e^{\otimes_L}$Then,

$$\begin{aligned} e^L(\neg p_1 \wedge \neg p_2 \wedge \neg p_3) &= 1 - \sum_{i=1}^{3} e(p_i) + \sum_{1 \leq i < j \leq 3} e^L(p_i \wedge p_j) - e^L(p_1 \wedge p_2 \wedge p_3) \\ &= 1 - 0.5 - 0.5 - 0.5 \\ &= -0.5. \end{aligned}$$

Note that

$$e^L(p_i \wedge p_j) = e(p_i) \otimes_L e(p_j) = 0.5 \otimes_L 0.5 = 0$$

and that $e^L(p_1 \wedge p_2 \wedge p_3) = 0$. Hence, we have showed that e^L is not a charge. ■

We shall show that in the case of Gödel's t-norm $x \otimes_G y = \min(x, y)$ and product t-norm $x \otimes_\Pi y = xy$, the corresponding extension $e^{\otimes}$ of an evaluation e will be a charge. In the case of Gödel's t-norm the corresponding extension of an evaluation e will be denoted by e^G, while e^Π corresponds to the case of the product t-norm.

An application to classification problems, particularly in the representation of the Grabisch's example [18], propositional letters will be used to code primary attributes (e.g., height, weight, grade etc.). Product measures e^Π correspond to one extreme situation: stochastic or probability independence of propositional letters. Gödel's measures e^G correspond to another kind of extreme situation: logical dependence of propositional letters. While stochastic independence is a measure-theoretic property and cannot be forced by some nontrivial logical conditions (see [19]), logical dependence is expressible in classical propositional calculus.

For example, the logical condition

$$p \to q$$

implies

$$\bar{e}(p \wedge q) = \min(e(p), e(q)).$$

On the other hand, charges are closed for linear convex combinations, so using e^G and e^Π we can construct a countable scale of charges

$$e^{(s)} = se^\Pi + (1-s)e^G, \quad s \in (0,1) \cap \mathbb{Q}.$$

From the uncertainty point of view, measures $e^{(s)}$ correspond to various degrees of dependence between propositional letters. From the fuzzyness point of view, measures $e^{(s)}$ provide countably many ways to extend initial evaluation e of propositional letters, which enables probability evaluations of fuzzy quantities. We will try to illustrate the intended meaning with the following example:

Example 6.3. Compounds $c_1, \ldots, c_5$ of the substances p and q have to be classified according to their harmfulness. It is known that both p and q are harmful and they neutralize each other. Concentrations of substances p and q in the given compounds are listed below:

compound	concentration of p	concentration of q
c_1	0.71	0.15
c_2	0.15	0.35
c_3	0.21	0.48
c_4	0.55	0.25
c_5	0.25	0.38

It is natural to consider p and q as propositional letters. Since substances p and q neutralize each other, the classification criteria is adequately formalized by $p \leftrightarrow q$. If e is any $[0,1]$-evaluation of p and q, then

$$\begin{aligned} e^{\Pi}(p \leftrightarrow q) &= e^{\Pi}((p \wedge q) \vee (\neg p \wedge \neg q)) \\ &= e^{\Pi}(p \wedge q) + e^{\Pi}(\neg p \wedge \neg q) \\ &= e(p) \cdot e(q) + e^{\Pi}(\neg p) - e^{\Pi}(\neg p \wedge q) \\ &= e(p) \cdot e(q) + 1 - e(p) - e(q) + e(p) \cdot e(q) \\ &= 1 - e(p) - e(q) + 2e(p) \cdot e(q). \end{aligned}$$

Similarly,

$$e^{G}(p \leftrightarrow q) = 1 - e(p) - e(q) + 2\min(e(p), e(q))$$

and

$$e^{(0.45)}(p \leftrightarrow q) = 0.45\, e^{\Pi}(p \leftrightarrow q) + 0.55\, e^{G}(p \leftrightarrow q).$$

Now we can interpret compounds c_i as $[0,1]$-evaluations of p and q and accordingly evaluate $p \leftrightarrow q$ using e^{Π}, e^{G} and $e^{(0.45)}$:

compound	$e(p)$	$e(q)$	$e^{\Pi}(p \leftrightarrow q)$	$e^{G}(p \leftrightarrow q)$	$e^{(0.45)}(p \leftrightarrow q)$
c_1	0.71	0.15	0.353	0.44	0.40085
c_2	0.15	0.35	0.605	0.8	0.71225
c_3	0.21	0.48	0.5116	0.73	0.63172
c_4	0.55	0.25	0.475	0.7	0.59875
c_5	0.25	0.38	0.56	0.87	0.7305

The corresponding classifications are given below:

e^{Π}: $c_1 < c_4 < c_3 < c_5 < c_2$;
e^{G}: $c_1 < c_4 < c_3 < c_2 < c_5$;
$e^{(0.45)}$: $c_1 < c_4 < c_3 < c_2 < c_5$.

Discrepancy emerged in classification of compounds c_2 and c_5. If we additionally evaluate $p \leftrightarrow q$ using $e^{(0.5)}$ (arithmetic mean of e^{Π} and e^{G}), we get the following results:

compound	$e^{\Pi}(p \leftrightarrow q)$	$e^{G}(p \leftrightarrow q)$	$e^{(0.5)}(p \leftrightarrow q)$
c_1	0.353	0.44	0.3965
c_2	0.605	0.8	0.7025
c_3	0.5116	0.73	0.6208
c_4	0.475	0.7	0.5875
c_5	0.56	0.87	0.715

In this case we have the following classification: $c_1 < c_4 < c_3 < c_2 < c_5$, which confirms the e^G classification. ■

6.2.1 Preliminaries on charges

Here we shall prove the following:

1. For any evaluation $e : Var \longrightarrow [0,1]$ and any t-norm $\otimes$, there is a unique $\mathbb{R}$-measure ($\mathbb{R}$-valued measure) $e^{\otimes}$ on For_C that extends e and that

$$e^{\otimes}(p_1 \wedge \cdots \wedge p_n) = \bigotimes_{i=1}^{n} e(p_i)$$

for pairwise distinct propositional letters $p_1, \ldots, p_n$;
2. In the case of Gödel's t-norm and product t-norm the corresponding extension $e^{\otimes}$ will be a charge.

We start with some basic notions. Recall that a literal is either a propositional letter or its negation, while an atom is a conjunction of literals. Let $Var(\alpha) = \{p_1, \ldots, p_n\}$. Then α determines unique n-ary Boolean function $\alpha^{\mathscr{B}} : \{0,1\}^n \longrightarrow \{0,1\}$ with respect to the usual truth tables of connectives. Formal definition of $\alpha^{\mathscr{B}}$ is given below:

- $p^{\mathscr{B}} = \mathrm{id}_{\{0,1\}}$, $p \in Var$, where $\mathrm{id}_{\{0,1\}}(x) = x$ for all $x \in \{0,1\}$
- $(\neg\alpha)^{\mathscr{B}} = 1 - \alpha^{\mathscr{B}}$
- $(\alpha \wedge \beta)^{\mathscr{B}} = \alpha^{\mathscr{B}}\beta^{\mathscr{B}}$
- $(\alpha \vee \beta)^{\mathscr{B}} = \alpha^{\mathscr{B}} + \beta^{\mathscr{B}} - \alpha^{\mathscr{B}}\beta^{\mathscr{B}}$
- $(\alpha \rightarrow \beta)^{\mathscr{B}} = 1 - \alpha^{\mathscr{B}} + \alpha^{\mathscr{B}}\beta^{\mathscr{B}}$
- $(\alpha \leftrightarrow \beta)^{\mathscr{B}} = 1 - \alpha^{\mathscr{B}} - \beta^{\mathscr{B}} + 2\alpha^{\mathscr{B}}\beta^{\mathscr{B}}$.

Definition 6.1. Suppose that $Var(\alpha) = \{p_1, \ldots, p_n\}$. By $\mathscr{D}_\alpha$ ($\mathscr{D}$ stands for disjunctive normal form) we will denote the set of all $A \subseteq \{1, \ldots, n\}$ such that

$$\alpha^{\mathscr{B}}(\chi_A(1), \ldots, \chi_A(n)) = 1,$$

where $\chi_A : \omega \longrightarrow \{0,1\}$ is the characteristic function of A. ■

In particular, $A \in \mathscr{D}_\alpha$ iff the atom

$$\bigwedge_{i\in A} p_i \wedge \bigwedge_{j\in\{1,\dots,n\}\setminus A} \neg p_j$$

appears in the complete disjunctive normal form of α with respect to the propositional letters $p_1,\dots,p_n$. If $A=\{1,\dots,n\}$ or $A=\emptyset$, the corresponding atoms are $p_1\wedge\cdots\wedge p_n$ and $\neg p_1\wedge\cdots\wedge\neg p_n$. If α is a contradiction, then $\alpha^{\mathscr{B}}$ is by definition the constant function 0.

An $\mathbb{R}$-measure is any function $\mu : For_C \longrightarrow \mathbb{R}$ with the following properties:

- $\mu(\alpha)=\mu(\alpha')$ whenever α and α' are equivalent;
- $\mu(\alpha\vee\beta)=\mu(\alpha)+\mu(\beta)$ for disjoint formulas α and β ($\alpha\wedge\beta$ is a contradiction);
- $\mu(\alpha)=1$ for any tautology α.

Note that an $\mathbb{R}$-measure μ is a charge iff $0\leqslant\mu(\alpha)\leqslant 1$ for all $\alpha\in For_C$. The following lemma can be proved straightforwardly.

Lemma 6.1. Suppose that $\mu : For_C \longrightarrow \mathbb{R}$ is an $\mathbb{R}$-measure. Then:
1. $\mu(\neg\alpha)=1-\mu(\alpha)$;
2. $\mu(\alpha)=0$ for any contradiction α;
3. $\mu(\alpha\wedge\neg\beta)=\mu(\alpha)-\mu(\alpha\wedge\beta)$. ■

In order to show that any evaluation of propositional variables can be extended to an $\mathbb{R}$-measure, we need the following lemma.

Lemma 6.2. Let $\otimes$ be any t-norm and let $\bigotimes_{i\in\emptyset} p_i =_{\text{def}} 1$. Then,

$$\sum_{A\in\mathscr{P}(\{1,\dots,n\})}\ \sum_{X\in\mathscr{P}(\{1,\dots,n\}\setminus A)} (-1)^{|X|} \bigotimes_{i\in A\cup X} p_i = 1.$$

Proof. Let $1\leqslant i_1<\cdots<i_k\leqslant n$. Then, the term

$$p_{i_1}\otimes\cdots\otimes p_{i_k}$$

appears in

$$\sum_{A\in\mathscr{P}(\{1,\dots,n\})}\ \sum_{X\in\mathscr{P}(\{1,\dots,n\}\setminus A)} (-1)^{|X|} \bigotimes_{i\in A\cup X} p_i$$

exactly $2^k=|\mathscr{P}(\{i_1,\dots,i_k\})|$ times. Furthermore, there are exactly 2^{k-1} subsets of $\{i_1,\dots,i_k\}$ of odd cardinalities, so there are exactly 2^{k-1} negative syntactical appearances of $p_{i_1}\otimes\cdots\otimes p_{i_k}$ in

$$\sum_{A\in\mathscr{P}(\{1,\dots,n\})}\ \sum_{X\in\mathscr{P}(\{1,\dots,n\}\setminus A)} (-1)^{|X|} \bigotimes_{i\in A\cup X} p_i.$$

Consequently, the term $p_{i_1}\otimes\cdots\otimes p_{i_k}$ vanishes in

$$\sum_{A\in\mathscr{P}(\{1,\dots,n\})}\ \sum_{X\in\mathscr{P}(\{1,\dots,n\}\setminus A)} (-1)^{|X|} \bigotimes_{i\in A\cup X} p_i.$$

This is true for all nonempty increasing subsequences of the sequence $1,\dots,n$, which concludes the proof of the lemma. ■

Theorem 6.1. Suppose that $\otimes$ is any t-norm and that e is any $[0,1]$-valued evaluation of the set *Var* of propositional letters. Then, there is the unique $\mathbb{R}$-measure $e^{\otimes}$ that extends e so that

$$e^{\otimes}(p_1 \wedge \cdots \wedge p_n) = \bigotimes_{i=1}^{n} e(p_i)$$

for any sequence of pairwise distinct propositional letters $p_1,\dots,p_n$.

Proof.

Uniqueness. Suppose that μ_1 and μ_2 are $\mathbb{R}$-measures such that $\mu_1(p) = \mu_2(p)$ for all $p \in Var$. Pick an arbitrary $\alpha \in For_C$. If α is a contradiction, then $\mu_1(\alpha) = \mu_2(\alpha) = 0$. Otherwise, α is equivalent to

$$\bigvee_{A \in \mathscr{D}_\alpha} \left(\bigwedge_{i \in A} p_i \wedge \bigwedge_{j \in \{1,\dots,n\} \setminus A} \neg p_j \right),$$

so by additivity of each μ_k and Lemma 6.1,

$$\begin{aligned}
\mu_k(\alpha) &= \sum_{A \in \mathscr{D}_\alpha} \mu_k \left(\bigwedge_{i \in A} p_i \wedge \bigwedge_{j \in \{1,\dots,n\} \setminus A} \neg p_j \right) \\
&= \sum_{A \in \mathscr{D}_\alpha} \left(\sum_{X \in \mathscr{P}(\{1,\dots,n\} \setminus A)} (-1)^{|X|} \mu_i \left(\bigwedge_{i \in A \cup X} p_i \right) \right) \\
&= \sum_{A \in \mathscr{D}_\alpha} \left(\sum_{X \in \mathscr{P}(\{1,\dots,n\} \setminus A)} (-1)^{|X|} \bigotimes_{i \in A \cup X} \mu_k(p_i) \right) \\
&= \sum_{A \in \mathscr{D}_\alpha} \left(\sum_{X \in \mathscr{P}(\{1,\dots,n\} \setminus A)} (-1)^{|X|} \bigotimes_{i \in A \cup X} e(p_i) \right),
\end{aligned}$$

so $\mu_1(\alpha) = \mu_2(\alpha)$. Note that

$$\sum_{X \in \mathscr{P}(\{1,\dots,n\} \setminus A)} (-1)^{|X|} \bigotimes_{i \in A \cup X} \mu_k(p_i)$$

is obtained from

$$\sum_{A \in \mathscr{D}_\alpha} \mu_k \left(\bigwedge_{i \in A} p_i \wedge \bigwedge_{j \in \{1,\dots,n\} \setminus A} \neg p_j \right)$$

by applications of $\mu(\beta \wedge \neg\gamma) = \mu(\beta) - \mu(\beta \wedge \gamma)$ and $\mu(\neg p) = 1 - \mu(p)$ for disjoint formulas until all negations are eliminated.

Existence. If α is a contradiction, then let $e^{\otimes}(\alpha) = 0$. Otherwise, $e^{\otimes}(\alpha)$ is defined as follows:

$$e^{\otimes}(\alpha) = \sum_{A \in \mathscr{D}_\alpha} \left(\sum_{X \in \mathscr{P}(\{1,\dots,n\} \setminus A)} (-1)^{|X|} \bigotimes_{i \in A \cup X} e(p_i) \right).$$

It remains to verify that $e^{\otimes}$ is an $\mathbb{R}$-measure. Note that by Lemma 6.2, we have $e^{\otimes}(\alpha) = 1$ for any tautology α.

Next we will show additivity of $e^{\otimes}$, i.e., that

$$e^{\otimes}(\alpha \vee \beta) = e^{\otimes}(\alpha) + e^{\otimes}(\beta)$$

for disjoint formulas α and β. Note that $X = Var(\alpha) \cap Var(\beta) \neq \emptyset$. Then, for any $A \in \mathscr{D}_{\alpha \vee \beta}$, either $A \cap Var(\alpha) \in \mathscr{D}_\alpha$ or $A \cap Var(\beta) \in \mathscr{D}_\beta$. Furthermore, if $A \cap Var(\alpha) \in \mathscr{D}_\alpha$, then

$$(X \setminus A) \cup (Var(\beta) \setminus X) \in \mathscr{D}_\beta$$

and vice versa. Now we can proceed similarly as in the proof of Lemma 6.2 to obtain additivity of $e^{\otimes}$.

As a consequence of additivity of $e^{\otimes}$ and the fact that $e^{\otimes}(\alpha) = 1$ for any tautology α we have that

$$e^{\otimes}(\neg\alpha) = 1 - e^{\otimes}(\alpha) \text{ and } e^{\otimes}(\alpha \wedge \neg\beta) = e^{\otimes}(\alpha) - e^{\otimes}(\alpha \wedge \beta).$$

Finally, equivalent formulas α and α' have the same complete disjunctive normal form with respect to the set of propositional variables $Var(\alpha) \cup Var(\alpha')$, so by definition of $e^{\otimes}$ we have that $e^{\otimes}(\alpha) = e^{\otimes}(\alpha')$. ■

Definition 6.2. Let e be any $[0,1]$-valued evaluation of propositional letters. Then:

- Gödel's measure e^G is an $\mathbb{R}$-measure $e^{\otimes}$ where $\otimes$ is the Gödel's t-norm;
- Product measure e^{Π} is an $\mathbb{R}$-measure $e^{\otimes}$ where $\otimes$ is the product t-norm;
- An intermediate measure $e^{(\xi)}$, $\xi \in [0,1]$ is defined by

$$e^{(xi)}(\alpha) = \xi e^{\Pi}(\alpha) + (1 - \xi) e^G(\alpha).$$

■

The next two lemmas, together with Lemma 6.2 will provide the fact that e^G and e^{Π} are probability measures for any evaluation e.

Lemma 6.3. Suppose that n is a positive integer, A is a subset of $\{1,\dots,n\}$, and $\langle x_1,\dots,x_n \rangle \in [0,1]^n$. Then,

$$0 \leqslant \sum_{X \in \mathscr{P}(\{1,\dots,n\} \setminus A)} (-1)^{|X|} \prod_{i \in A \cup X} x_i \leqslant 1,$$

where $\prod_{i \in \emptyset} x_i =_{\text{def}} 1$.

Proof. A consequence of the distributivity of multiplication over addition of reals. ∎

Lemma 6.4. Suppose that n is a positive integer, A is a subset of $\{1,\ldots,n\}$, and $\langle x_1,\ldots,x_n\rangle \in [0,1]^n$. Then,

$$0 \leqslant \sum_{X\in\mathscr{P}(\{1,\ldots,n\}\setminus A)} (-1)^{|X|} \min_{i\in A\cup X} x_i \leqslant 1,$$

where $\min_{i\in\emptyset} x_i =_{\text{def}} 1$.

Proof. Let $\xi = \min_{i\in\{1,\ldots,n\}} x_i$. If $\xi \in A$, then

$$\sum_{X\in\mathscr{P}(\{1,\ldots,n\})\setminus A} (-1)^{|X|} \min_{i\in A\cup X} x_i = \xi \cdot \sum_{X\in\mathscr{P}(\{1,\ldots,n\}\setminus A)} (-1)^{|X|}.$$

Since there exist exactly $2^{n-|A|}$ subsets of $\{1,\ldots,n\}\setminus A$, we have the following two possibilities for the value of $\sum_{X\in\mathscr{P}(\{1,\ldots,n\}\setminus A)}(-1)^{|X|}$:

- 1, in the case $n = |A|$;
- 0, in the case $n > |A|$. Namely, the number of subsets of $\{1,\ldots,n\}\setminus A$ with odd cardinalities is equal to the number of subsets of $\{1,\ldots,n\}\setminus A$ that have even cardinality, and that number is $2^{n-|A|-1}$.

Either way, the lemma in this case holds.

Let $0 \leqslant \xi_1 < \cdots < \xi_k < \min_{i\in A} x_i$, where $\xi_1 = \min\{x_1,\ldots,x_n\}$, $\xi_2 = \min(\{1,\ldots,n\}\setminus\{\xi_1\})$ etc. For each $i \in \{1,\ldots,k\}$, let I_i be the set of all $X \subseteq \{1,\ldots,n\}\setminus A$ such that

$$\xi_i = \min_{j\in A\cup X} x_j.$$

Note that $|I_1| > \cdots > |I_k|$. Since there are exactly $2^{|I_i|}$ subsets X of $\{1,\ldots,n\}\setminus A$ such that $\xi_i = \min_{j\in A\cup X} x_j$, we have that

$$\sum_{X\subseteq I_i} (-1)^{|X|} \min_{j\in A\cup X} x_j = \xi_i \cdot \sum_{X\subseteq I_i} (-1)^{|X|} = 0$$

for every $i < k$. If $|I_k| > 0$, then $\sum_{X\subseteq I_k}(-1)^{|X|}\min_{j\in A\cup X} x_j = 0$ as well, so we can proceed in the same way as in the case $\min_{i\in\{1,\ldots,n\}} x_i = \min_{i\in A} x_i$ and conclude that the lemma also holds in this case.

It remains to see what happens if $|I_k| = 0$. There are exactly three possibilities for the value of $\sum_{X\in\mathscr{P}(\{1,\ldots,n\}\setminus A)}(-1)^{|X|}\min_{i\in A\cup X} x_i$:

- $1-\xi_k$, if $A = \emptyset$;
- ξ_k, if $|\{X \subseteq \{1,\ldots,n\}\setminus A \mid \min_{i\in A\cup X} x_i = \min_{i\in A} x_i\}| > 0$;
- $\min_{i\in A} x_i - \xi_k$ otherwise. ∎

As a consequence, we have the following theorem.

Theorem 6.2. Suppose that e is a $[0,1]$-valued evaluation of the set Var of propositional letters. Then, e^G, e^Π and $e^{(\xi)}$, $\xi \in [0,1]$ are charges. ■

Suppose that $Var(\alpha) = \{p_1,\dots,p_n\}$. Let the $2n$-ary functions

$$f^{\Pi}_{\mathscr{D}_\alpha}, F^{\Pi}_{\mathscr{D}_\alpha}, f^{G}_{\mathscr{D}_\alpha}, F^{G}_{\mathscr{D}_\alpha}, f^{(s)}_{\mathscr{D}_\alpha}, F^{(s)}_{\mathscr{D}_\alpha} : [0,1]^{2n} \longrightarrow [0,1]$$

be defined as follows:

- $f^{\Pi}_{\mathscr{D}_\alpha}(\bar{x},\bar{y}) = \sum\limits_{A\in\mathscr{D}_\alpha}\left(\sum\limits_{X\in\mathscr{P}(\{1,\dots,n\}\setminus A)}(-1)^{|X|}\prod\limits_{i\in A\cup X}\left(\frac{1+(-1)^{|X|}}{2}x_i+\frac{1-(-1)^{|X|}}{2}y_i\right)\right)$
- $F^{\Pi}_{\mathscr{D}_\alpha}(\bar{x},\bar{y}) = \sum\limits_{A\in\mathscr{D}_\alpha}\left(\sum\limits_{X\in\mathscr{P}(\{1,\dots,n\}\setminus A)}(-1)^{|X|}\prod\limits_{i\in A\cup X}\left(\frac{1+(-1)^{|X|}}{2}y_i+\frac{1-(-1)^{|X|}}{2}x_i\right)\right)$
- $f^{G}_{\mathscr{D}_\alpha}(\bar{x},\bar{y}) = \sum\limits_{A\in\mathscr{D}_\alpha}\left(\sum\limits_{X\in\mathscr{P}(\{1,\dots,n\}\setminus A)}(-1)^{|X|}\min\limits_{i\in A\cup X}\left(\frac{1+(-1)^{|X|}}{2}x_i+\frac{1-(-1)^{|X|}}{2}y_i\right)\right)$
- $F^{G}_{\mathscr{D}_\alpha}(\bar{x},\bar{y}) = \sum\limits_{A\in\mathscr{D}_\alpha}\left(\sum\limits_{X\in\mathscr{P}(\{1,\dots,n\}\setminus A)}(-1)^{|X|}\min\limits_{i\in A\cup X}\left(\frac{1+(-1)^{|X|}}{2}y_i+\frac{1-(-1)^{|X|}}{2}x_i\right)\right)$
- $f^{(s)}_{\mathscr{D}_\alpha}(\bar{x},\bar{y}) = sf^{\Pi}_{\mathscr{D}_\alpha}(\bar{x},\bar{y}) + (1-s)f^{G}_{\mathscr{D}_\alpha}(\bar{x},\bar{y})$
- $F^{(s)}_{\mathscr{D}_\alpha}(\bar{x},\bar{y}) = sF^{\Pi}_{\mathscr{D}_\alpha}(\bar{x},\bar{y}) + (1-s)F^{G}_{\mathscr{D}_\alpha}(\bar{x},\bar{y})$.

Here $\bar{x}$ and $\bar{y}$ are lower and upper bounds for values of $p_1,\dots,p_n$. If α is a contradiction, then we adopt the convention that all of the above defined function are a constant function 0.

Observe that

$$f^{\Pi}_{\mathscr{D}_\alpha}(\bar{x},\bar{x}) = F^{\Pi}_{\mathscr{D}_\alpha}(\bar{x},\bar{x}) = \sum_{A\in\mathscr{D}_\alpha}\left(\sum_{X\in\mathscr{P}(\{1,\dots,n\}\setminus A)}(-1)^{|X|}\prod_{i\in A\cup X}x_i\right)$$

and

$$f^{\Pi}_{\mathscr{D}_\alpha}(\bar{x},\bar{x}) = F^{\Pi}_{\mathscr{D}_\alpha}(\bar{x},\bar{x}) = \sum_{A\in\mathscr{D}_\alpha}\left(\sum_{X\in\mathscr{P}(\{1,\dots,n\}\setminus A)}(-1)^{|X|}\min_{i\in A\cup X}x_i\right).$$

Some straightforward statements about $f^{*}_{\mathscr{D}_\alpha}$ and $F^{*}_{\mathscr{D}_\alpha}$ are given in the next theorem.

Theorem 6.3. Suppose that e is a $[0,1]$-valued evaluation of Var. If $\alpha \in For_C$ is not a contradiction and $Var(\alpha) = \{p_1,\dots,p_n\}$, then, for any choice of $x_1,\dots,x_n,y_1,\dots y_n \in [0,1]$ such that $x_i \leqslant e(p_i) \leqslant y_i$ for all $i \in \{1,\dots,n\}$, the following statements hold:

1. $\max(0,f^{\Pi}_{\mathscr{D}_\alpha}(\bar{x},\bar{y})) \leqslant e^{\Pi}(\alpha) \leqslant \min(1,F^{\Pi}_{\mathscr{D}_\alpha}(\bar{x},\bar{y}))$. If $x_i = e(p_i) = y_i$ for all i, then

$$f^{\Pi}_{\mathscr{D}_\alpha}(\bar{x},\bar{y}) = e^{\Pi}(\alpha) = F^{\Pi}_{\mathscr{D}_\alpha}(\bar{x},\bar{y});$$

2. $\max(0, f^G_{\mathscr{D}\alpha}(\bar{x},\bar{y})) \leqslant e^G(\alpha) \leqslant \min(1, F^G_{\mathscr{D}\alpha}(\bar{x},\bar{y}))$. If $x_i = e(p_i) = y_i$ for all i, then

$$f^G_{\mathscr{D}\alpha}(\bar{x},\bar{y}) = e^G(\alpha) = F^G_{\mathscr{D}\alpha}(\bar{x},\bar{y});$$

3. $\max(0, f^{(s)}_{\mathscr{D}\alpha}(\bar{x},\bar{y})) \leqslant e^{(s)}(\alpha) \leqslant \min(1, F^{(s)}_{\mathscr{D}\alpha}(\bar{x},\bar{y}))$, for all $s \in (0,1) \cap \mathbb{Q}$. If $x_i = e(p_i) = y_i$ for all i, then

$$f^{(s)}_{\mathscr{D}\alpha}(\bar{x},\bar{y}) = e^{(s)}(\alpha) = F^{(s)}_{\mathscr{D}\alpha}(\bar{x},\bar{y}).$$

■

6.2.2 Classification via $e^{(s)}$

By classification of the finite set of evaluations $\{e_0,\ldots,e_k\}$, $e_i : Var \longrightarrow [0,1]$ with respect to the set of attributes (propositional letters) $\{p_1,\ldots,p_n\}$ and the criterium function $f : [0,1]^n \longrightarrow [0,1]$ we assume the partial ordering $<$ defined by

$$e_i < e_j \text{ iff } f(e_i(p_1),\ldots,e_i(p_n)) < f(e_j(p_1),\ldots,e_j(p_n)).$$

In practice, we can apply the measures $e_i^{(s)}$ in any classification problem where at least one part of the computation of the criterion function f involves computation of the truth value of certain formula $\alpha(p_1,\ldots,p_n)$.

Example 6.4. Suppose that we want to develop a fuzzy relational database for automated trade of furniture, where database entries are evaluations of predefined quality attributes. User's queries should be stated in the form of propositional formulas over the quality attributes. The resolution process we will illustrate on the example of the query "find me a sturdy but light wooden chair that is not too expensive":

- Prompt the user to choose rational $s \in [0,1]$. Here s represents user's estimation of dependence between quality attributes in the query. 1 represents stochastic independence, while 0 represents logical dependence;
- Compute
$$e^{(s)}(\text{sturdy} \wedge \text{light} \wedge \text{wooden} \wedge \neg(\text{too expensive}))$$
for all relevant database entries e;
- Return to the user all relevant database entries e with maximal $e^{(s)}$-values. ■

Product measures correspond to the one extreme - stochastic or probability independence of attributes (propositional letters), while Gödel's measures correspond to the other extreme - logical dependence of attributes. The standard statistical techniques, such as linear or nonlinear regression, can be applied to measure stochastic independence of fuzzy attributes.

Intermediate measures $e^{(s)}$ are particularly useful in the cases where both e^G and e^{Π} do not classify observed objects.

Example 6.5. Suppose that an evaluation e and a formula α induce the following:

object	$e^{\Pi}(\alpha)$	$e^{G}(\alpha)$	$e^{(0.5)}$
A	0.3	0.4	0.35
B	0.3	0.6	0.45
C	0.5	0.6	0.55

Note that neither Gödel's nor product measure provide total classification of objects A, B and C according to the classification criteria α, while arithmetic mean $e^{(0.5)}$ provides a classification - linear ordering $A < B < C$ that is sound with both partial orderings induced by e^{Π} and e^{G}. In this example, e^{Π} induces partial ordering $A < C$ and $B < C$, while e^{G} induces partial ordering $A < B$ and $A < C$. ■

6.2.3 Grabisch's example

The more interesting example of the classification problem is the well known Grabisch classification problem unsolvable by the standard aggregation via discrete Choquet integral given in [18]. Following [40], instead of using bicapacities in the process of aggregation, we can obtain the intended classification using the product measure e^{Π} in the aggregation process.

We begin with the formulation of the problem. Objects A, B, C and D are described by quality attributes p_1, p_2 and p_3, whose values are given below:

object	p_1	p_2	p_3
A	0.75	0.9	0.3
B	0.75	0.8	0.4
C	0.3	0.65	0.1
D	0.3	0.55	0.2

Objects A, B, C and D should be classified according to the following criteria:

ϕ_1: The average value of quality attributes;

ϕ_2: If the analyzed object is good with respect to p_1, then p_3 is more important than p_2. Otherwise, p_2 is more important than p_3.

The obvious solution is: $D < C < A < B$.

As the first step towards the decision mechanism, we will formalize ϕ_1 and ϕ_2:

- Criteria ϕ_1 does not give us any room for different interpretations: $p_1 \wedge p_2 \wedge p_3$ must be evaluated by the arithmetic mean of the values of p_1, p_2 and p_3;
- Criteria ϕ_2 is formally representable by the formula $(p_1 \wedge p_3) \vee (\neg p_1 \wedge p_2)$.

We will separately evaluate ϕ_1 and ϕ_2, and then aggregate the obtained evaluations. Since there is no information about significance of ϕ_1 and ϕ_2, we will assume that they are equally important, so the aggregation coefficient would be equal to $\frac{1}{2}$. In other words, the final evaluation would be equal to the arithmetic mean of evaluations of ϕ_1 and ϕ_2.

Evaluation of ϕ_1. The arithmetic means of values of p_1, p_2 and p_3 for the objects A, B, C and D are displayed in the following table:

object	p_1	p_2	p_3	arithmetic mean
A	0.75	0.9	0.3	0.65
B	0.75	0.8	0.4	0.65
C	0.3	0.65	0.1	0.35
D	0.3	0.55	0.2	0.35

Evaluation of ϕ_2. Since ϕ_2 can be represented by $(p_1 \wedge p_3) \vee (\neg p_1 \wedge p_2)$, its e^{Π}-value can be computed in the following way:

$$\begin{aligned} e^{\Pi}((p_1 \wedge p_3) \vee (\neg p_1 \wedge p_2)) &= e^{\Pi}(p_1 \wedge p_3) + e^{\Pi}(\neg p_1 \wedge p_2) \\ &= e(p_1)e(p_3) + e(p_2) - e(p_1)e(p_2). \end{aligned}$$

The corresponding results for the objects A, B, C and D are displayed in the following table:

object	p_1	p_2	p_3	$e^{\Pi}((p_1 \wedge p_3) \vee (\neg p_1 \wedge p_2))$
A	0.75	0.9	0.3	0.45
B	0.75	0.8	0.4	0.5
C	0.3	0.65	0.1	0.458
D	0.3	0.55	0.2	0.445

Aggregation of ϕ_1 *and* ϕ_2. As we have sad before, the aggregation will be carried out by computing the arithmetic mean of the evaluations of ϕ_1 and ϕ_2, which gives the correct classification $D < C < A < B$.

object	ϕ_1	ϕ_2	aggregation
A	0.65	0.45	0.55
B	0.65	0.5	0.575
C	0.35	0.458	0.4175
D	0.35	0.445	0.3725

6.2.4 Probabilistic formalization

The probabilistic formalization presented in this section is a slight modification of the formal system for reasoning about qualitative probability given in [32]. The only difference is that we need more probability operators. The set of all classical formulas For_C is the same as before, while the set of probabilistic formulas For_P is the smallest set of formulas closed for finite Boolean combinations containing the basic probabilistic formulas

$$P^{i,\Pi}_{\geqslant r}\alpha,\ P^{i,G}_{\geqslant r}\alpha,\ P^{i,(s)}_{\geqslant r}\alpha \text{ and } P^{i,*_1}(\alpha) \leqslant P^{j,*_2}(\beta),$$

where $r \in [0,1] \cap \mathbb{Q}$, $s \in (0,1) \cap \mathbb{Q}$, $*_1, *_2 \in \{\Pi, G\} \cup \{(s) \mid s \in (0,1) \cap \mathbb{Q}\}$ and i and j are arbitrary nonnegative integers. The intended meanings are:

- $P^{i,\Pi}_{\geqslant r}\alpha$: $e^{\Pi}_i(\alpha) \geqslant r$, where $e_i : Var \longrightarrow [0,1]$;
- $P^{i,G}_{\geqslant r}\alpha$: $e^{G}_i(\alpha) \geqslant r$
- $P^{i,s}_{\geqslant r}\alpha$: $e^{(s)}_i(\alpha) \geqslant r$
- $P^{i,*_1}(\alpha) \leqslant P^{j,*_2}(\beta)$: $e^{*_1}_i(\alpha) \leqslant e^{*_2}_j(\beta)$.

Formulas are either classical propositional formulas, or probabilistic formulas. Mixing of those two types of formulas is not allowed. The class of models is the same as for the LPP_2-logic, see Chapter 1 or [34]. Let $* \in \{\Pi, G\} \cup \{(s) \mid s \in (0,1) \cap \mathbb{Q}\}$. The satisfiability relation is defined in the following way:

- $M \models \alpha$ if $[\alpha] = W$
- $M \models P^{*}_{i \geqslant r}\alpha$ if $\mu_i^*([\alpha]) \geqslant r$, where $\mu_i^* : H \longrightarrow [0,1]$ is defined by $\mu_i^*([\alpha]) = e_i^*(\alpha)$;
- $M \models \phi \wedge \psi$ if $M \models \phi$ and $M \models \psi$
- $M \models \neg\phi$ if $M \not\models \phi$.

The notions of satisfiability, validity and semantical consequence are the same as for the LPP_2-logic. A strongly complete axiomatization is given below:

Axioms

A1 Substitutional instances of tautologies;

A2 $P^{j,*}_{\geqslant 0}\alpha$, $* \in \{\Pi, G\} \cup \{(s) \mid s \in (0,1) \cap \mathbb{Q}\}$

A3 $P^{i,*_1}_{\geqslant r} p \leftrightarrow P^{j,*_2}_{\geqslant r} p$, $p \in Var$

A4 $P^{j,*}_{\geqslant r}\alpha \rightarrow P^{j,*}_{> t}\alpha$, $t < r$

A5 $(P^{j,*}_{\geqslant r}\alpha \wedge P^{j,*}_{\geqslant t}\beta \wedge P^{j,*}_{=0}(\alpha \wedge \beta)) \rightarrow P^{j,*}_{\geqslant r+t}(\alpha \vee \beta)$

A6 $(P^{j,*}_{< r}\alpha \wedge P^{j,*}_{\leqslant t}\beta) \rightarrow P^{j,*}_{\leqslant r+t}(\alpha \vee \beta)$, $r + t \leqslant 1$

A7 $\bigwedge\limits_{n=1}^{n} \left(P^{j,\Pi}_{\geqslant r_i} p_i \wedge P^{j,\Pi}_{\leqslant t_i} p_i\right) \rightarrow \left(P^{j,\Pi}_{\geqslant \prod_{i=1}^{n} r_i}\left(\bigwedge\limits_{i=1}^{n} p_i\right) \wedge P^{j,\Pi}_{\leqslant \prod_{i=1}^{n} t_i}\left(\bigwedge\limits_{i=1}^{n} p_i\right)\right)$

A8 $\bigwedge\limits_{n=1}^{n} \left(P^{j,G}_{\geqslant r_i} p_i \wedge P^{j,G}_{\leqslant t_i} p_i\right) \rightarrow$

$\left(P^{j,G}_{\geqslant \min\limits_{i \in \{1,\dots,n\}} r_i}\left(\bigwedge\limits_{i=1}^{n} p_i\right) \wedge P^{j,G}_{\leqslant \min\limits_{i \in \{1,\dots,n\}} t_i}\left(\bigwedge\limits_{i=1}^{n} p_i\right)\right)$

A9 $\bigwedge\limits_{n=1}^{n} \left(P^{j,\Pi}_{\geqslant r_i} p_i \wedge P^{j,\Pi}_{\leqslant t_i} p_i\right) \rightarrow$

$P^{j,s}_{\geqslant s\prod_{i=1}^{n} r_i + (1-s)\min\limits_{i \in \{1,\dots,n\}} r_i}\left(\bigwedge\limits_{i=1}^{n} p_i\right) \wedge P^{j,s}_{\leqslant s\prod_{i=1}^{n} t_i + (1-s)\min\limits_{i \in \{1,\dots,n\}} t_i}\left(\bigwedge\limits_{i=1}^{n} p_i\right)$

A10 $P^{i,*}(\alpha) \leqslant P^{i,*}(\alpha)$

A11 $P^{i,*_1}(\alpha) \leqslant P^{j,*_2}(\beta) \vee P^{j,*_2}(\beta) \leqslant P^{i,*_1}(\alpha)$

A12 $(P^{i,*_1}(\alpha) \leqslant P^{j,*_2}(\beta) \wedge P^{j,*_2}(\beta) \leqslant P^{k,*_3}(\gamma)) \rightarrow P^{i,*_1}(\alpha) \leqslant P^{k,*_3}(\gamma)$

A13 $P^{i,*_1}(\alpha) \leqslant P^{j,*_2}(\beta) \rightarrow (P^{i,*_1}_{\geqslant r}\alpha \rightarrow P^{j,*_2}_{\geqslant r}\beta)$

A14 $(P^{i,*_1}_{\leqslant r}\alpha \wedge P^{j,*_2}_{\geqslant r}\beta) \rightarrow P^{i,*_1}(\alpha) \leqslant P^{j,*_2}(\beta)$.

Inference rules

R1 Modus ponens for propositional formulas and modus ponens for probabilistic formulas;
R2 (Necessitation) From α infer $P^{j,*}_{=1}\alpha$;
R3 (Archimedean rule) From the set of premises

$$\left\{\phi \rightarrow P^{j,*}_{\geqslant r-\frac{1}{n}}\alpha \mid n > \frac{1}{r}\right\}$$

infer $\phi \rightarrow P^{j,*}_{\geqslant r}\alpha$;
R4 From the set of premises

$$\{\phi \rightarrow (P^{i,*_1}_{\geqslant r}\alpha \rightarrow P^{j,*_2}_{\geqslant r}\beta) \mid r \in [0,1] \cap \mathbb{Q}\}$$

infer $\phi \rightarrow P^{i,*_1}(\alpha) \leqslant P^{j,*_2}(\beta)$.

The notions of inference, theorem, consistency etc., are defined as usual. Proofs of the corresponding strong completeness theorem and decidability of the PSAT problem can be found in [39].

Though Grabisch's example from subsection 6.2.4 cannot be formalized within probability logic introduced here since arithmetic mean of the values of quality attributes cannot be syntactically expressed, nor it can be aggregated with the propositional criteria, we will illustrate the formalization process on the propositional part of this example.

The propositional part of the problem would be to classify the objects A, B, C and D using just ϕ_2. Let $T = \{\phi_A, \phi_B, \phi_C, \phi_D\}$ and $\alpha = (p_1 \wedge p_3) \vee (\neg p_1 \wedge p_2)$, where:

- $\phi_A = P^{0,\Pi}_{=0.75}p_1 \wedge P^{0,\Pi}_{=0.9}p_2 \wedge P^{0,\Pi}_{=0.3}p_3$
- $\phi_B = P^{1,\Pi}_{=0.75}p_1 \wedge P^{1,\Pi}_{=0.8}p_2 \wedge P^{1,\Pi}_{=0.4}p_3$
- $\phi_C = P^{2,\Pi}_{=0.3}p_1 \wedge P^{2,\Pi}_{=0.65}p_2 \wedge P^{2,\Pi}_{=0.1}p_3$
- $\phi_D = P^{3,\Pi}_{=0.3}p_1 \wedge P^{3,\Pi}_{=0.55}p_2 \wedge P^{3,\Pi}_{=0.2}p_3$.

Clearly, T is satisfiable. Let M be any model of T. Then,

$$M \models P^{0,\Pi}_{=0.45}\alpha \wedge P^{1,\Pi}_{=0.5}\alpha \wedge P^{2,\Pi}_{=0.458}\alpha \wedge P^{3,\Pi}_{=0.455}\alpha,$$

so, by the completeness theorem,

$$T \vdash P^{0,\Pi}_{=0.45}\alpha \wedge P^{1,\Pi}_{=0.5}\alpha \wedge P^{2,\Pi}_{=0.458}\alpha \wedge P^{3,\Pi}_{=0.455}\alpha.$$

Now

$$T \vdash P^{0,\Pi}(\alpha) < P^{3,\Pi}(\alpha) < P^{2,\Pi}(\alpha) < P^{1,\Pi}(\alpha),$$

which is a formal representation of the ordering

$$A < D < C < B$$

induced by ϕ_2. The full formalization of this problem is carried out in the $\text{Ł}\Pi\frac{1}{2}$-logic, see [39].

6.3 On weighted logics

In [35] we have presented a general propositional framework for description of weighted logics with countable object languages. An $\mathbb{F}$-valued weight is any mapping $w : For_C \longrightarrow \mathbb{F}$. On the semantical level we can think of weighted logics as coherent assignments of weights to formulas. The prime examples are fuzzy logics, probability logics, possibility and necessity logics.

In case of fuzzy logics, an arbitrary evaluation $e : Var \longrightarrow [0,1]$ (the initial weights) are extended using the corresponding truth functions for conjunction (t-norms) and implication (residuated implications of t-norms). In case of probability logics, probability distributions determine the assignment of weights.

Additionally, weights range over some recursive first-order structure. Hence, a general framework for description of weighted logics needs to be expressive enough to represent not only particular weights $w(\alpha)$, $\alpha \in For_C$, but also to express connections between weights. For example, in product fuzzy logic,

$$w(\alpha \vee \beta) = w(\alpha) + w(\beta) - w(\alpha)w(\beta),$$

while in any probability logic we have that

$$w(\alpha \vee \beta) = w(\alpha) + w(\beta) - w(\alpha \wedge \beta).$$

The adequate formal language is formed by Σ_0-sentences of a suitable first-order language L. Recall that a Σ_0-sentence is any L-formula without quantifiers and variables. For example, if c_0, c_1 and c_2 are constant symbols and $+$ is a binary function symbol, then

$$c_0 = c_1 + c_2 \ \rightarrow \ c_1 = c_2$$

is a Σ_0-sentence.

We extend such languages with the countable set C of names for the elements of the range of weights.

6.3.1 Axiomatization

We assume that the reader is familiar with the notion of a first-order language. For the given language L, $Const(L)$ is the set of symbols for constants, $Fun(L)$ is the set of symbols for functions, and $Rel(L)$ is the set of symbols for relations. Those three sets are pairwise disjoint. Additionally, we assume that $C \cap Const(L) = \emptyset$. Recall that the countable set C contains names for the elements of the range of weights.

The set of terms $Term$ is defined as follows:

$$c \mid F(f_1, \ldots, f_{ar(F)}).$$

Here c ranges over $C \cup Const(L)$, F ranges over $Fun(L)$, and $ar(F)$ is the arity (number of argument places) of F. Variables for terms are f, g and h, indexed if necessary.

The set of formulas For is defined in the following way:

$$f = g \mid R(f_1, \ldots, f_{ar(R)}) \mid \neg\phi \mid \phi \wedge \psi.$$

Here R ranges over the set $Rel(L)$. In other words, formulas are Boolean combinations of variable-free atomic formulas.

Our models are C-valued L-structures, i.e., , L-structures of the form $\mathbb{A} = (A, \ldots)$, where each $a \in A$ is an interpretation of some $c \in C$. Our satisfiability relation is the restriction of the classical first-order satisfiability relation on Σ_0-sentences.

The axiom system $\mathscr{A}$ is defined as follows:

Propositional axioms

A1 Tautology instances;

Equality axioms

A2 $f = f$;
A3 $f = g \rightarrow (\phi(\ldots, f, \ldots) \rightarrow \phi(\ldots, g, \ldots)$;

Inference rules

R1 Modus ponens: from ϕ and $\phi \rightarrow \psi$ infer ψ;
R2 f-rule: from the set of premises $\{\phi \rightarrow f \neq c \,:\, c \in C\}$ infer $\phi \rightarrow f \neq f$.

The purpose of the infinitary rule R2 is to ensure that value of each term is in the set of interpretations of C. This rule allows us to modify Henkin's completion technique and to prove the strong completeness theorem.

The notion of formal deduction (inference) is defined similarly as in the case of the LPP_2-logic. Some basic properties like soundness theorem and deduction theorem can be obtained by the standard argument, see [34, 41].

Theorem 6.4. The compactness theorem fails for $\mathscr{A}$.

Proof. Let $d \in Const(L) \setminus C$. We define theory T as the union of theories

$$T_1 = \{d \neq c \,:\, c \in C\}$$

and

$$T_2 = \{c_1 \neq c_2 \,:\, c_1, c_2 \in C \text{ and } c_1 \neq c_2\}.$$

Let $\mathbb{A} = (A, \ldots)$ be any C-valued L-structure that satisfies T_2. Since A coincides with the interpretation of C and the interpretation of d is an element of A, we have that $\mathbb{A} \not\models T_1$.

On the other hand, each finite subset of T is satisfiable. Indeed, let $C = \{c_n \,:\, n \in \omega\}$. For the given positive integer m, a model $\mathbb{A} = (\omega, \ldots)$ such that interpretation of each c_n is n and that interpretation of d is m satisfies T_2 and the sentence

$$d \neq c_0 \wedge \cdots \wedge d \neq c_{m-1}.$$

Thus, T is finitely satisfiable. ■

The next technical lemma is needed for the proof of the Lindenbaum's theorem for $\mathscr{A}$.

Lemma 6.5. Suppose that T is a consistent theory and that f is an arbitrary term. Then, there is $c \in C$ such that $T, f = c$ is consistent.

Proof. Suppose that $T, f = c$ is inconsistent for all $c \in C$. By the Deduction theorem,

$$T \vdash f \neq c$$

for all $c \in C$, so by A1, A2 and MP we have that

$$T \vdash f = f \to f \neq c$$

for all $c \in C$. By f-rule, we have that

$$T \vdash f = f \to f \neq f.$$

On the other hand, $T \vdash f = f$, so by MP, $T \vdash f \neq f$, which contradicts the assumed consistency of T. ■

Theorem 6.5 (Lindenbaum's theorem for $\mathscr{A}$). *Each consistent theory can be extended to the maximal consistent theory.*

Proof. Fix a consistent L-theory T. Let $For(L) = \{\phi_n : n \in \omega\}$ and $Term(L) = \{f_n : n \in \omega\}$. We define an ω-sequence $\langle T_n : n \in \omega \rangle$ of theories inductively as follows:

1. $T_0 = T$
2. $T_{2n+1} = \begin{cases} T_{2n} \cup \{\phi_n\}, & T_{2n} \cup \{\phi_n\} \text{ is consistent} \\ T_{2n} \cup \{\neg \phi_n\}, & \text{otherwise} \end{cases}$
3. $T_{2n+2} = T_{2n+1} \cup \{f_n = c_m\}$, where m is the least nonnegative integer such that T_{2n+1} is consistent (the existence of such m is provided by Lemma 6.5).

Let $T_\omega = \bigcup_{n\in\omega} T_n$. By induction on the length of inference we can show that T_ω is deductively closed, which with consistence of all T_n's implies consistency of T_ω. Indeed, if $\bot \in T_\omega$, then by definition there is $n \in \omega$ such that $\bot \in T_n$, which contradicts consistency of T_n. ■

Theorem 6.6. Every consistent theory is satisfiable.

Proof. By Theorem 6.5 it is sufficient to prove the statement for a complete (maximal consistent) L-theory T.

Following the Henkin's construction, we define a binary relation $\sim$ on C by

$$c_1 \sim c_2 \Longleftrightarrow_{\text{def}} T \vdash c_1 = c_2.$$

The domain A of the canonical model $\mathbb{A}$ is defined as the quotient set $C_{/\sim}$. Furthermore, the corresponding interpretation I of the language L is defined as follows:

1. $I(c) = c_{/\sim}$, $c \in C$;
2. Let $d \in Const(L) \setminus C$. By Lemma 6.5, there is $c \in C$ such that $T, d = c$ is consistent. Since T is complete, $T \vdash d = c$. We define $I(d)$ by
$$I(d) = I(c);$$
3. Let f be an n-ary function symbol. Similarly as above, for all $c_1, \ldots, c_n \in C$ there is $c \in C$ such that $T \vdash f(c_1, \ldots, c_n) = c$. We define $I(f)$ by
$$I(f)(I(c_1), \ldots, I(c_n)) = I(c);$$
Let R be n-ary relation symbol. We define $I(R)$ by
$$(I(c_1), \ldots, I(c_n) \in I(R)) \Longleftrightarrow_{\text{def}} T \vdash R(c_1, \ldots, c_n).$$

The fact that $\sim$ is a congruence compatible with introduced relations is a consequence of equality axioms. It remains to show that $\mathbb{A} \models T$. The only nontrivial step is to verify that $\mathbb{A}$ satisfies atomic formulas appearing in T. Without the loss of generality, we shall consider only unary function and relation symbols.

Let $T \vdash f(u) = g(v)$. By Lemma 6.5 and the completeness of T, there are $c_1, c_2 \in C$ so that
$$T \vdash f(u) = c_1 \wedge g(v) = c_2.$$
By equality axioms, $T \vdash c_1 = c_2$, so
$$I(f)(I(u)) = I(c_1) = I(c_2) = I(g)(I(v)).$$
In other words, $\mathbb{A} \models f(u) = g(v)$.

Let $T \vdash R(u)$. Similarly as above, there is $c \in C$ such that $T \vdash u = c$. Hence, $T \vdash R(c)$, so
$$c_{/\sim} = I(u) \in I(R),$$
i.e., $\mathbb{A} \models R(u)$. ∎

6.3.2 Applications

We conclude this section by showing how probability and possibility weights can be formalized by $\mathscr{A}$-theories. The same is done for the Gödel's propositional fuzzy logic.

Example 6.6 (probabilistic weights). We start with the logic for reasoning about simple probabilities presented in [14] and further modified in [36, 37]. The language L is the union of the language of ordered division groups $L = \{+, \leq\} \cup \mathbb{Q}$ and the set of names for weights

$$Wgh = \{w(\alpha) : \alpha \in For_C\}.$$

We assume that $\mathbb{Q} \subseteq C$ and that $C \cap Wgh = \emptyset$. In order to assure that elements of Wgh behave like probability weights we define the following $\mathscr{A}$-theory:

1. Diagram[1] of the ordered field of rational numbers;
2. $f + g = g + f$;
3. $f + (g + h) = (f + g) + h$;
4. $f + 0 = f$;
5. $f - f = 0$;
6. $f \leq f$;
7. $f \leq g \wedge g \leq h \to f \leq h$;
8. $f \leq g \wedge g \leq f \to f = g$;
9. $f \leq g \vee g \leq f$;
10. $f \leq g \to f + h \leq g + h$;
11. $0 \leq w(\alpha) \leq 1$;
12. $w(\alpha) = 1$, α is a tautology;
13. $w(\neg\alpha) = 1 - w(\alpha)$;
14. $w(\alpha \vee \beta) = w(\alpha) + w(\beta) - w(\alpha \wedge \beta)$;
15. $w(\alpha) = w(\beta)$, $\alpha \leftrightarrow \beta$ is a tautology. ■

Example 6.7 (possibility and necessity functions). A binary relation $\leq_\Pi$ on For_C is a qualitative possibility relation iff it satisfies the following conditions:

1. $\leq_\Pi$ is a nontrivial weak order (linear, transitive and $\bot <_\Pi \top$);
2. $\leq_\Pi$ is compatible with the equivalence of propositional formulas. More precisely, if $\alpha \leq_\Pi \beta$, $\alpha \Leftrightarrow \alpha'$ and $\beta \Leftrightarrow \beta'$, then $\alpha' \leq_\Pi \beta'$;
3. $\bot \leq_\Pi \alpha$;
4. Disjunctive stability: if $\beta \leq_\Pi \gamma$, then, for all α, $\alpha \vee \beta \leq_\Pi \alpha \vee \gamma$.

The dual notion is the qualitative necessity relation $\leq_N$. The first two properties are the same as for the qualitative possibility relation, while the remaining two are $\alpha \leq_N \top$ and conjunctive stability:

If $\beta \leq_N \gamma$, then, for all α, $\alpha \wedge \beta \leq_N \alpha \wedge \gamma$.

Note that, for the given qualitative relation $\leq_\Pi$, the relation $\leq_N^\Pi$ defined by

$$\alpha \leq_N^\Pi \beta \Longleftrightarrow_{\text{def}} \neg\beta \leq_\Pi \neg\alpha$$

is a qualitative necessity relation. Similarly, relation $\leq_\Pi^N$ defined by

$$\alpha \leq_\Pi^N \beta \Longleftrightarrow_{\text{def}} \neg\beta \leq_N \neg\alpha$$

[1] A diagram is a list of all tuples describing relations and operations of the given structure.

is a qualitative possibility relations. Hence, it is naturally to consider this two types of relations as dual notions.

Moreover, for any possibility relation $\leq_\Pi$, the quotient structure $(For_{C/\sim}, \leq_{\Pi/\sim})$ ($\alpha \sim \beta$ iff $\alpha \leq_\Pi \beta$ and $\beta \leq_\Pi \alpha$) is a countable linear ordering with endpoints, so it can be embedded into $([0,1]_\mathbb{Q}, \leq)$. Such embeddings are called possibility functions or possibility distributions.

If π is a possibility function of $\leq_\Pi$, then it satisfies the maxitivity condition

$$\pi(\alpha \vee \beta) = \max(\pi(\alpha), \pi(\beta)).$$

On the other hand, any function $\pi : For_C \longrightarrow [0,1]_\mathbb{Q}$ satisfying maxitivity condition that additionally satisfy $\pi(\alpha) = \pi(\beta)$ whenever α and β are equivalent generates the qualitative possibility relation by

$$\alpha \leq_\Pi \beta \Leftrightarrow_{\text{def}} \pi(\alpha) \leq \pi(\beta).$$

Now, let us describe the interpretation as an $\mathscr{A}$-theory. The underlying language has names for weights (same as in the previous example), names for rational numbers (the set C) and the binary relation symbol $\leq$. The corresponding theory is defined as follows:

1. Diagram of $([0,1]_\mathbb{Q}, \leq)$;
2. $f \leq f$;
3. $f \leq g \wedge g \leq f$;
4. $f \leq g \wedge g \leq h \rightarrow f \leq h$;
5. $f \leq g \vee g \leq f$;
6. $w(\alpha) = w(\beta)$, $\models \alpha \leftrightarrow \beta$;
7. $w(\bot) = 0$;
8. $w(\top) = 1$;
9. $w(\alpha) \leq w(\beta) \rightarrow w(\alpha \vee \beta) = w(\beta)$. ■

Example 6.8. Formally, Gödel's logic is an extension of the basic *BL* propositional fuzzy logic that adds idempotency of conjunction. As a consequence, conjunction and disjunction in Gödel's fuzzy logic behave like min and max. Truth functions of implication and negation are given by

$$x \Rightarrow y = \begin{cases} 1\ , x \leq y \\ y\ , y < x \end{cases}$$

and

$$\neg x = \begin{cases} 1\ ,\ x = 0 \\ 0\ ,\ x > 0. \end{cases}$$

Let us proceed with the formalization of the Gödel weights, i.e., weights induced by evaluations in Gödel's logic. The formal language L will be the same as in the previous example. The axioms are given below:

1. Diagram of $([0,1]_\mathbb{Q}, \leq)$;

2. $f \leq f$;
3. $f \leq g \wedge g \leq f$;
4. $f \leq g \wedge g \leq h \rightarrow f \leq h$;
5. $f \leq g \vee g \leq f$;
6. $0 < w(\alpha) \rightarrow w(\neg\alpha) = 0$;
7. $w(\alpha) = 0 \rightarrow w(\neg\alpha) = 1$;
8. $w(\alpha) \leq w(\beta) \rightarrow w(\alpha \wedge \beta) = w(\alpha)$;
9. $w(\alpha) \leq w(\beta) \rightarrow w(\alpha \vee \beta) = w(\beta)$;
10. $w(\alpha) \leq w(\beta) \rightarrow w(\alpha \rightarrow \beta) = 1$;
11. $w(\alpha) > w(\beta) \rightarrow w(\alpha \rightarrow \beta) = w(\beta)$. ■

6.4 Big-stepped probabilities

In [6] we have presented a formalization of the concept of big-stepped probabilities, i.e., discrete probability distributions (defied on finitely many elementary events) that can emulate infinitesimals.

More precisely, let At be the set of all atoms over the set of propositional letters $\{p_1,\dots,p_n\}$. A probability distribution $\mu : At \longrightarrow [0,1]$ is called a big-stepped probability distribution iff it satisfied the following conditions:

1. $\mu(\alpha) > 0$ for all $\alpha \in At$;
2. $\mu(\alpha) \neq \mu(\beta)$ for $\alpha \neq \beta$;
3. Suppose that $At = \{\alpha_i \ : i \in \omega,\ i \leq 2^n - 1\}$ and that $\mu(\alpha_i) < \mu(\alpha_j)$ for $i < j$. Then, for all positive $j < 2^n$,
$$\mu(\alpha_0) + \cdots + \mu(\alpha_{j-1}) < \mu(\alpha_j).$$

Example 6.9. Consider two distinct propositional letters p and q. The corresponding set of atoms is given by

$$At = \{p \wedge q, p \wedge \neg q, \neg p \wedge q, \neg p \wedge \neg q\}.$$

The probability distribution given by

$$\begin{pmatrix} p \wedge q & p \wedge \neg q & \neg p \wedge q & \neg p \wedge \neg q \\ \frac{1}{100} & \frac{1}{50} & \frac{1}{25} & \frac{93}{100} \end{pmatrix}$$

is an example of a big-stepped probability distribution. ■

Example 6.10. Let $Var = \{p_1,\dots,p_n\}$ and $At = \{\alpha_1,\dots,\alpha_{2^n}\}$. Since

$$\frac{1}{2^k} = \sum_{i=k+1}^{\infty} \frac{1}{2^i},$$

the probability distribution given by

$$\begin{pmatrix} \alpha_1 & \alpha_2 & \alpha_3 & \alpha_4 & \dots & \alpha_{2^n-1} & \alpha_{2^n} \\ \frac{2^n-1}{2^n} & \frac{1}{2} & \frac{1}{4} & \frac{1}{8} & \dots & \frac{1}{2^n-1} & \frac{1}{2^n} \end{pmatrix}$$

is also an example of a big-stepped probability distribution. ■

A big-stepped probability on For_C, where $Var = \{p_1, \dots, p_n\}$, is any probability measure on For_C generated by some big-stepped probability distribution.

Recall that a binary relation $\mid\!\sim$ on For_C is called a preference relation (see [26]) iff it satisfies the following conditions:

1. Reflexivity: $\phi \mid\!\sim \phi$;
2. Left logical equivalence: if ϕ and ψ are equivalent and $\phi \mid\!\sim \theta$, then $\psi \mid\!\sim \theta$;
3. Right weakening: if ψ implies θ and $\phi \mid\!\sim \psi$, then $\phi \mid\!\sim \theta$;
4. And: if $\phi \mid\!\sim \psi$ and $\phi \mid\!\sim \theta$, then $\phi \mid\!\sim \psi \wedge \theta$;
5. Or: if $\phi \mid\!\sim \theta$ and $\psi \mid\!\sim \theta$, then $\phi \vee \psi \mid\!\sim \theta$;
6. Cautious monotonicity: if $\phi \mid\!\sim \psi$ and $\phi \mid\!\sim \theta$, then $\phi \wedge \psi \mid\!\sim \theta$.

In [2] is shown that if *Var* is a finite set, then all preferential relations on For_C are generated by big-stepped probabilities. More precisely, if $\mid\!\sim$ is a preferential relation on For_C and *Var* is finite, then there is a big-stepped probability measure $\mu : For_C \longrightarrow [0,1]$ such that

$$\phi \mid\!\sim \psi \text{ iff } \mu(\psi|\phi) > \frac{1}{2} \text{ or } \mu(\phi) = 0.$$

6.4.1 Syntax and semantics

As a base for formalization of reasoning about big-stepped probabilities we used the $LPP_{2,\leq}$-logic with the qualitative probability operator presented in [32]. The only difference is in the fact that the set of propositional variables *Var* is finite, and that we need three additional axioms.

More precisely, for the given positive integer n, the syntax of the logic $LPBS_n$ is defined as follows:

- $Var\{p_1, \dots, p_n\}$;
- *At* is the set of all atoms over *Var*, i.e., all formulas of the form

$$\pm p_1 \wedge \dots \wedge \pm p_n,$$

  where $+p_i =_{\text{def}} p_i$ and $-p_i =_{\text{def}} \neg p_i$. Note that $|At| = 2^n$. Atoms will be denoted by α, β and γ, indexed if necessary;
- For_C is the set of all classical propositional formulas over *Var*. Variables for For_C are ϕ, ψ and θ, indexed if necessary;

- Probabilistic operators are unary operators $P_{\geq r}$ (the probability is at least r) , $r \in [0,1]_{\mathbb{Q}}$, and a binary qualitative probability operator $\leq$ (the probability is lesser or equal);
- Basic (atomic) probabilistic formulas are $P_{\geq r}\phi$ and $\phi \leq \psi$. Probabilistic formulas are finite Boolean combinations of basic probabilistic formulas. As before, the set of all probabilistic formulas is denoted by For_P. Variables for For_P are χ, ξ and ζ, indexed if necessary.

Probability operators $P_{\leq r}, P_{>r}, P_{<r}, P_{=r}$ are defined as before. Binary probability operators $<, >$ and $=$ are defined in the obvious way. For example, $\phi = \psi$ is the formula $\psi \leq \phi \wedge \phi \leq \psi$, while $\phi < \psi$ is the formula $\phi \leq \psi \wedge \neg(\psi \leq \phi)$.

A model is a structure $M = (W, H, v, \mu)$ whit the following properties:

- $W \neq \emptyset$ is the set of worlds (states);
- $v : W \times Var \longrightarrow [0,1]$ is the evaluation that determines satisfiability of propositional letters in worlds. For $\phi \in For_C$, by $[\phi]$ is denoted the set of all worlds that satisfy ϕ. Note that $[p] = \{w \in W : v(w,p) = 1\}$, $[\neg\phi] = W \setminus [\phi]$, $[\phi \wedge \psi] = [\phi] \cap [\psi]$ etc;
- $H = \{[\phi] : \phi \in For_C\}$. Note that H is a finite Boolean algebra and that $|H| = 2^{2^n}$;
- $\mu : H \longrightarrow [0,1]$ is a big-stepped probability measure.

The satisfiability relation $\models$ is defined as follows:

- $M \models \phi$ iff $[\phi] = W$;
- $M \models P_{\geq r}\phi$ iff $\mu([\phi]) \geq r$;
- $M \models \phi \leq \psi$ iff $\mu([\phi]) \leq \mu([\psi])$;
- $M \models \neg\chi$ iff $M \not\models \chi$;
- $M \models \chi \wedge \xi$ iff $M \models \chi$ and $M \models \xi$.

Basic model theoretical notions such as satisfiability, validity and semantical consequence are defined in the usual way.

6.4.2 Axiomatization

Axioms:

A1 Tautology instances;
A2 $P_{\geq 0}\phi$;
A3 $P_{=1} \bigvee_{\alpha \in At} \alpha$;
A4 $P_{>0}\alpha$ for all $\alpha \in At$;
A5 $P_{\geq r}\phi \to P_{>s}\phi, s < r$;
A6 $P_{>r}\phi \to P_{\geq r}\phi$;
A7 $P_{=1}(\phi \to \psi) \to (P_{\geq r}\phi \to P_{\geq r}\psi)$;
A8 $(P_{\geq r}\phi \wedge P_{\geq s}\psi \wedge P_{=0}(\phi \wedge \psi)) \to P_{\geq r+s}(\phi \vee \psi), r+s \leq 1$;
A9 $(P_{\leq r}\phi \wedge P_{\leq s}\psi) \to P_{\leq r+s}(\phi \vee \psi), r+s \leq 1$;

A10 $(\phi \leq \psi \wedge P_{\geq r}\phi) \rightarrow P_{\geq r}\psi$;
A11 $\bigwedge_{i=1}^{k-1} (\alpha_i < \alpha_k) \rightarrow (\alpha_1 \vee \cdots \vee \alpha_{k-1}) < \alpha_k$ for all pairwise distinct atoms $\alpha_1, \ldots, \alpha_k \in At$;
A12 $\alpha < \beta \vee \beta < \alpha$ for $\alpha, \beta \in At$, $\alpha \neq \beta$.

Inference rules:

R1 (Modus Ponens) from ϕ and $\phi \rightarrow \psi$ infer ψ;
R2 (Archimedean rule) from the set of premises

$$\{\chi \rightarrow P_{\geq r}\phi \ : \ r < s\}$$

infer $\chi \rightarrow P_{\geq s}\phi$;
R3 ($\leq$ rule) from the set of premises

$$\{\chi \rightarrow (P_{\geq r}\phi \rightarrow P_{\geq r}\psi) \ : \ r \in [0,1]_{\mathbb{Q}}\}$$

infer $\chi \rightarrow (\phi \leq \psi)$.

The standard proof theoretical notions such as the notion of the inference, syntactical consequence, consistency and maximal consistent theory are defined similarly as in the case of probability logics with infinitary inference rules studied in the previous chapters.

The proof of the corresponding strong completeness theorem can be carried out in the way described in the previous chapters, so it will be omitted. Similarly we omit the proof of decidability of the corresponding satisfiability problem.

Note that the formula

$$P_{=0}\phi \vee (\phi \wedge \neg\psi < \phi \wedge \psi)$$

models $\phi \mid\!\sim \psi$. Indeed, let μ be any big-stepped probability measure such that

$$\mu(\phi \wedge \neg\psi) < \mu(\phi \wedge \psi).$$

Since

$$\mu(\phi \wedge \neg\psi) = \mu(\phi) - \mu(\phi \wedge \psi),$$

we have that

$$\mu(\phi) < 2\mu(\phi \wedge \psi),$$

which is equivalent to $\mu(\psi|\phi) > \frac{1}{2}$.

Example 6.11. A trivial consequence of the completeness theorem is the fact that the formula χ of the form

$$(P_{\leq r}\phi \wedge P_{\geq r}\psi) \rightarrow \phi \leq \psi$$

is valid. Let us provide a direct proof that χ is a theorem. By the Deduction theorem, we need to prove that

$$P_{\leq r}\phi \wedge P_{\geq r}\psi \vdash \phi \leq \psi.$$

Since $(P_{\leq r}\phi \wedge P_{\geq r}\psi) \to P_{\leq r}\phi$, $(P_{\leq r}\phi \wedge P_{\geq r}\psi) \to P_{\geq r}\psi$ and

$$(P_{\leq r}\phi \to (P_{\geq r}\psi \to (P_{\leq r}\phi \wedge P_{\geq r}\psi))$$

are tautology instances, the initial problem is equivalently reducible to verification of the sequent

$$P_{\leq r}\phi, P_{\geq r}\psi \vdash \phi \leq \psi.$$

Let $s > r$. Since $P_{\leq r}\phi \vdash \neg P_{\geq s}\phi$ and $\neg P_{\geq s}\phi \to (P_{\geq s}\phi \to P_{\geq s}\psi)$ is a tautology instance, we have that

$$P_{\leq r}\phi, P_{\geq r}\psi \vdash P_{\geq s}\phi \to P_{\geq s}\psi$$

for all $s > r$. Furthermore, let $s \leq r$. Then $P_{\geq r}\psi \vdash P_{\geq s}\phi \to P_{\geq s}\psi$, so

$$P_{\leq r}\phi, P_{\geq r}\psi \vdash P_{\geq s}\phi \to P_{\geq s}\psi$$

for all $s \leq r$. Hence, by $\leq$ rule, we obtain that $\phi \leq \psi$ is a consequence of $P_{\leq r}\phi$ and $P_{\geq r}\psi$. ∎

References

[1] C. Beierle and G. Kern-Isberner. The relationship of the logic of big-stepped probabilities to standard probabilistic logics. In *International Symposium on Foundations of Information and Knowledge Systems*, pages 191–210. Springer, 2010.

[2] S. Benferhat, D. Dubois, and H. Prade. Possibilistic and standard probabilistic semantics of conditional knowledge bases. *Journal of Logic and Computation*, 9(6):873–895, 1999.

[3] H. Bezzazi, D. Makinson, and P. P. Pérez. Beyond rational monotony: some strong non-Horn rules for nonmonotonic inference relations. *Journal of Logic and Computation*, 7(5):605–631, 1997.

[4] K.P.S. Bhaskara Rao and M. Bhaskara Rao. *Theory of charges*. Academic Press, 1983.

[5] C. Boutilier. Modal logics for qualitative possibility theory. *International Journal of Approximate Reasoning*, 10(2):173–201, 1994.

[6] D. Doder. A logic with big-stepped probabilities that can model nonmonotonic reasoning of system P. *Publications de l'Institut Mathématique*, Ns. 90(104):13–22, 2011.

[7] D. Doder, B. Marinković, P. Maksimović, and A. Perović. A Logic with Conditional Probability Operators. *Publications de l'Institut Mathématique*, Ns. 87(101):85–96, 2010.

[8] D. Doder, Z. Ognjanović, Z. Marković, A. Perović, and M. Rašković. A probabilistic temporal logic that can model reasoning about evidence. In *Founda-*

tions of Information and Knowledge Systems, 6th International Symposium, FoIKS 2010, Sofia, Bulgaria, February 15-19, 2010. Proceedings, pages 9–24. Springer, 2010.
[9] D. Doder, A. Perović, and Z. Ognjanović. Probabilistic Approach to Nonmonotonic Consequence Relations. In *11th European Conference on Symbolic and Quantitative Approaches to Reasoning with Uncertainty, ECSQARU 2011, Belfast, United Kingdom, 29 June-1 July 2011*, volume 6717 of *Lecture Notes in Computer Science*, pages 459–471. Springer, 2011.
[10] D. Dubois. Belief structures, possibility theory and decomposable confidence measures on finite sets. *Computers and Artificial Intelligence*, 5(5):403–416, 1986.
[11] D. Dubois and H. Prade. *Conditional objects, possibility theory and default rules*. Oxford University Press, 1996.
[12] D. Dubois and H. Prade. Qualitative possibility functions and integrals. In *Handbook of measure theory*, pages 1469–1522. Elsevier, 2002.
[13] F. Esteva, L. Godo, and F. Montagna. The $L\Pi$ and $L\Pi\frac{1}{2}$ logics: two complete fuzzy logics joining Łukasiewicz and product logic. *Archive for Mathematical Logic*, 40:39–67, 2001.
[14] R. Fagin, J.Y. Halpern, and N. Megiddo. A logic for reasoning about probabilities. *Information and Computation*, 87:78–128, 1990.
[15] L. Farifias del Cerro and A. Herzig. A modal analysis of possibility theory. *Symbolic and Qualitative Approaches to Uncertainty*, pages 58–62, 1991.
[16] T. Flaminio. Strong non-standard completeness for fuzzy logics. *Soft Computing*, 12(4):321–333, 2008.
[17] L. Godo and E. Marchioni. Coherent conditional probability in a fuzzy logic setting. *Logic Journal of IGPL*, 14 (3):457–481, 2006.
[18] M. Grabisch and C. Labreuche. Bi-capacities for decision making on bipolar scales. In *EUROFUSE Workshop in Information Systems*, pages 185–190, 2002.
[19] T. Hailperin. *Sentential Probability Logic. Origins, development, current status and technical applications*. Lehigh University Press, 1996.
[20] P. Hajek. *Metamathematics offuzzy logic*. Kluwer Academic Publishers, 1998.
[21] P Hajek, F. Esteva, and L. Godo. Fuzzy logic and probability. In *Proceedings of the 11th conference "Uncertainty in AI"*, pages 237–244, 1995.
[22] D. Hoover. Probability logic. *Annals of mathematical logic*, 14:287–313, 1978.
[23] N. Ikodinović and Z. Ognjanović. A logic with coherent conditional probabilities. In L. Godo, editor, *Proceedings of the 8th European Conference Symbolic and Quantitative Approaches to Reasoning with Uncertainty, ECSQARU 2005; Barcelona; Spain; 6-8 July 2005*, volume 3571 of *Lecture Notes in Computer Science*, pages 726–736. Springer, 2005.
[24] H.J. Keisler. Hyperfinite model theory. In R.O. Gandy and J.M.E. Hyland, editors, *Logic Colloquium 76*, pages 5–110. North-Holland, 1977.
[25] H.J. Keisler. Probability quantifiers. In J. Barwise and S. Feferman, editors, *Model Theoretic Logic*, pages 509–556. Springer, Berlin, 1985.

[26] S. Kraus, D. Lehmann, and M. Magidor. Nonmonotonic reasoning, preferential models and cumulative logics. *Artificial Intelligence*, 44 (1-2):167–207, 1990.

[27] D. Lehmann and M. Magidor. What does a conditional knowledge base entail? *Artificial Intelligence*, 55:1–60, 1992.

[28] E. Marchioni and L. Godo. A logic for reasoning about coherent conditional probability: a modal fuzzy logic approach. In *JELIA'04*, volume 3229 of *Lecture Notes in Computer Science*, pages 213–225, 2004.

[29] E. Marchioni and F. Montagna. On triangular norms and uninorms definable in $l\pi\frac{1}{2}$. *International Journal of Approximate Reasoning*, 47(2):179–201, 2008.

[30] Z. Ognjanović and N. Ikodinović. A logic with higher order conditional probabilities. *Publications de l'Institut Mathématique*, Ns. 82(96):141–154, 2007.

[31] Z. Ognjanović, Z. Marković, M. Rašković, D. Doder, and A. Perović. A propositional probabilistic logic with discrete linear time for reasoning about evidence. *Annals of Mathematics and Artificial Intelligence*, 65(2-3):217–243, 2012.

[32] Z. Ognjanović, A. Perović, and M. Rašković. Logics with the Qualitative Probability Operator. *Logic Journal of IGPL*, 16(2):105–120, 2008.

[33] Z. Ognjanović and M. Rašković. Some probability logics with new types of probability operators. *Journal of Logic and Computation*, 9(2):181–195, 1999.

[34] Z. Ognjanović, M. Rašković, and Z. Marković. *Probability logics. Probability-Based Formalization of Uncertain Reasoning*. Springer, Cham, Switzerland, 2016.

[35] A. Perović, D. Doder, Z. Ognjanović, and M. Rašković. On evaluations of propositional formulas in countable structures. *Filomat*, 30(1):1–13, 2016.

[36] A. Perović, Z. Ognjanović, M. Rašković, and Z. Marković. A Probabilistic Logic with Polynomial Weight Formulas. In L.S. Hartmann and G. Kern-Isberner, editors, *Proceedings of the 5th International Symposium Foundations of Information and Knowledge Systems FoIKS 2008, Pisa, Italy, February 11-15, 2008*, volume 4932 of *Lecture Notes in Computer Science*, pages 239–252, Berlin, Heidelberg, 2008. Springer.

[37] A. Perović, Z. Ognjanović, M. Rašković, and Z. Marković. How to Restore Compactness into Probabilistic Logics? In S. Hölldobler, C. Lutz, and H. Wansing, editors, *Proceedings of the 11th European Conference, JELIA 2008, Dresden, Germany, September 28-October 1, 2008*, volume 5293 of *Lecture Notes in Computer Science*, pages 338–348, Berlin, Heidelberg, 2008. Springer.

[38] A. Perović, Z. Ognjanović, M. Rašković, and Z. Marković. Qualitative possibilities and necessities. In *Proceedings of the 10th European Conference on Symbolic and Quantitative Approaches to Reasoning and Uncertainty, ECSQARU 2009, Verona, Italy, 1-3 July 2009*, volume 5590 of *Lecture Notes in Computer Science*, pages 651–662. Springer, 2009.

[39] A. Perović, Z. Ognjanović, M. Rašković, and D. Radojević. Finitely additive probability measures on classical propositional formulas definable by Gödel's t-norm and product t-norm. *Fuzzy Sets and Systems*, 169:65–90, 2011.

[40] D. Radojević, A. Perović, Z. Ognjanović, and M. Rašković. Interpolative Boolean Logic. In *Proceedings of the 13th conference Artificial Intelligence: Methodology, Systems and Applications AIMSA 2008*, pages 209–219, 2008.

[41] M. Rašković, Z. Marković, and Z. Ognjanović. A logic with approximate conditional probabilities that can model default reasoning. *International Journal of Approximate Reasoning*, 49(1):52–66, 2008.

Chapter 7
Probabilized Sequent Calculus and Natural Deduction System for Classical Logic

Marija Boričić

Abstract By combining, on one side Carnap–Popper–Leblance and Suppes concepts of sentence probability, and Gentzen's sequent calculus **LK** and natural deduction system **NK** for classical propositional logic, on the other, we obtain their probabilistic versions. Through an introductory review, we briefly present Gentzen's calculi, Carnap–Popper–Leblance and Suppes probability semantics. Afterwards we introduce systems **LKprob**, **LKprob**(ε) and **NKprob**, through axioms and inference rules and define the corresponding probabilistic models followed by examples of derivations in these systems. We define the notion of "probabilized sequent" $\Gamma \vdash_a^b \Delta$ with the intended meaning that "the probability of truthfulness of $\Gamma \vdash \Delta$ is into the interval $[a,b]$", and in a similar way the notion of "probabilized formula" $A[a,b]$. The soundness and completeness theorems are proved for all of the presented systems with respect to defined models.

7.1 Sequent calculus, natural deductions and probability

In this section, we give a brief overview of sequent calculi, natural deduction and probabilistic treatment of logical inference. We follow Gentzen's original approach (see [8, 19, 24, 25]). Sequent calculus can be considered a redefining of propositional logic presented in Hilbert style based on a new predicate $\vdash$ playing the role of deduction relation.

Definition 7.1. For arbitrary finite (possibly empty) words Γ and Δ consisting of propositional formulas, expression $\Gamma \vdash \Delta$ is called a sequent. Γ and Δ are the antecedent and succedent, respectively. ■

Marija Boričić
Faculty of Organizational Sciences, University of Belgrade, Jove Ilića 154, Belgrade, Serbia, e-mail: marija.boricic@fon.bg.ac.rs

Z. Ognjanović (ed.), *Probabilistic Extensions of Various Logical Systems*,
https://doi.org/10.1007/978-3-030-52954-3_7

A sequent $A_1 \ldots A_n \vdash B_1 \ldots B_m$, where $m, n \geq 1$, can be interpreted as the following formula $A_1 \wedge \cdots \wedge A_n \to B_1 \vee \cdots \vee B_m$.

The system of classical propositional logic presented as sequent calculus **LK** consists of an axiom $p \vdash p$, where p is any propositional letter, and the inference rules are indicated by the figures below.

Structural rules:

$$\text{permutation:} \quad \frac{\Gamma AB\Pi \vdash \Delta}{\Gamma BA\Pi \vdash \Delta} (P\vdash) \qquad \frac{\Gamma \vdash \Delta AB\Lambda}{\Gamma \vdash \Delta BA\Lambda} (\vdash P)$$

$$\text{contraction:} \quad \frac{\Gamma AA \vdash \Delta}{\Gamma A \vdash \Delta} (C\vdash) \qquad \frac{\Gamma \vdash AA\Delta}{\Gamma \vdash A\Delta} (\vdash C)$$

$$\text{weakening:} \quad \frac{\Gamma \vdash \Delta}{\Gamma A \vdash \Delta} (W\vdash) \qquad \frac{\Gamma \vdash \Delta}{\Gamma \vdash A\Delta} (\vdash W)$$

$$\text{cut rule:} \quad \frac{\Gamma \vdash A\Delta \quad A\Pi \vdash \Lambda}{\Gamma\Pi \vdash \Delta\Lambda} (\text{cut})$$

and *logical rules* treating

$$\text{negation:} \quad \frac{\Gamma \vdash A\Delta}{\Gamma \neg A \vdash \Delta} (\neg\vdash) \qquad \frac{\Gamma A \vdash \Delta}{\Gamma \vdash \neg A\Delta} (\vdash\neg)$$

$$\text{conjunction:} \quad \frac{\Gamma A \vdash \Delta}{\Gamma A \wedge B \vdash \Delta}, \frac{\Gamma B \vdash \Delta}{\Gamma A \wedge B \vdash \Delta} (\wedge\vdash) \qquad \frac{\Gamma \vdash A\Delta \quad \Gamma \vdash B\Delta}{\Gamma \vdash A \wedge B\Delta} (\vdash\wedge)$$

$$\text{disjunction:} \quad \frac{\Gamma A \vdash \Delta \quad \Gamma B \vdash \Delta}{\Gamma A \vee B \vdash \Delta} (\vee\vdash) \qquad \frac{\Gamma \vdash A\Delta}{\Gamma \vdash A \vee B\Delta}, \frac{\Gamma \vdash B\Delta}{\Gamma \vdash A \vee B\Delta} (\vee\vdash)$$

$$\text{implication:} \quad \frac{\Gamma \vdash A\Delta \quad \Pi B \vdash \Lambda}{\Gamma\Pi A \to B \vdash \Delta\Lambda} (\to\vdash) \qquad \frac{\Gamma A \vdash B\Delta}{\Gamma \vdash A \to B\Delta} (\vdash\to)$$

On the other side the natural deduction system for classical propositional logic **NK** is based on the inference rules consisting of at least one introduction and one elimination rule for each logical connective:

$$\text{implication:} \quad \frac{\begin{array}{c}[A]\\ \vdots \\ B\end{array}}{A \to B} (I\to) \qquad \frac{A \quad A \to B}{B} (E\to)$$

$$\text{conjunction:} \quad \frac{A \quad B}{A \wedge B} (I\wedge) \qquad \frac{A \wedge B}{A}, \frac{A \wedge B}{B} (E\wedge)$$

$$\text{disjunction:} \quad \frac{A}{A \vee B}, \frac{B}{A \vee B} (I\vee) \qquad \frac{A \vee B \quad \begin{array}{c}[A]\\ \vdots \\ C\end{array} \quad \begin{array}{c}[B]\\ \vdots \\ C\end{array}}{C} (E\vee)$$

and *rules for the absurdity constant* $\bot$:

$$\frac{\bot}{A} (\bot) \qquad \frac{\begin{array}{c}[\neg A]\\ \vdots \\ \bot\end{array}}{A} (RAA)$$

Negation is defined as usual: $\neg A = A \to \bot$.

Let us explain the rules $(I \to)$, $(E\vee)$ and (RAA): each appearance of $[A]$, formula A in brackets, means that it is possible to cancel A in the set of hypotheses. More accurately, in this case, the set of hypotheses is obtained by striking out some (or none, or all) occurrences, if any, of formula A on the top of a derivation tree (see [19]).

The following statement describes the connection between the sequent calculus **LK**, the system of natural deduction **NK**, and the traditional Hilbert style classical propositional logic (see [8, 19, 20, 25]).

Theorem 7.1. *(a) Sequent $\Gamma \vdash A$ is provable in* **LK** *iff it is possible to infer a conclusion A from the set of hypotheses Γ in* **NK**.

(b) Sequent $\Gamma \vdash A$ is provable in **LK** *iff $\bigwedge \Gamma \to A$ is a classical logic tautology, where $\bigwedge \Gamma$ denotes the conjunction of all formulas in Γ.*

The idea of probabilistic nature of a proposition was expressed in Boole's works (see [1]) and later treated by Keynes (see [12]), but without serious influence on its later development. Actually, Carnap's and Popper's approach to proposition probability opened a new era in investigation of this phenomenon (see [17, 22]). Our semantics is based on the following properties of proposition probability defined in works by Carnap, Popper and Leblanc (see [6, 13–15, 18]):

(1) For any tautology A, $P(A) = 1$;
(2) For any contradiction A, $P(A) = 0$;
(3) For any A, $P(\neg A)$=1-P(A);
(4) For any A and B, if $A \wedge B$ is a contradiction, then $P(A \vee B) = P(A) + P(B)$;
(5) For any A and B, $P(A \vee B) + P(A \wedge B) = P(A) + P(B)$;
(6) For any A and B, if $A \leftrightarrow B$ is a tautology, then $P(A) = P(B)$;

where we recognize the general algebraic properties of probability function P usually defined over a set of random events.

7.2 Natural deductions probabilized – NKprob

Our first step will be a probabilization of classical logic natural deduction system which will enable us to follow the inferences of probabilized formulas (see [2, 3]).

Let A be any propositional formula and I a finite subset of reals $[0,1]$ containing 0 and 1, closed under addition, also meaning that, for instance, a denotes the $\min\{a,1\}$, and b denotes the $\max\{0,b\}$. Formulas in **NKprob** are of the form $A[a,b]$, for $a,b \in I$, with the intended meaning that "the probability of truthfulness of a sentence A belongs to the interval $[a,b]$" and the system is defined over the set of all probabilized formulas $A[a,b]$.

We will introduce the axioms and inference rules of **NKprob**. Axioms of the system **NKprob** are the probabilized formulae $A[1,1]$, for each propositional formula A provable in classical logic.

The inference rules of system **NKprob** are the corresponding introduction (I-) and elimination (E-) rules, for each propositional connective, as well as, additivity, monotonicity and rules regarding inconsistency:
conjunction:

$$\frac{A[a,b] \quad B[c,d]}{(A\wedge B)[a+c-1,\min(b,d)]}(I\wedge) \qquad \frac{A[a,b] \quad (A\wedge B)[c,d]}{B[c,1+d-a]}(E\wedge)$$

disjunction:

$$\frac{A[a,b] \quad B[c,d]}{(A\vee B)[\max(a,c),b+d]}(I\vee) \qquad \frac{A[a,b] \quad (A\vee B)[c,d]}{B[c-b,d]}(E\vee)$$

implication:

$$\frac{A[a,b] \quad B[c,d]}{(A\rightarrow B)[\max(1-b,c),1-a+d]}(I\rightarrow) \qquad \frac{A[a,b] \quad (A\rightarrow B)[c,d]}{B[a+c-1,d]}(E_1\rightarrow)$$

$$\frac{B[a,b] \quad (A\rightarrow B)[c,d]}{A[1-d,1-c+b]}(E_2\rightarrow)$$

negation:

$$\frac{A[a,b]}{(\neg A)[1-b,1-a]}(I\neg) \qquad \frac{(\neg A)[a,b]}{A[1-b,1-a]}(E\neg)$$

additivity rule:

$$\frac{A[a,b] \qquad B[c,d] \qquad (A\wedge B)[e,f]}{(A\vee B)[a+c-f,b+d-e]}(ADD),$$

two *monotonicity rules:*

$$\frac{A[a,b] \quad A[c,d]}{A[\max(a,c),\min(b,d)]}(M\downarrow) \qquad \frac{A[a,b]}{A[c,d]}(M\uparrow)$$

where, for $(M\uparrow)$ it is supposed that $[a,b]\subseteq[c,d]$, and, finally, two rules regarding *inconsistency:*

$$\frac{\dfrac{A[c_1,c_1]}{A\emptyset} \quad \dfrac{A[c_2,c_2]}{A\emptyset} \quad \dots \quad \dfrac{A[c_m,c_m]}{A\emptyset}}{A\emptyset}(I\emptyset) \qquad \frac{A\emptyset}{B[a,b]}(E\emptyset)$$

for any propositional formulas A and B, and any $a,b\in I=\{c_1,c_2,\dots c_m\}$, where $A\emptyset$ stands for $A[a,b]$ when $a>b$, and the double line is an abbreviation for several steps in a derivation. Also, note that the additivity rule can be considered an additional case treating $(E\wedge)$ and $(I\vee)$ rules.

We provide an example of a derivation in our system.

Example 7.1. Let $A[0.8,1]$ present that the probability of selling the product A to a certain customer lies in interval $[0.8,1]$. Similarly, suppose that we have the following information about products A, B, C and D: $(A \to B)[0.3, 0.3]$, $C[0.6, 0.8]$ and $D[0.1, 0.4]$. We are interested in the probability interval of $B \to C \vee D$. Using derivations in our system we have:

$$\frac{A[0.8,1] \quad (A \to B)[0.3,0.3]}{B[0.1,0.3]}(E_1 \to)$$

$$\dfrac{B[0.1,0.3] \qquad \dfrac{C[0.6,0.8] \quad D[0.1,0.4]}{(C \vee D)[0.6,1]}(I\vee)}{(B \to C \vee D)[0.7,1]}(I \to)$$

concluding that, under the hypotheses given above, we infer $(B \to C \vee D)[0.7, 1]$. ■

Note that if $A \leftrightarrow B$ is provable in classical logic, where A and B are any propositional formulas, and $A[a, b]$ is provable in **NKprob**, then $B[a, b]$ is also provable in **NKprob**. Having in mind that $(A \to B)[1, 1]$ and $(B \to A)[1, 1]$ are the axioms of **NKprob**, we derive $B[a, b]$ as follows:

$$\dfrac{\dfrac{A[a,b] \quad (A \to B)[1,1]}{B[a,1]}(E1 \to) \qquad \dfrac{A[a,b] \quad (B \to A)[1,1]}{B[0,b]}(E2 \to)}{B[a,b]}(M\downarrow)$$

Models for **NKprob** and satisfiability relation are defined as follows.

Definition 7.2. Let For be the set of all propositional formulas. Then any mapping $p : \text{For} \to I$ will be **NKprob**-model if it satisfies the following axioms:

(i) $p(\top) = 1$ and $p(\bot) = 0$;
(ii) if $p(A \wedge B) = 0$, then $p(A \vee B) = p(A) + p(B)$;
(iii) if $A \leftrightarrow B$ in classical logic, then $p(A) = p(B)$.

■

Let us note that the above axioms of mappings p roughly correspond to Carnap's and Popper's sentence probability axioms (see [6, 18]).

Definition 7.3. A formula $A[a, b]$ is satisfied in a model p, denoted by $\models_p A[a, b]$, iff $a \le p(A) \le b$. A formula $A[a, b]$ is valid iff it is satisfied in each model, and this will be denoted by $\models A[a, b]$. ■

The following statements justifies some of the introduced inference rules, using basic characteristics of mappings which can be easily derived from Definition 7.2 (see [2]).

Lemma 7.1. *(a) If* $a \le p(A) \le b$ *and* $c \le p(A \wedge B) \le d$*, then* $c \le p(B) \le d + 1 - a$.
(b) If $a \le p(A) \le b$ *and* $c \le p(A \vee B) \le d$*, then* $c - b \le p(B) \le d$.
(c) If $a \le p(A) \le b$ *and* $c \le p(A \to B) \le d$*, then* $a + c - 1 \le p(B) \le d$.

(d) If $a \leq p(\neg B) \leq b$ and $c \leq p(A \to B) \leq d$, then $a+c-1 \leq p(\neg A) \leq d$.
The bounds in (a), (b), and (c) are the best possible.

Proof. The proof is given in [2]. For instance we will present the proof of (d): From the axioms in Definition 7.2 we can infer $p(\neg A) = p(\neg A \wedge B) + p(\neg A \vee B) - p(B) = p(\neg A \wedge B) + p(\neg A \vee B) - 1 + p(\neg B) \geq c - 1 + a$. Also, $p(\neg A) \leq p(\neg A) + p(A \wedge B) = p(\neg A \vee (A \wedge B)) = p(\neg A \vee B) \leq d$. The upper bound and the lower are reached when $p(\neg A \wedge B) = 0$ and $p(A \wedge B) = 0$, respectively, meaning that the given bounds are the best possible.

Lemma 7.2. *(a) From $A[a,a]$, $B[b,b]$ and $C[c,c]$ in* **NKprob** *we can infer:*

$$(A \to C)[\max(1-a,b) + \max(1-a,1-b,c) - 1, \min(1,1-b+c) + 1 - a].$$

(b) From $A[a,a]$, $C[c,c]$, $(A \to B)[r,r]$ and $(B \to C)[s,s]$ in **NKprob** *we can infer:*

$$(A \to C)[\max(r-a, r+s-1), \min(s+1-a, r+c)].$$

(c) From $(A \to B)[a,b]$ and $(B \to C)[c,d]$ in **NKprob** *we can infer:*

$$(A \to C)[\max(0, a+c-1), \min(b+d, 1)].$$

In the sequel we define the notion of a theory, consistency and a maximal consistent theory in **NKprob**.

Definition 7.4. We say that formula $A[a,b]$ can be derived from the set of hypotheses Γ in **NKprob** if there is a finite sequence of probabilized formulas ending with $A[a,b]$, such that each its formula is an axiom, it belongs to Γ or it is obtained by an **NKprob**-rule applied to some previous formulae of this sequence. ■

Definition 7.5. By **NKprob**-theory (or theory) we mean a set of formulas which are derivable from the set of hypothese $\{A_1[a_1,b_1], \ldots, A_n[a_n,b_n]\}$ in **NKprob**, denoted by **NKprob**$(A_1[a_1,b_1], \ldots, A_n[a_n,b_n])$. We say that theory T is inconsistent if there is a proposition A such that both $A[a,b]$ and $A[c,d]$ are contained in T, and $[a,b] \cap [c,d] = \emptyset$; otherwise, we say that it is consistent. A consistent theory is called a maximal consistent theory if each of its proper extensions is inconsistent. ■

We emphasize that inconsistency means that, bearing in mind the presence of monotonicity rules in **NKprob**, each probabilized formula $A[a,b]$, is provable.

Now, we are following the usual justification of the soundness and completeness theorem (see [2]).

Lemma 7.3. *Each consistent theory can be extended to a maximal consistent theory.*

Theorem 7.2. *If* **NKprob**-*theory has a model, then it is consistent.*

The previous statement presents the soundness theorem and it can be proved by induction on the length of the proof for any formula $A[a,b]$ in **NKprob** using Lemma 7.1.

In order to prove the completeness part, we define the notion of canonical model. Let $\mathrm{C}(\mathbf{NKprob}(A_1[a_1,b_1],\ldots,A_n[a_n,b_n])$ be the class of all maximal consistent extensions of all $\mathbf{NKprob}(A_1[a_1,b_1],\ldots,A_n[a_n,b_n])$-provable formulas, existing by Lemma 7.3.

Definition 7.6. For any $X \in \mathrm{C}(\mathbf{NKprob}(\sigma_1,\ldots,\sigma_n))$ we define

$$\models_{p^X} A[a,b] \text{ iff } a \leq \max\{c|A[c,1] \in X\} \text{ and } b \geq \min\{c|A[0,c] \in X\}.$$

p^X is called a canonical model. ■

Lemma 7.4. *The canonical model is a model.*

Lemma 7.5. $\models_{p^X} A[a,b]$ *iff* $A[a,b] \in X$.

As a consequence of Lemmata 7.3, 7.4 and 7.5 we obtain the completeness theorem:

Theorem 7.3. *Each consistent* **NKprob**-*theory has a model.*

7.3 Sequent calculus probabilized – LKprob

In this section we will present a type of probabilization of Gentzen's sequent calculus **LK** for classical propositional logic – **LKprob**, accompanied by soundness and completeness theorems. In **LKprob** we manipulate with propositions of the form "the probability of $\Gamma \vdash \Delta$ belongs to the interval $[a,b]$" (see [3, 5]).

For any sequent $\Gamma \vdash \Delta$ of **LK** and any $a,b \in I$ we suppose that, $\Gamma \vdash_a^b \Delta$ is a sequent of **LKprob**. The sequent $\Gamma \vdash_a^b \Delta$ is interpreted as "there exists $c \in [a,b]$ such that the probability of $\Gamma \vdash \Delta$ equals c". In cases when $a=1$ or $b=0$, for $\Gamma \vdash_a^b \Delta$ and $a \leq b$ we use the following abbreviations $\Gamma \vdash_1 \Delta$ and $\Gamma \vdash^0 \Delta$, respectively, and in the case of $a > b$ we treat $\Gamma \vdash^{\emptyset} \Delta$ as a pure contradiction.

The axioms of **LKprob** are :$\Gamma \vdash_0^1 \Delta$, $\vdash^0$, $A \vdash_1 A$, for any words Γ and Δ, and any formula A. The second axiom could be understood that the probability of absurdity (empty sequent) equals to zero.

For any words Γ, Δ, Π and Λ and any formulas A and B, the *structural rules* of **LKprob** are as follows:

permutation: $$\frac{\Gamma AB\Pi \vdash_a^b \Delta}{\Gamma BA\Pi \vdash_a^b \Delta}(P\vdash_a^b) \qquad \frac{\Gamma \vdash_a^b \Delta AB\Lambda}{\Gamma \vdash_a^b \Delta BA\Lambda}(\vdash_a^b P)$$

contraction: $$\frac{\Gamma AA \vdash_a^b \Delta}{\Gamma A \vdash_a^b \Delta}(C\vdash_a^b) \qquad \frac{\Gamma \vdash_a^b AA\Delta}{\Gamma \vdash_a^b A\Delta}(\vdash_a^b C)$$

for any $a,b \in I$, *the cut rule:*

$$\frac{\Gamma \vdash_a^b A\Delta \quad \Pi A \vdash_c^d \Lambda}{\Gamma\Pi \vdash_{\max(0,a+c-1)}^{\min(b+d,1)} \Delta\Lambda}(\mathrm{cut}^{[a,b][c,d]})$$

for any $a,b,c,d \in I$, and the following *specific structural rules:*

weakening: $$\frac{\Gamma \vdash_a^b \Delta \quad \vdash_c^d A}{\Gamma A \vdash_{\max(a,1-d)}^{\min(1,b+1-c)} \Delta}(W\vdash_a^b) \qquad \frac{\Gamma \vdash_a^b \Delta \quad \vdash_c^d A}{\Gamma \vdash_{\max(a,c)}^{\min(1,b+d)} A\Delta}(\vdash_a^b W)$$

for any $a,b,c,d \in I$,

monotonicity: $$\frac{\Gamma \vdash_a^b \Delta}{\Gamma \vdash_c^d \Delta}(M\uparrow) \qquad \frac{\Gamma \vdash_a^b \Delta \quad \Gamma \vdash_c^d \Delta}{\Gamma \vdash_{\max(a,c)}^{\min(b,d)} \Delta}(M\downarrow)$$

for any $[a,b] \subseteq [c,d]$, and any $a \leq b$ and $c \leq d$, respectively, for $(M\uparrow)$ and $(M\downarrow)$, and the following rule regarding *additivity:*

$$\frac{AB \vdash_1 \quad \vdash_a^b A \quad \vdash_c^d B}{\vdash_{\min(1,a+c)}^{\min(1,b+d)} AB}(ADD)$$

The rule regarding *inconsistency:*

$$\frac{\Gamma \vdash^{\emptyset} \Delta}{\Pi \vdash^{\emptyset} \Lambda}(\bot)$$

is inspired by Frisch and Haddawy (see [7]).

The *logical rules* of **LKprob** are as follows:

negation: $$\frac{\Gamma \vdash_a^b A\Delta}{\Gamma \neg A \vdash_a^b \Delta}(\neg\vdash_a^b) \qquad \frac{\Gamma A \vdash_a^b \Delta}{\Gamma \vdash_a^b \neg A\Delta}(\vdash_a^b \neg)$$

conjunction: $$\frac{\Gamma AB \vdash_a^b \Delta}{\Gamma A\wedge B \vdash_a^b \Delta}(\wedge\vdash_a^b) \qquad \frac{\Gamma \vdash_a^b A\Delta \quad \Gamma \vdash_c^d B\Delta}{\Gamma \vdash_{\max(0,a+c-1)}^{\min(b,d)} A\wedge B\Delta}(\vdash_a^b \wedge)$$

disjunction: $$\frac{\Gamma A \vdash_a^b \Delta \quad \Gamma B \vdash_c^d \Delta}{\Gamma A\vee B \vdash_{\max(0,a+c-1)}^{\min(b,d)} \Delta}(\vee\vdash_a^b) \qquad \frac{\Gamma \vdash_a^b AB\Delta}{\Gamma \vdash_a^b A\vee B\Delta}(\vdash_a^b \vee)$$

implication: $$\frac{\Gamma \vdash_a^b A\Delta \quad \Gamma B \vdash_c^d \Delta}{\Gamma A\rightarrow B \vdash_{\max(0,a+c-1)}^{\min(b,d)} \Delta}(\rightarrow\vdash_a^b) \qquad \frac{\Gamma A \vdash_a^b B\Delta}{\Gamma \vdash_a^b A\rightarrow B\Delta}(\vdash_a^b \rightarrow)$$

Also, we have the following specific inference rule

$$\frac{\dfrac{[\Gamma \vdash_{c_1}^{c_1} \Delta]}{\Gamma \vdash^{\emptyset} \Delta} \quad \dfrac{[\Gamma \vdash_{c_2}^{c_2} \Delta]}{\Gamma \vdash^{\emptyset} \Delta} \quad \cdots \quad \dfrac{[\Gamma \vdash_{c_m}^{c_m} \Delta]}{\Gamma \vdash^{\emptyset} \Delta}}{\Gamma \vdash^{\emptyset} \Delta}(\vdash \emptyset)$$

for any words Γ and Δ, where $I = \{c_1, c_2, \ldots c_m\}$, as an alternative to the axiom $\Gamma \vdash_0^1 \Delta$.

The following example illustrates a derivation in **LKprob**.

Example 7.2. Let the additional axioms of **LKprob** be $AB \vdash^1$, $\vdash^{.8}_{.5} A$, $\vdash^{.8}_{.4} B$, $\vdash^{.8}_{.6} C$, $\Pi A \vdash^{.7}_{.7} \Lambda$ and $\Pi B \vdash^{.9}_{.6} \Lambda$. We will show how to derive a truthfulness probability interval for sequent $C\Pi \vdash \Lambda$ in **LKprob**. From the following derivation tree

$$\dfrac{\dfrac{\vdash^{.8}_{.6} C \qquad \dfrac{AB \vdash^1 \quad \vdash^{.8}_{.5} A \quad \vdash^{.8}_{.4} B}{\vdash^1_{.9} A \vee B}(ADD, \vdash \vee)}{C \vdash^1_{.9} A \vee B}(W \vdash) \qquad \dfrac{\Pi A \vdash^{.7}_{.7} \Lambda \quad \Pi B \vdash^{.9}_{.6} \Lambda}{\Pi A \vee B \vdash^{.7}_{.3} \Lambda}(\vee \vdash)}{C\Pi \vdash^1_{.2} \Lambda}(\text{cut})$$

we conclude $C\Pi \vdash^1_{.2} \Lambda$. ■

The basic characteristics of **LKprob** are given by the following statements.

Lemma 7.6. *(a) If the sequent* $\vdash^b_a A$ *is provable in* **LKprob**, *then* $\vdash^{1-a}_{1-b} \neg A$ *is provable in* **LKprob**; *if the sequent* $A \vdash^b_a$ *is provable in* **LKprob**, *then* $\neg A \vdash^{1-a}_{1-b}$ *is provable in* **LKprob**.

(b) If the sequents $\vdash^0 A \wedge B$, $\vdash^b_a A$ *and* $\vdash^d_c B$ *are provable in* **LKprob**, *then* $\vdash^{\min(1,b+d)}_{\min(1,a+c)} A \vee B$ *is provable in* **LKprob**.

(c) If $\Gamma \vdash \Delta$ *is provable in* **LK**, *then* $\Gamma \vdash_1 \Delta$ *is provable in* **LKprob**.

(d) If $\vdash^c_c A$ *and* $A \vdash_1 B$ *are provable in* **LKprob**, *then* $\vdash^1_c B$ *is provable in* **LKprob**.

(e) Let suppose that $\vdash^{\emptyset} A$ *and* $\vdash^{\emptyset} B$ *are not provable. If* $\vdash^c_c A$, $\vdash^d_d B$, $A \vdash_1 B$ *and* $B \vdash_1 A$ *are provable in* **LKprob**, *then* $c = d$.

Proof. (d) Directly by the cut rule: $\dfrac{\vdash^c_c A \quad A \vdash_1 B}{\vdash^1_c B}(\text{cut})$

(e) Using part (d) and the fact that sequents $\vdash^c_c A$ are $A \vdash_1 B$ provable, we have that $\vdash^1_c B$ is also provable. Having in mind that $\vdash^{\emptyset} B$ is not provable, and both $\vdash^1_c B$ and $\vdash^d_d B$ are provable, we conclude that $d \geq c$. In a similar way, from the fact that $\vdash^d_d B$ and $B \vdash_1 A$ are provable, we infer $c \geq d$, which means that $c = d$. ■

Now we describe models in **LKprob** and the satisfiability relation.

Definition 7.7. T he mapping $p : \text{Seq} \to I$ will be a model, if it satisfies the following conditions:

(i) $p(A \vdash A) = 1$, for any formula A;

(ii) if $p(AB \vdash) = 1$, then $p(\vdash AB) = p(\vdash A) + p(\vdash B)$, for any formulas A and B;

(iii) if sequents $\Gamma \vdash \Delta$ and $\Pi \vdash \Lambda$ are equivalent in **LK**, in sense that there are proofs for both sequents $\bigwedge \Gamma \to \bigvee \Delta \vdash \bigwedge \Pi \to \bigvee \Lambda$ and $\bigwedge \Pi \to \bigvee \Lambda \vdash \bigwedge \Gamma \to \bigvee \Delta$ in **LK**, then $p(\Gamma \vdash \Delta) = p(\Pi \vdash \Lambda)$,

where Seq is the set of all unlabelled sequents, i.e. of sequents of form $\Gamma \vdash \Delta$. ■

Definition 7.8. The probabilized sequent $\Gamma \vdash^b_a \Delta$ is satisfied in the model p, denoted by $\models_p \Gamma \vdash^b_a \Delta$ iff $a \leq p(\Gamma \vdash \Delta) \leq b$. Sequent $\Gamma \vdash^b_a \Delta$ is valid iff it is satisfied in each model, and this is denoted by $\models \Gamma \vdash^b_a \Delta$. ■

The soundness of **LKprob** with respect to the given models is an immediate consequence of the following lemmas.

Lemma 7.7. *For any formula A and each sequent $\Gamma \vdash \Delta$, we have:*
(a) $\models \Gamma \vdash_0^1 \Delta$;
(b) $\models \vdash^0$;
(c) $\models A \vdash_1 A$.

Lemma 7.8. *For any formulas A and B, and any words Γ, Δ, Π and Λ:*
(a) $p(\Gamma AB\Pi \vdash \Delta) = p(\Gamma BA\Pi \vdash \Delta)$;
(b) $p(\Gamma \vdash \Delta AB\Lambda) = p(\Gamma \vdash \Delta BA\Lambda)$;
(c) $p(\Gamma A\Pi \vdash \Delta) = p(\Gamma AA\Pi \vdash \Delta)$;
(d) $p(\Gamma \vdash \Delta A\Lambda) = p(\Gamma \vdash \Delta AA\Lambda)$;
(e) $p(\Gamma \vdash A\Delta) = p(\Gamma \neg A \vdash \Delta)$;
(f) $p(\Gamma A \vdash \Delta) = p(\Gamma \vdash \neg A\Delta)$;
(g) $p(\Gamma AB \vdash \Delta) = p(\Gamma A \wedge B \vdash \Delta)$;
(h) $p(\Gamma \vdash AB\Delta) = p(\Gamma \vdash A \vee B\Delta)$;
(k) $p(\Gamma A \vdash B\Delta) = p(\Gamma \vdash A \to B\Delta)$.

Lemma 7.9. *For any formulas A and B, and any words Γ, Δ, Π and Λ, we have:*
(a) if $a \leq p(\Gamma \vdash \Delta) \leq b$ *and* $c \leq p(\vdash A) \leq d$, *then*

$$\max(a, 1-d) \leq p(\Gamma A \vdash \Delta) \leq \min(1, b+1-c);$$

(b) if $a \leq p(\Gamma \vdash \Delta) \leq b$ *and* $c \leq p(\vdash A) \leq d$, *then*

$$\max(a, c) \leq p(\Gamma \vdash A\Delta) \leq \min(1, b+d);$$

(c) if $a \leq p(\Gamma \vdash A\Delta) \leq b$ *and* $c \leq p(\Gamma \vdash B\Delta) \leq d$, *then*

$$\max(0, a+c-1) \leq p(\Gamma \vdash A \wedge B\Delta) \leq \min(b, d);$$

(d) if $a \leq p(\Gamma A \vdash \Delta) \leq b$ *and* $c \leq p(\Gamma B \vdash \Delta) \leq d$, *then*

$$\max(0, a+c-1) \leq p(\Gamma A \vee B \vdash \Delta) \leq \min(b, d);$$

(e) if $a \leq p(\Gamma \vdash A\Delta) \leq b$ *and* $c \leq p(\Gamma B \vdash \Delta) \leq d$, *then*

$$\max(0, a+c-1) \leq p(\Gamma A \to B \vdash \Delta) \leq \min(b, d).$$

(f) (hypothetical syllogism rule probabilized)

$$\max(1-a, b) + \max(1-a, 1-b, c) - 1 \leq p(A \vdash C) \leq 2-a-b+c$$

(g) (hypothetical syllogism rule probabilized)

$$\max(r-a, r+s-1) \leq p(A \vdash C) \leq \min(s+1-a, r+c)$$

The bounds in (a), (b), (c) and (d) are the best possible.

Proof. (b) From the given assumptions we have

$$\begin{aligned} p(\Gamma \vdash A\Delta) &= p(\vdash \neg(\bigwedge\Gamma)A\Delta) \\ &= p(\vdash \neg(\bigwedge\Gamma)\Delta) + p(\vdash A) - p(\vdash (\neg(\bigwedge\Gamma)\Delta) \wedge A) \\ &= p(\Gamma \vdash \Delta) + p(\vdash A) - p(\vdash (\neg(\bigwedge\Gamma)\Delta) \wedge A) \end{aligned}$$

i.e. $p(\Gamma \vdash A\Delta) \in [\max(a,c), \min(1, b+d)]$.

(c) Suppose that $p(\Gamma \vdash A\Delta) \in [a,b]$ and $p(\Gamma \vdash B\Delta) \in [c,d]$. We have that

$$\begin{aligned} p(\Gamma \vdash (A\wedge B)\Delta) &= p(\vdash (A\wedge B)\Delta\neg(\bigwedge\Gamma)) \\ &= p(\vdash (A\vee\Delta\vee\neg(\bigwedge\Gamma))\wedge(B\vee\Delta\vee\neg(\bigwedge\Gamma))) \\ &= p(\vdash A\Delta\neg(\bigwedge\Gamma)) + p(\vdash B\Delta\neg(\bigwedge\Gamma)) - p(\vdash AB\Delta\neg(\bigwedge\Gamma)) \\ &= p(\Gamma \vdash A\Delta) + p(\Gamma \vdash B\Delta) - p(\vdash AB\Delta\neg(\bigwedge\Gamma)) \end{aligned}$$

Therefore, $p(\Gamma \vdash (A\wedge B)\Delta) \in [\max(0, a+c-1), \min(b,d)]$. ■

The best possible bounds for the probability logical functions were considered by Hailperin (see [9–11]).

LKprob-theory over the sequents $\Gamma_1 \vdash_{a_1}^{b_1} \Delta_1, \ldots, \Gamma_n \vdash_{a_n}^{b_n} \Delta_n$, for $a_i, b_i \in I$ $(1 \le i \le n)$, denoted by $\mathbf{LKprob}(\Gamma_1 \vdash_{a_1}^{b_1} \Delta_1, \ldots, \Gamma_n \vdash_{a_n}^{b_n} \Delta_n)$, presents a deductive closure of the system **LKprob** extended by the list of sequents $\Gamma_1 \vdash_{a_1}^{b_1} \Delta_1, \ldots, \Gamma_n \vdash_{a_n}^{b_n} \Delta_n$ as additional axioms.

Definition 7.9. Theory T is inconsistent if there are two sequents $\Gamma \vdash_a^b \Delta$ and $\Gamma \vdash_c^d \Delta$, both provable in T such that $[a,b] \cap [c,d] = \emptyset$; otherwise, T is consistent. ■

A sequent $\Gamma \vdash_a^b \Delta$ is said to be consistent with respect to $\mathbf{LKprob}(\Gamma_1 \vdash_{a_1}^{b_1} \Delta_1, \ldots, \Gamma_n \vdash_{a_n}^{b_n} \Delta_n)$ if $\mathbf{LKprob}(\Gamma_1 \vdash_{a_1}^{b_1} \Delta_1, \ldots, \Gamma_n \vdash_{a_n}^{b_n} \Delta_n, \Gamma \vdash_a^b \Delta)$ is consistent. By a proper extension of a theory we mean any its deductively closed proper superset. A consistent theory is called a maximal consistent theory if each of its proper extensions is inconsistent.

Lemma 7.10. *Each consistent theory can be extended to a maximal consistent theory.*

Proof. Let T be a consistent theory and let sequence (T_n) of theories be defined inductively as follows: $T_0 = T$, and $T_{n+1} = T_n \cup \{\Gamma_n \vdash_{c_1}^{c_1} \Delta_n\}$, if $\Gamma_n \vdash_{c_1}^{c_1} \Delta_n$ is consistent with respect to T_n, but if it is not consistent, then: $T_{n+1} = T_n \cup \{\Gamma_n \vdash_{c_2}^{c_2} \Delta_n\}$, if $\Gamma_n \vdash_{c_2}^{c_2} \Delta_n$ is consistent with respect to T_n, but if it is not, then ... $T_{n+1} = T_n \cup \{\Gamma_n \vdash_{c_{m-1}}^{c_{m-1}} \Delta_n\}$, if $\Gamma_n \vdash_{c_{m-1}}^{c_{m-1}} \Delta_n\}$ is consistent with respect to T_n, and finally, $T_{n+1} = T_n \cup \{\Gamma_n \vdash_{c_m}^{c_m} \Delta_n\}$, otherwise; where $I = \{c_1, c_2, \ldots, c_m\}$. Let us note that the final result of this construction depends on the order of points $c_1, c_2, \ldots, c_m$ of set I. Let

$$T' = \cup_{n\in\omega} T_n$$

By induction on n we will prove that T' is a maximal consistent extension of T. Note that if T_n is consistent, then T_{n+1} is consistent, beacause of the following.

The only nontrivial case is when $T_{n+1} = T_n \cup \{\Gamma_n \vdash_{cm}^{cm} \Delta_n\}$. Suppose that T_{n+1} is inconsistent, i.e. that the sequent $\Gamma_n \vdash_{cm}^{cm} \Delta_n$ is not consistent with respect to T_n. Then there exists an interval $[a,b] \subset [0,1]$ such that $c_m \notin [a,b]$ and $\Gamma_n \vdash_a^b \Delta_n$ is provable in T_n, which is impossible because the theory $T_n \cup \{\Gamma_n \vdash_{c_j}^{c_j} \Delta_n\}$ is inconsistent for each j $(1 \leq j \leq m-1)$. In order to prove that T' is a *maximal* consistent extension of T we extend T' by the sequent $\Gamma_k \vdash_a^b \Delta_k$. In case that this is a proper extension, we already have that theory $T_{k+1} \subset T'$ contains $\Gamma_k \vdash_c^c \Delta_k$ for some $c \notin [a,b]$, and, consequently, this extension will be inconsistent. ■

We define the notion of canonical model as a mapping $p^X : \text{Seq} \to I \subseteq [0,1]$ as follows. Let $\text{ConExt}(T)$ be the class of all maximal consistent extensions of theory T. Then, for any $X \in \text{ConExt}(T)$ we define $\models_{p^X} \Gamma \vdash_a^b \Delta$ iff $a \leq \max\{c | \Gamma \vdash_c^1 \Delta \in X\}$ and $b \geq \min\{c | \Gamma \vdash_0^c \Delta \in X\}$. Obviously, such a definition provides that mapping p^X, depending on X, always has an adequate value, meaning that $p^X(\Gamma \vdash \Delta) \in [a,b]$. The canonical model is essentially included in the proof of the completeness theorem.

Lemma 7.11. $\models_{p^X} \Gamma \vdash_a^b \Delta$ *iff* $\Gamma \vdash_a^b \Delta \in X$.

Lemma 7.12. *Any canonical model is a model.*

As a consequence we obtain soundness and completeness theorem (see [5]).

Theorem 7.4. *If* **LKprob**-*theory has a model, then it is consistent. Each consistent* **LKprob**-*theory has a model.*

7.4 Suppes' style sequent calculus – LKprob(ε)

As we announced at the beginning of this text we will introduce another probabilization of **LKprob**, denoted by **LKprob**(ε).

Let ε be a number of the form $\varepsilon = \frac{1}{k}$, where k is a natural number. Every sequent $\Gamma \vdash \Delta$ of **LK** defines sequents $\Gamma \vdash^n \Delta$ of **LKprob**(ε), where n is a natural number, meaning that "the probability of $\Gamma \vdash \Delta$ is greater than or equal to $1 - n\varepsilon$" (see [23, 26]). For any words Γ and Δ, and any natural number n such that $[1 - n\varepsilon, 1] \subseteq [0,1]$, $\Gamma \vdash^n \Delta$ is a sequent of **LKprob**(ε). In case $n\varepsilon \geq 1$, we take 0 as a lower bound.

The *axioms* of **LKprob**(ε) are (see [3, 4]):

$$A \vdash^0 A$$

$$\Gamma \vdash^k \Delta,$$

for any words Γ and Δ, and any formula A. Note that the second axiom is equivalent to the axiom $\Gamma \vdash_0^1 \Delta$ of the system **LKprob**.

The structural rules of **LKprob**(ε) are as follows:

$$\textit{permutation:}\quad \frac{\Gamma AB\Pi \vdash^n \Delta}{\Gamma BA\Pi \vdash^n \Delta}(P\vdash) \qquad \frac{\Gamma \vdash^n \Delta AB\Lambda}{\Gamma \vdash^n \Delta BA\Lambda}(\vdash P)$$

$$\textit{contraction:}\quad \frac{\Gamma AA \vdash^n \Delta}{\Gamma A \vdash^n \Delta}(C\vdash) \qquad \frac{\Gamma \vdash^n AA\Delta}{\Gamma \vdash^n A\Delta}(\vdash C)$$

$$\textit{weakening:}\quad \frac{\Gamma \vdash^n \Delta}{\Gamma A \vdash^n \Delta}(W\vdash) \qquad \frac{\Gamma \vdash^n \Delta}{\Gamma \vdash^n A\Delta}(\vdash W)$$

the cut rule:

$$\frac{\Gamma \vdash^n A\Delta \quad \Pi A \vdash^m \Lambda}{\Gamma\Pi \vdash^{m+n} \Delta\Lambda}(\text{cut})$$

and the following two *specific structural rules* treating *monotonicity:*

$$\frac{\Gamma \vdash^n \Delta}{\Gamma \vdash^m \Delta}(M\uparrow)$$

for any m $(m \geq n)$, and *additivity:*

$$\frac{AB \vdash^0 \quad \vdash^m A \quad \vdash^n B}{\vdash^{m+n-k} AB}(ADD)$$

where $k\varepsilon = 1$.

The logical rules of **LKprob**(ε) are as follows:

$$\frac{\Gamma \vdash^n A\Delta}{\Gamma\neg A \vdash^n \Delta}(\neg\vdash) \qquad \frac{\Gamma A \vdash^n \Delta}{\Gamma \vdash^n \neg A\Delta}(\vdash\neg)$$

$$\frac{\Gamma AB \vdash^n \Delta}{\Gamma A\wedge B \vdash^n \Delta}(\wedge\vdash) \qquad \frac{\Gamma \vdash^n A\Delta \quad \Gamma \vdash^m B\Delta}{\Gamma \vdash^{m+n} A\wedge B\Delta}(\vdash\wedge)$$

$$\frac{\Gamma A \vdash^n \Delta \quad \Gamma B \vdash^m \Delta}{\Gamma A\vee B \vdash^{m+n} \Delta}(\vee\vdash) \qquad \frac{\Gamma \vdash^n AB\Delta}{\Gamma \vdash^n A\vee B\Delta}(\vdash\vee)$$

$$\frac{\Gamma \vdash^n A\Delta \quad \Pi B \vdash^m \Lambda}{\Gamma\Pi A\rightarrow B \vdash^{m+n} \Delta\Lambda}(\rightarrow\vdash) \qquad \frac{\Gamma A \vdash^n B\Delta}{\Gamma \vdash^n A\rightarrow B\Delta}(\vdash\rightarrow)$$

These rules can be justified by using well-known Boole's and Bonferroni's inequalities. Let us note that rule $(M\uparrow)$ is also known as "the interval expansion inference rule" (see [7]).

We provide an example of a derivation in our system.

Example 7.3. In this example we present the derivation of $D \vdash^{65} (A\vee B)\wedge C$ in **LKprob**(10^{-2}), from the following axioms: $AB \vdash^0$, $\vdash^{65} A$, $\vdash^{80} B$, $D \vdash^{20} C$. For instance, $\vdash^{65} A$ is interpreted as "the probabilty of A is greater than $1 - 0.65\cdot 10^{-2} = 0.35$".

$$\cfrac{\cfrac{\cfrac{AB \vdash^0 \quad \vdash^{65} A \quad \vdash^{80} B}{\vdash^{45} AB}(ADD)}{\vdash^{45} (A\vee B)}(\vdash\vee)}{D \vdash^{45} (A\vee B)}(W\vdash)$$

$$\frac{D \vdash^{45} (A \vee B) \quad D \vdash^{20} C}{D \vdash^{65} (A \vee B) \wedge C} (\vdash \wedge)$$

■

Models in **LKprob**(ε) are defined as in **LKprob**, i. e. mapping $p : \text{Seq} \to I$ will be a model, if it satisfies the conditions from definition 7.7. We say that sequent $\Gamma \vdash^n \Delta$ is satisfied in a model p, denoted by $\models_p \Gamma \vdash^n \Delta$, if and only if $p(\Gamma \vdash \Delta) \geq 1 - n\varepsilon$. Sequent $\Gamma \vdash^n \Delta$ is valid if and only if it is satisfied in each model, denoted by $\models \Gamma \vdash^n \Delta$. Mapping p has the same characteristics mentioned in the previous section (see [4]).

The notion of consistency in **LKprob**(ε) has to be defined in a different way than that in **LKprob**.

Definition 7.10. Theory **LKprob**(ε)($\Gamma_1 \vdash^{c_1} \Delta_1, \ldots, \Gamma_n \vdash^{c_n} \Delta_n$) is said to be consistent iff there exists a sequent $\Gamma \vdash^0 \Delta$ which is unprovable in **LKprob**(ε)($\Gamma_1 \vdash^{c_1} \Delta_1, \ldots, \Gamma_n \vdash^{c_n} \Delta_n$). Otherwise, we say that it is inconsistent. A consistent theory is called a maximal consistent theory if each its proper extension is inconsistent. ■

Note that in an inconsistent theory every sequent is provable.

In **LKprob**(ε), also holds:

Lemma 7.13. *Each consistent theory can be extended to a maximal consistent theory.*

Proof. Let T be a consistent theory, and $\Gamma_n \vdash^k \Delta_n$ be the sequence of **LKprob**(ε), where $k \in \mathbf{N}$ and $1 \leq k \leq \min\{p \in \mathbf{N} |$ and $1 - p\varepsilon \leq 0\}$. Let's define theories T_n, $n \in \mathbf{N}$ as follows: $T_0 = T$, and $T_{n+1} = T_n \cup \{\Gamma_n \vdash^0 \Delta_n\}$ if it is consistent, else, $T_{n+1} = T_n \cup \{\Gamma_n \vdash^1 \Delta_n\}$ if it is consistent, else, ..., $T_{n+1} = T_n \cup \{\Gamma_n \vdash^{m-1} \Delta_n\}$, else, $T_{n+1} = T_n \cup \{\Gamma_n \vdash^m \Delta_n\}$. Theory $T' = \cup_{n \in \mathbf{N}} T_n$ is consistent and it is a maximal consistent extension of T. Suppose that we extend T' by sequent $\Gamma_k \vdash^t \Delta_k$. If sequent $\Gamma_k \vdash^s \Delta_k$, for some $s < t$, is already in T', than by monotonicity inference rule $(M \uparrow)$, we infer $\Gamma_k \vdash^t \Delta_k$, which means that $T' \cup \{\Gamma_k \vdash^t \Delta_k\}$ is not a proper extension of T'. Else, if any of sequents $\Gamma_k \vdash^s \Delta_k$, for $s < t$, are not in T', there is sequent $\Gamma_k \vdash^l \Delta_k$ in T', for some $l > t$. But that is impossible, because that means that theory $T_k \cup \{\Gamma_k \vdash^t \Delta_k\}$ is inconsistent, and consequently, $T' \cup \{\Gamma_k \vdash^t \Delta_k\}$ is inconsistent as well. ■

Now we will prove the soundness and completeness of **LKprob**(ε)-theory with respect to defined models.

Theorem 7.5. **LKprob**(ε)*-theory has a model iff it is consistent.*

Proof. Soundness is justified directly by the basic characteristics of the model (see [4]). To prove the completeness part we have to construct a canonical model first. Suppose that theory $T =$ **LKprob**(ε)($\Gamma_1 \vdash^{c_1} \Delta_1, \ldots, \Gamma_n \vdash^{c_n} \Delta_n$) is consistent. Let CE($T$) be the class of all maximal consistent extensions of the set of all **LKprob**(ε)($\Gamma_1 \vdash^{c_1} \Delta_1, \ldots, \Gamma_n \vdash^{c_n} \Delta_n$)−provable sequents. For any $X \in \text{CE}(T)$ we define $\models_{p^X} \Gamma \vdash^m \Delta$ iff $1 - m\varepsilon \leq p_X(\Gamma \vdash \Delta)$, where $p_X(\Gamma \vdash \Delta) = 1 - \varepsilon \min\{n | \Gamma \vdash^n$

$\Delta \in X\}$, meaning that $\models_{p^X} \Gamma \vdash^m \Delta$ iff $m \geq \min\{n | \Gamma \vdash^n \Delta \in X\}$. From this definition and the monotonicity rule $(M\uparrow)$ we can conclude that $\models_{p^X} \Gamma \vdash^m \Delta$ iff $\Gamma \vdash^m \Delta \in X$. Also, note that the constructed canonical model is a model (see [4]). Finally, since X is a maximal consistent set and every theory can be extended to a maximal consistent theory, there is a model for **LKprob**(ε). ∎

7.5 End view on NKprob, LKprob and LKprob(ε)

Here we will discuss some relationships between systems presented in this chapter and give some ideas for further research.

Our first comment regarding sequent calculi **LKprob** and **LKprob**(ε) is that **LKprob**(ε) obviously can be treated as a proper subsystem of **LKprob**. On the other side, Suppes' approach provides a nice opportunity to define a system of derivations which is general enough and extremely elegant. Subtle differences between definitions of **LKprob**-theories and **LKprob**(ε)-theories, as well as between completeness proofs, show that it is more comfortable to work within the Suppes' style systems where probabilities are connected only to natural numbers and a fixed small real positive ε, recommending **LKprob**(ε) as easier for applications.

The system **NKprob** can be considered a natural counterpart of **LKprob**, bearing in mind the parallelism with original Gentzen's systems of natural deductions **NK** and sequent calculus **LK** for classical logic. Here we see a possibility to define a new Suppes' style system of natural deductions **NKprob**(ε), in order to complete a puzzle of four probabilistic versions of sequent calculi and natural deduction systems for classical logic.

Regarding possibilities for further research, the first quite natural problem is to investigate the relationships between **NKprob** and the corresponding fragments of classical probability logic with the pure probabilistic formulae defined in the literature such as in [21]. The next one is the construction of **NK** + **NKprob**, the extension of original Gentzen's system by our probabilistic one, and investigation of its relationship with systems developed in [21] and similar calculi.

A very ambitious project is to introduce intuitionistic versions of sequent calculi **LJprob** and **LJprob**(ε), and natural deduction systems **NJprob** and **NJprob**(ε). The good starting point for such research can be found in [16].

All that ideas of combining different concepts and making new structures can be compared with the concept of hybridization in agriculture, chemistry or genetics. For instance, starting from natural deduction concept **N**, classical **K** and intuitionistic logic **J**, Gentzen made his systems **NK** and **NJ**. Similarly he has done with the sequent calculi **LK** and **LJ**. In our work we add two new elements, probabilistic **prob** and Suppes' probabilistic one **prob**(ε), to natural deductions **N** and sequents **L** in order to obtain **NKprob**, **NKprob**(ε), **NJprob**, **LJprob** etc. It is important that, in this process, such "hybrids" are deeply founded on basic philosophical principles of probability and logic, applicable, and justified by theorems like soundness and completeness.

References

[1] G. Boole. *An investigation of the laws of thought on which are founded the mathematical theories of logic and probabilities.* London, 1854.

[2] M. Boričić. Inference rules for probability logic. *Publications de l'Institut Mathématique*, Ns. 100(114):77–86, 2016.

[3] M. Boričić. *Probability sequent calculi and entropy based non-classical logics classification.* PhD thesis, University of Belgrade, Belgrade, 2016.

[4] M. Boričić. Suppes-style sequent calculus for probability logic. *Journal of Logic and Computation*, 27 (4):1157–1168, 2017.

[5] M. Boričić. Sequent calculus for classical logic probabilized. *Archive for Mathematical Logic*, 58:119–138, 2019.

[6] R. Carnap. *Logical Foundations of Probability.* The University of Chicago Press, 1962. 2nd edition, 1st edition 1950.

[7] A. Frisch and P. Haddawy. Anytime deduction for probabilistic logic. *Artificial Intelligence*, 69:93–122, 1994.

[8] G. Gentzen. Die Widerspruchsfreiheit der reinen Zahlentheorie. *Mathematische Annalen*, 112:493–565, 1936.

[9] T. Hailperin. Best Possible Inequalities for the Probability of a Logical Function of Events. *The American Mathematical Monthly*, 72(4):343–359, 1965.

[10] T. Hailperin. Probability logic. *Notre Dame Journal of Formal Logic*, 25:198–212, 1984.

[11] T. Hailperin. *Boole's logic and probability.* North-Holland, 1986.

[12] J.M. Keynes. *Treatise on Probability.* Macmillan & Co, London, 1921.

[13] H. Leblanc. Popper's 1955 axiomatization of absolute probability. *Pacific Philosophical Quarterly*, 44:133–145, 1982.

[14] H. Leblanc. Probability functions and their assumption sets - the singulary case. *Journal of Philosophical Logic*, 12:382–402, 1983.

[15] H. Leblanc and B. C. van Fraassen. On Carnap and Popper probability functions. *The Journal of Symbolic Logic*, 44:369–373, 1979.

[16] Z. Marković, Z. Ognjanović, and M. Rašković. An intuitionistic logic with probabilistic operators. *Publications de l'Institut Mathématique*, Ns. 73 (87):31–38, 2003.

[17] Z. Ognjanović, M. Rašković, and Z. Marković. Probability logics. In Z. Ognjanović, editor, *Zbornik radova, subseries Logic in computer science*, volume 12(20), pages 35–111, Beograd, Serbia, 2009. Matematički institut SANU.

[18] K. Popper. Two autonomous axiom systems for the calculus of probabilities. *The British Journal for the Philosophy of Science*, 6:51–57, 1955.

[19] D. Prawitz. *Natural Deduction. A Proof-theoretical Study.* Almquist and Wiksell, Stockholm, 1965.

[20] J.G. Raftery. Correspondence between Gentzen and Hilbert systems. *The Journal of Symbolic Logic*, 71:903–957, 2006.

[21] M. Rašković. Classical logic with some probability operators. *Publications de l'Institut Mathématique*, Ns. 53(67):1–3, 1993.

[22] M. Rašković and R. Đorđević. *Probability Quantifiers and Operators*. VESTA, Beograd, 1996.
[23] P. Suppes. Probabilistic inference and the concept of total evidence. In J. Hintikka and P. Suppes, editors, *Aspects of inductive logic*, pages 49–65. North-Holland, Amsterdam, 1966.
[24] M.E. Szabo. *Algebra of Proofs*. North-Holland, Amsterdam, 1975.
[25] G. Takeuti. *Proof Theory*. North-Holland, Amsterdam, 1975.
[26] C.G. Wagner. Modus tollens probabilized. *British Journal for the Philosophy of Science*, 54(4):747–753, 2004.

Chapter 8
Justification Logics with Probability Operators*

Ioannis Kokkinis, Nenad Savić, and Thomas Studer

Abstract In this chapter we present a formal system that results from the combination of two well known formalisms for knowledge representation: probabilistic logic and justification logic. This framework, called probabilistic justification logic, allows the analysis of epistemic situations with incomplete information. We present two sound and strongly complete probabilistic justification logics, which are defined by adding probability operators to the minimal justification logic J. The first logic does not allow nesting of the probability operators and can be used to express statements like "t is a justification for A with probability at least 30%". The second logic allows iterations of the probability operators and can be used to express statements like "I am uncertain of the fact that t is a justification for a coin being counterfeit" or to describe more complex epistemic situations like Kyburg's Lottery Paradox. We also present tight complexity bounds for the satisfiability problem in the aforementioned logics which are obtained with the help of the theory of linear programming and by applying a tableau procedure. Finally, we present two more extensions of the logic J.

8.1 Introduction

In this section we briefly discuss how the frameworks of justification logic and probabilistic logic were introduced. We also explain why the system of probabilistic

Ioannis Kokkinis
Department of Mathematics, Aristotle University of Thessaloniki, Thessaloniki, Greece
e-mail: ykokkinis@gmail.com

Nenad Savić, Thomas Studer
Institute of Computer Science, University of Bern, Bern, Switzerland,
e-mail: {nenad.savic,thomas.studer}@inf.unibe.ch

* This work was supported by the Swiss National Science Foundation grant 200020_184625.

Z. Ognjanović (ed.), *Probabilistic Extensions of Various Logical Systems*,
https://doi.org/10.1007/978-3-030-52954-3_8

justification logic is necessary and discuss some related work. For all the aforementioned systems we present examples that illustrate their expressive power.

Justification logic

The description of knowledge as "justified true belief" is usually attributed to Plato. While traditional modal epistemic logic [7] uses formulas of the form $\Box\alpha$ to express that an agent believes/knows α, the language of justification logic [6] "unfolds" the $\Box$-modality into a family of so-called *justification terms*, which are used to represent evidence for the agent's belief/knowledge. Hence, instead of $\Box\alpha$, justification logic includes formulas of the form $t : \alpha$ (which are usually called justification assertions) meaning

the agent believes α for reason t.

Artemov developed the first justification logic, the Logic of Proofs (usually abbreviated to LP), to provide intuitionistic logic with a classical provability semantics [2, 3, 29, 30]. There, justification terms represent formal proofs in Peano Arithmetic. However, justification terms can be used to represent evidence of a more informal nature. This more general reading of terms lead to the development of justification logics for various purposes and applications [9–11, 27, 28].

Melvin Fitting [15] introduced the use of Kripke models in justification logic in order to place justification logics in the broader family of modal logics. However, semantics to justification logics can also be given by so-called basic modular models. Artemov [4] initially proposed these models to provide an ontologically transparent semantics for justifications. Kuznets and Studer [27] further developed basic modular models so that they can provide semantics to many different justification logics. Note that basic modular models are mathematically equivalent to appropriate adaptations of Mkrtychev models [36] which were introduced earlier.

It is interesting that a famous correspondence between modal logics and justification logics has been established: the so called *realization theorem*. Artemov [3] proved that any theorem in LP can be translated into a theorem in the modal logic S4 by replacing any justification term by the modal operator $\Box$ and that any theorem in S4 can be translated into a theorem in LP by replacing any occurrence of a $\Box$ by an appropriate justification term. So, we say that LP realizes S4, or that LP is the explicit counterpart of S4. In the same way explicit counterparts for many famous modal logics were found [8]. For example, the minimal modal logic K corresponds to the basic justification logic J.

An overview on decidability and complexity results for justification logic can be found in Kuznets' PhD thesis [25]. The problem of decidability for systems with negative introspection has been solved later in [41].

Probabilistic logic

Chapter 1 describes several probability logics motivated by Nilsson's paper [37]. The logic LPP_2 adds (non-nested) probability operators of the form $P_{\geq s}$ (where s is a rational number) to the language of classical propositional logic, and the logic LPP_1 allows iterations (nesting) of the probability operators. In both logics the formula $P_{\geq s}\alpha$ intuitively means "the probability of truthfulness of the classical propositional formula α is at least s". Having iterations of the probability operators, we can describe more complicated situations, e.g., if c is a coin and p denotes the event "c lands tails", then, $P_{\geq 0.8}P_{\geq 0.6}p$, which reads as "it is at least 80% certain that the probability of c landing tails is at least 60%", can express our uncertainty about the fact of c being counterfeit.

Already in Boole's "Laws of Thought" a procedure for reducing sets of probabilistic constraints to systems of linear (in)equalities was provided. That problem, denoted PSAT, can be stated in the following form: "assume that we are given a formula in conjunctive normal form and a probability for each clause. Is there a probability distribution (over the set of all possible truth assignments of the variables appearing in the clauses) that satisfies all the clauses?" The papers [13, 17] proved NP-completeness of PSAT for logics without iterations of probability operators, while [12] mentioned (without giving a complete formal proof) that complexity bounds for the satisfiability problem in a probabilistic logic that allows nesting of the probability operators (like in LPP_1) can be obtained by employing an algorithm based on a tableau construction as in classical modal logic [20].

Probabilistic justification logic

In everyday life we often have to deal with incomplete evidence which naturally leads to vague justifications. So, it seems necessary to combine reasoning about knowledge and uncertainty. Let us consider the following example.

> **Motivating Example**
> Anna receives a phone call from Bethany. Bethany tells Anna that tax rates will increase. Anna reads in the New York Times that tax rates will increase. Anna considers the New York Times to be a much more reliable source than Bethany.

In order to describe the situation in the Motivating Example, Anna needs a framework that allows reasoning about justifications and uncertainty together. Anna for example needs to say that "the probability of the fact that tax rates will increase, because Bethany said so, is 30%" or that "the probability of the fact that tax rates will increase, because it is written in the New York times, is 80%". This kind of statements can be nicely expressed in probabilistic justification logic, which is a framework that allows reasoning about the probabilities of justified statements.

In Section 8.3 we describe the probabilistic justification logic PJ [22], which is a probabilistic logic over the basic justification logic J that allows formulas of the form "$P_{\geq s}(t : \alpha)$", meaning that

the probability that t justifies α is at least s.

For instance, we can study the formula:

$$P_{\geq r}(u : (\alpha \to \beta)) \to (P_{\geq s}(v : \alpha) \to P_{\geq r \cdot s}(u \cdot v : \beta)) , \tag{8.1}$$

which states that the probability of the conclusion of an application axiom is greater than or equal to the product of the probabilities of its premises. We will see later that this, of course, only holds in models where the premises are independent.

The semantics of PJ consists of a set of possible worlds, each a model of justification logic, and a probability measure $\mu(\cdot)$ on sets of possible worlds. We assign a probability to a formula α of justification logic as follows. We first determine the set $[\alpha]$ of possible worlds that satisfy α. Then we obtain the probability of α as $\mu([\alpha])$, i.e. by applying the measure function to the set $[\alpha]$. Hence our logic relies on the usual model of probability. This makes it possible, e.g., to explore the role of independence and to investigate formulas like (8.1) in full generality.

In Section 8.4 we present the logic PPJ [23] that allows iterations of the probability operators and also application of justification terms to probability operators and vice versa (as we will see later this is the property that makes finding complexity bounds for the satisfiability problem in PPJ a challenging task). So, continuing our example with the counterfeit coin c, if t is some explicit reason to believe that c is counterfeit, then in PPJ we could have the formula $P_{\geq 0.8}(t : P_{\geq 0.6}p)$, meaning "I am uncertain for a particular justification of c being counterfeit, e.g. because this coin looks similar to a counterfeit coin I have seen some time ago". A more interesting and complicated application of PPJ is that it can be used to analyze Kyburg's famous lottery paradox [31].

We conclude that probabilistic justification logics are intended for comparing different sources of information. Thus the key idea behind the introduction of logics PJ and PPJ is that:

$$\begin{array}{c}\text{different kinds of evidence for } \alpha \\ \text{lead to different degrees of belief in } \alpha.\end{array} \tag{8.2}$$

Also, in Section 8.6 we explain how tight complexity bounds for the probabilistic justification logics PJ and PPJ can be obtained. In the case of PJ the complexity bounds are obtained via a small model property and in the case of PPJ the bounds are obtained by applying a tableau procedure [21].

In the last Section of this chapter, 8.7, we briefly discuss two more extensions of the basic justification logic J.

Related work

In the past 5 years there have been several attempts to develop a framework that models uncertain reasoning in justification logic. Very closely related to our approach are Milnikel's proposal [35] for a system with uncertain justifications, Ghari's fuzzy justification logics [18, 19], the possibilistic justification logic introduced by Su, Fan and Liau in [14, 42] and a justification logic based on both probabilistic and possibilistic logic proposed by Lurie [34].

Milnikel introduces formulas of the form $t :_q \alpha$, which correspond to our

$$P_{\geq q}(t : \alpha).$$

However, there are two important differences between our work and Milnikel's.

The first difference concerns the semantics. Whereas our models are probability spaces, Milnikel uses a variation of Kripke-Fitting models. He also uses a special kind of semantics that allow him to avoid infinitary rules. To every world w, term t and formula α, Milnikel assigns an interval $E(w,t,\alpha)$ that is equal to $[0,r)$ or $[0,r]$ for some rational r from $[0,1]$. Then the justification assertion $t :_q \alpha$ is true at a world w iff $q \in E(w,t,\alpha)$ and also α is true in all worlds accessible from w.

The second difference concerns independence. As will be evident later, in our setting formula (8.1) holds only in models that satisfy certain conditions. However Milnikel accepts this formula as an axiom, which implies that he assumes that various pieces of evidence are independent.

Ghari combines justification logic with uncertain reasoning in a different way. He presents several justification logics where the classical base is replaced with Hájek's rational Pavelka logic. He proves that all the principles that hold in Milnikel's logic of uncertain justifications also hold in his framework and also presents an analysis of the famous sorites paradox.

The logic of Su, Fan and Liau includes formulas $t :_r A$ to express that *according to evidence t, A is believed with certainty at least r*. However, the following principle holds in their logic:

$$s :_r A \wedge t :_q A \rightarrow s :_{\max(r,q)} A.$$

Hence all justifications for a belief yield the same (strongest) certainty, which is not in accordance with our guiding idea (8.2).

Lurie introduced the justification logic pr-LP which incorporates the essential features of both a fuzzy logic and a probabilistic logic. He shows that any model for pr-LP is a model for Kolmogorov probability and also discusses several arguments which show that pr-LP is the most appropriate model for evidentialist justification.

The combination of evidenced-based reasoning and reasoning under uncertainty has also been studied by Artemov in [5] and Schechter in [40]. Schechter combined features from justification logics and logics of plausibility based beliefs to build a normal modal logic of explicit beliefs, where each agent can explicitly state which is their justification for believing in a given sentence. Artemov studied a justification logic to formalize aggregated probabilistic evidence. His approach can handle conflicting and inconsistent data and positive and negative evidence for the same proposition as well.

8.2 The basic justification logic J

In this section we present the syntax and semantics and recall some fundamental properties of the minimal justification logic J.

Justification terms are built from countably many constants and countably many variables according to the following grammar:

$$t ::= c \mid x \mid (t \cdot t) \mid (t + t) \mid !t \,,$$

where c is a constant and x is a variable. Tm denotes the set of all terms and Con denotes the set of all constants. For $c \in$ Con and $n \in \omega$ we define

$$!^0 c := c \qquad \text{and} \qquad !^{n+1} c := !\,(!^n c)\,.$$

The operators $\cdot$ and $+$ are assumed to be left-associative. The intended meaning of the connectives used to construct terms will be clear when we present the deductive system for J.

Let Prop be a countable set of propositional letters. Formulas of the language $\mathscr{L}_{\mathsf{J}}$ (justification formulas) are built according to the following grammar:

$$\alpha ::= p \mid t : \alpha \mid \neg\alpha \mid \alpha \wedge \alpha \,,$$

where $t \in$ Tm and $p \in$ Prop.

The deductive system for J is the Hilbert system presented in Table 8.1. Axiom (J) is also called the *application axiom* and is the explicit version of the distribution axiom in modal logic. It states that we can combine a justification for $\alpha \to \beta$ and a justification for α in order to obtain a justification for β. Axiom (+), which is also called the *monotonicity axiom*, states that if s or t is a justification for α then the term $s + t$ is also a justification for α. This operator can model monotone reasoning like proofs in some formal system of mathematics: if I already have a proof t for a formula α, then t remains a proof for α if a few more lines are added to it. Rule (AN!) states that any constant can be used to justify any axiom and also that we can use the operator ! to express positive introspection: if c justifies axiom instance α, then $!c$ justifies $c : \alpha$, $!!c$ justifies $!c : c : \alpha$ and so on. The previous situation is the explicit analogue of the positive iteration of modalities in traditional modal logic: I know α, I know that I know α and so on. The operator ! is also called proof checker or proof verifier. This is because we can think that α is a problem given to a student, c is the solution (or the proof) given by the student and $!c$ is the verification of correctness for the proof given by the tutor. So justification logic can model the following situation:

Student: I have a proof for α (i.e. $c : \alpha$).

Tutor: I can verify your proof for α (i.e. $!c : c : \alpha$).

In justification logic it is common to assume that only some axioms are justified by constants (see the notion of *constant specification* in [6]). However, in our approach we assume that every constant justifies every axiom (this assumption corresponds to the notion of a *total constant specification* [6]).

Axioms:	
(P)	finite set of axiom schemata axiomatizing classical propositional logic in the language $\mathscr{L}_{\mathsf{J}}$
(J)	$\vdash s:(\alpha \to \beta) \to (t:\alpha \to s\cdot t:\beta)$
(+)	$\vdash (s:\alpha \vee t:\alpha) \to s+t:\alpha$
Rules:	
(MP)	if $T \vdash \alpha$ and $T \vdash \alpha \to \beta$ then $T \vdash \beta$
(AN!)	$\vdash !^n c : !^{n-1} c : \cdots : !c : c : \alpha$, where c is a constant, α is an instance of (P), (J) or (+) and $n \in \omega$

Table 8.1 The Deductive System J

In order to illustrate the usage of axioms and rules in J we present the following example:

Example 8.1. Let $a, b \in \mathsf{Con}$, $\alpha, \beta \in \mathscr{L}_{\mathsf{J}}$ and x, y be variables. Then we have the following:

$$\vdash_{\mathsf{J}} (x:\alpha \vee y:\beta) \to a\cdot x + b\cdot y : (\alpha \vee \beta).$$

Proof. Since $\alpha \to \alpha \vee \beta$ and $\beta \to \alpha \vee \beta$ are instances of (P), we can use (AN!) to obtain

$$\vdash_{\mathsf{J}} a : (\alpha \to \alpha \vee \beta)$$

and

$$\vdash_{\mathsf{J}} b : (\beta \to \alpha \vee \beta).$$

Using (J) and (MP) we obtain

$$\vdash_{\mathsf{J}} x:\alpha \to a\cdot x : (\alpha \vee \beta)$$

and

$$\vdash_{\mathsf{J}} y:\beta \to b\cdot y : (\alpha \vee \beta).$$

Using (+) and propositional reasoning we obtain

$$\vdash_{\mathsf{J}} x:\alpha \to a\cdot x + b\cdot y : (\alpha \vee \beta)$$

and

$$\vdash_{\mathsf{J}} y:\beta \to a\cdot x + b\cdot y : (\alpha \vee \beta).$$

We can now obtain the desired result by applying propositional reasoning. ∎

Logic J also enjoys the *internalization property*, which is presented in the following theorem. Internalization states that the logic internalizes its own notion of proof. The version without premises is an explicit form of the necessitation rule of modal logic.

Theorem 8.1 (Internalization, [27]). *For any formulas α, $\beta_1, \ldots, \beta_n \in \mathscr{L}_\mathsf{J}$ and terms $t_1, \ldots, t_n \in \mathsf{Tm}$, if*

$$\beta_1, \ \ldots \ , \beta_n \vdash_\mathsf{J} \alpha$$

then there exists a term t such that

$$t_1 : \beta_1, \ \ldots \ , t_n : \beta_n \vdash_\mathsf{J} t : \alpha \,. \qquad \blacksquare$$

The models for J which we are going to use are called M-models and were introduced by Mkrtychev [36] for the logic LP. Later Kuznets [24] adapted these models for other justification logics (including J) and proved the corresponding soundness and completeness theorems. An M-model consists of an evaluation for propositional atoms and an evidence function that assigns justifications to formulas. Formally, we have the following:

Definition 8.1 (**M-Model).** An M-model is a pair $\langle v, \mathscr{E} \rangle$, where $v : \mathsf{Prop} \to \{\mathsf{T}, \mathsf{F}\}$ and $\mathscr{E} : \mathsf{Tm} \to \mathbb{P}(\mathscr{L}_\mathsf{J})$ such that for every $s, t \in \mathsf{Tm}$, for $c \in \mathsf{Con}$ and $\alpha, \beta \in \mathscr{L}_\mathsf{J}$, for γ being an axiom instance of J and $n \in \omega$ we have

1. $\big(\alpha \to \beta \in \mathscr{E}(s)$ and $\alpha \in \mathscr{E}(t)\big) \Longrightarrow \beta \in \mathscr{E}(s \cdot t)$;
2. $\mathscr{E}(s) \cup \mathscr{E}(t) \subseteq \mathscr{E}(s+t)$;
3. $!^{n-1}c : !^{n-2}c : \cdots :!c : c : \gamma \in \mathscr{E}(!^n c)$. $\blacksquare$

Definition 8.2 (Truth in an M-model). We define what it means for an $\mathscr{L}_\mathsf{J}$-formula to hold in the M-model $M = \langle v, \mathscr{E} \rangle$ inductively as follows (the connectives $\neg$ and $\wedge$ are treated classically):

$$\begin{aligned} M \models p &\Longleftrightarrow v(p) = \mathsf{T} \qquad \text{for } p \in \mathsf{Prop}\,; \\ M \models t : \alpha &\Longleftrightarrow \alpha \in \mathscr{E}(t)\,. \end{aligned} \qquad \blacksquare$$

Last but not least, we have soundness and completeness of J with respect to M-models [4, 27] and some tight complexity bounds for the satisfiability problem[2]

Theorem 8.2 (Soundness and Completeness of J**).** *Let $\alpha \in \mathscr{L}_\mathsf{J}$. Then we have:*

$$\vdash_\mathsf{J} \alpha \quad \Longleftrightarrow \quad \models_\mathsf{M} \alpha \,,$$

where $\models_\mathsf{M} \alpha$ means that α holds in any M*-model.* $\blacksquare$

Theorem 8.3. *The* J*-satisfiability problem is Σ_2^p-complete.* $\blacksquare$

---

[2] The satisfiability problem for some logic L is defined as usual: given a formula A in the language of L, is there an L-model for this formula?

The upper bound in the previous theorem was shown in [24] and the lower bound in [1]. It is interesting that the complexity gap between the satisfiability problems in the minimal modal logic K and its explicit analogue, which is the justification logic J, is huge: the former is PSPACE-complete [20] while the latter is only Σ_2^p-complete.

8.3 Non-iterated probabilistic justification logic

In this section we present the syntax, semantics and some interesting properties of the non-iterated probabilistic logic PJ , which is defined over the basic justification logic J.

Syntax and semantics

Let $\mathbb{S}$ denote the set of all rational numbers from the interval $[0,1]$. The formulas of the language $\mathcal{L}_{\mathsf{PJ}}$ are built according to the following grammar:

$$A ::= P_{\geq s}\alpha \mid \neg A \mid A \wedge A$$

where $s \in \mathbb{S}$, and $\alpha \in \mathcal{L}_{\mathsf{J}}$. The intended meaning of the formula $P_{\geq s}\alpha$ is that "the probability of truthfulness for the justification formula α" is at least s.

We also use the following syntactical abbreviations:

$$P_{<s}\alpha \equiv \neg P_{\geq s}\alpha \qquad P_{>s}\alpha \equiv \neg P_{\leq s}\alpha$$
$$P_{\leq s}\alpha \equiv P_{\geq 1-s}\neg\alpha \qquad P_{=s}\alpha \equiv P_{\geq s}\alpha \wedge P_{\leq s}\alpha \,.$$

We use capital Latin letters like A, B, C, ... for members of $\mathcal{L}_{\mathsf{PJ}}$ possibly primed or with subscripts.

The system PJ is the deductive system presented in Table 8.2. Axiom (NN) corresponds to the fact that the probability of truthfulness for every formula is at least 0 (the acronym (NN) stands for non-negative). Observe that by substituting $\neg A$ for A in (NN), we have $P_{\geq 0}\neg A$, which by our syntactical abbreviations is $P_{\leq 1}A$. Hence axiom (NN) also corresponds to the fact that the probability of truthfulness for every formula is at most 1. Axioms (L1) and (L2) describe some properties of inequalities (the L in (L1) and (L2) stands for less). Axioms (Add1) and (Add2) correspond to the additivity of probabilities for disjoint events (the Add in (Add1) and (Add2) stands for additivity).

Rule (PN) is the probabilistic analogue of the necessitation rule in modal logics (hence the acronym (PN) stands for probabilistic necessitation): if a justification formula is valid, then it has probability 1. Rule (ST) intuitively states that if the probability of a formula is arbitrary close to s, then it is at least s. Observe that the rule (ST) is infinitary in the sense that it has an infinite number of premises. The acronym (ST) stands for strengthening, since the statement of the result is stronger than the statement of the premises. The reader should pay attention to the premises of the 3 rules. Whereas for (MP) and (ST), the formula in the premises can have a proof of any length, in the case of (PN) the formula has to be a theorem. This is an important restriction. Without it we are unable to prove the Deduction Theorem in PJ, and as a consequence we do not have strong completeness either.

A proof of an $\mathscr{L}_{\mathsf{PJ}}$-formula A from a set T of $\mathscr{L}_{\mathsf{PJ}}$-formulas is a sequence of formulas A_k indexed by countable ordinal numbers such that the last formula is A, and each formula in the sequence is an axiom, or a formula from T, or it is derived from the preceding formulas by a PJ-rule of inference.

Axioms:	
(P)	finitely many axiom schemata axiomatizing classical propositional logic in the language $\mathscr{L}_{\mathsf{PJ}}$
(NN)	$\vdash P_{\geq 0} A$
(L1)	$\vdash P_{\leq r} A \to P_{<s} A$, where $s > r$
(L2)	$\vdash P_{<s} A \to P_{\leq s} A$
(Add1)	$\vdash P_{\geq r} A \wedge P_{\geq s} B \wedge P_{\geq 1} \neg(A \wedge B) \to P_{\geq \min(1, r+s)}(A \vee B)$
(Add2)	$\vdash P_{\leq r} A \wedge P_{<s} B \to P_{<r+s}(A \vee B)$, where $r + s \leq 1$
Rules:	
(MP)	if $T \vdash A$ and $T \vdash A \to B$ then $T \vdash B$
(PN)	if $\vdash_{\mathsf{J}} \alpha$ then $\vdash_{\mathsf{PJ}} P_{\geq 1} \alpha$
(ST)	if $T \vdash A \to P_{\geq s - \frac{1}{k}} \alpha$ for every integer $k \geq \frac{1}{s}$ and $s > 0$ then $T \vdash A \to P_{\geq s} \alpha$

Table 8.2 System PJ

A model for PJ is a probability space. The universe of the probability space is a set of models for the logic J. In order to determine the probability of a justification formula α in such a probability space we have to find the measure of the set containing all the M-models that satisfy α. The following definitions formalize the notion of a PJ-model and the notion of satisfiability in a PJ-model (the notions of algebras, finitely additive measures and probability spaces are introduced in Definition 1.1).

Definition 8.3 (PJ-Model). A model for PJ or simply a PJ-*model* is a structure $M = \langle W, H, \mu, v \rangle$ where:

- $\langle W, H, \mu \rangle$ is a probability space ;
- v is a function from W to the set of all M-models, i.e. $v(w)$ is an M-model for each world $w \in W$. We will usually write v_w instead of $v(w)$.
- for every $\alpha \in \mathscr{L}_{\mathsf{J}}$, $[\alpha]_M \in H$ holds, where

$$[\alpha]_M = \{ w \in W \mid v_w \models \alpha \} .$$

We will omit the subscript M, i.e. we will simply write $[\alpha]$, if M is clear from the context. ■

Definition 8.4 (Truth in a PJ**-model).** Let $M = \langle W, H, \mu, v\rangle$ be a PJ-model. We define what it means for an $\mathcal{L}_{\mathsf{PJ}}$-formula to hold in M inductively as follows:

$$M \models P_{\geq s}\alpha \Longleftrightarrow \mu([\alpha]_M) \geq s \; ;$$
$$M \models \neg A \Longleftrightarrow M \not\models A \; ;$$
$$M \models A \wedge B \Longleftrightarrow (M \models A \text{ and } M \models B) \; .$$
■

Properties

In this part we establish some useful properties about the logic PJ. The following lemma states that if $\alpha \to \beta$ is a theorem of J, then PJ proves that β is at least as probable as α. It is interesting to observe that this property resembles the distribution axiom in modal logic (and of course also the rule modus ponens). The lemma also states some monotonicity properties of inequalities that can be proved in PJ.

Lemma 8.1 ([22, 38]). *If* $\vdash_{\mathsf{J}} \alpha \to \beta$ *then* $\vdash_{\mathsf{PJ}} P_{\geq s}\alpha \to P_{\geq s}\beta$. ■

The next property we are going to present is a probabilistic version of internalization. Many forms of probabilistic internalization can be proved for the logic PJ. Theorem 8.4 states two of them. Item 1 of Theorem 8.4 states that if we have uncertainty for the conjunction of the premises, this uncertainty is passed to the result, whereas item 2 of Theorem 8.4 states that uncertainty in a single premise is again passed to the result.

Theorem 8.4 (Probabilistic Internalization, [22]). *For any* $\alpha, \beta_1, \ldots, \beta_n \in \mathcal{L}_{\mathsf{J}}$, $t_1, \ldots, t_n \in \mathsf{Tm}$ *and* $s \in \mathbb{S}$, *if:*

$$\beta_1, \ldots, \beta_n \vdash_{\mathsf{J}} \alpha$$

then there exists a term t such that:

1. $P_{\geq s}(t_1 : \beta_1 \wedge \ldots \wedge t_n : \beta_n) \vdash_{\mathsf{PJ}} P_{\geq s}(t : \alpha)$ *;*
2. for every $i \in \{1, \ldots, n\}$*:*

$$\{ P_{\geq 1}(t_j : \beta_j) \mid j \neq i \}, P_{\geq s}(t_i : \beta_i) \vdash_{\mathsf{PJ}} P_{\geq s}(t : \alpha) \; .$$
■

Remark 8.1. If we consider the formulation of probabilistic internalization without premises, then we obtain that:

$$\vdash_{\mathsf{J}} \alpha \quad \text{implies} \quad \vdash_{\mathsf{PJ}} P_{\geq 1}(t : \alpha) \quad \text{for some term } t.$$

The above rule contains a combination of constructive and probabilistic necessitation. ■

We close this section by presenting a semantical characterization of independence in the system PJ.

The notion of independent set in a model is defined as usual.

Definition 8.5 (Independent Sets in a PJ-Model). Let $M = \langle W, H, \mu, v \rangle$ be a model for PJ and let $U, V \in H$. U and V will be called *independent* in M iff the following holds:

$$\mu(U \cap V) = \mu(U) \cdot \mu(V) . \qquad \blacksquare$$

Theorem 8.5. *Let $u, v \in \mathsf{Tm}$, let $\alpha, \beta \in \mathscr{L}_\mathsf{J}$ and let M be a PJ-model. Assume that $[u : (\alpha \to \beta)]_M$ and $[v : \alpha]_M$ are independent in M. Then for any $r, s \in \mathbb{S}$ we have:*

$$M \models P_{\geq r}(u : (\alpha \to \beta)) \to \big(P_{\geq s}(v : \alpha) \to P_{\geq r \cdot s}(u \cdot v : \beta)\big) . \qquad \blacksquare$$

So, as we promised in Section 8.1 we have shown that equation (8.1) holds in models where the premises are independent. It seems that a syntactical characterization of independence is impossible in PJ, unless we assume that all pieces of evidence are independent as in the work of Milnikel [35]. Of course this would imply that we have to abandon our natural semantics which is based on the standard model for probability.

8.4 Iterated probabilistic justification logic

In this section, we present the syntax and semantics of the iterated probabilistic justification logic PPJ [23]. The logic PPJ follows the design of LPP_1 [38] and allows formulas of the form $t : (P_{\geq s}A)$ as well as $P_{\geq r}(P_{\geq s}A)$. This explains the name PPJ: the two P's refer to iterated P-operators. We also show how Kyburg's [31] famous lottery paradox can be formalized in the language of PPJ.

Syntax and semantics

The language $\mathscr{L}_\mathsf{PPJ}$ is defined by the following grammar:

$$A ::= p \mid P_{\geq s}A \mid \neg A \mid A \wedge A \mid t : A$$

where $t \in \mathsf{Tm}$, $s \in \mathbb{S}$ and $p \in \mathsf{Prop}$. For the language $\mathscr{L}_\mathsf{PPJ}$ we assume the same abbreviations as for the language $\mathscr{L}_\mathsf{PJ}$.

This system is presented in Table 8.3. As we can see, the axiomatization for PPJ simply consists of the axiomatization for PJ and the axiomatization for J that we have already presented. There is only one difference: in PJ rule (PN) can only be applied if α is a theorem of J (and not PJ), whereas in the case of PPJ, (PN) can be applied if A is a theorem of PPJ.

Axioms:	
(P)	finitely many axiom schemata axiomatizing classical propositional logic in the language $\mathcal{L}_{\mathsf{PPJ}}$
(NN)	$\vdash P_{\geq 0}A$
(L1)	$\vdash P_{\leq r}A \rightarrow P_{<s}A$, where $s > r$
(L2)	$\vdash P_{<s}A \rightarrow P_{\leq s}A$
(Add1)	$\vdash P_{\geq r}A \wedge P_{\geq s}B \wedge P_{\geq 1}\neg(A \wedge B) \rightarrow P_{\geq \min(1,r+s)}(A \vee B)$
(Add2)	$\vdash P_{\leq r}A \wedge P_{<s}B \rightarrow P_{<r+s}(A \vee B)$, where $r+s \leq 1$
(J)	$\vdash s:(A \rightarrow B) \rightarrow (t:A \rightarrow s \cdot t:B)$
(+)	$\vdash (s:A \vee t:A) \rightarrow s+t:A$
Rules:	
(MP)	if $T \vdash A$ and $T \vdash A \rightarrow B$ then $T \vdash B$
(PN)	if $\vdash A$ then $\vdash P_{\geq 1}A$
(ST)	if $T \vdash A \rightarrow P_{\geq s-\frac{1}{k}}B$ for every integer $k \geq \frac{1}{s}$ and $s > 0$ then $T \vdash A \rightarrow P_{\geq s}B$
(AN!)	$\vdash !^n c : !^{n-1} c : \cdots : !c : c : A$, where $c \in \mathsf{Con}$, A is an instance of some PPJ-axiom and $n \in \omega$

Table 8.3 The Deductive System PPJ

A PPJ-model is a combination of a PJ-model and a J-model. It consists of a probability space, where to every possible world an M-model as well as a probability space is assigned. This way we can deal with iterated probabilities and justifications over probabilities. The formal defining of a PPJ-model follows:

Definition 8.6 (PPJ-Model and Truth in a PPJ-Model). Let $M = \langle U, W, H, \mu, v \rangle$ where:

1. U is a nonempty set of objects called worlds;
2. W, H, μ and v are functions, which have U as their domain, such that for every $w \in U$:
   - $\langle W(w), H(w), \mu(w) \rangle$ is a probability space with $W(w) \subseteq U$;
   - v_w is an M-model[3].

Truth in M is defined as follows:

[3] We will usually write v_w instead of $v(w)$.

$$M,w \models p \iff p^v_w = \mathsf{T} \quad \text{for } p \in \mathsf{Prop}\,;$$

$$M,w \models P_{\geq s}B \iff \big([B]_{M,w} \in H(w) \text{ and } \mu(w)([B]_{M,w}) \geq s\big)$$
$$\text{where } [B]_{M,w} = \{x \in W(w) \mid M,x \models B\}\,;$$

$$M,w \models \neg B \iff M,w \not\models B\,;$$

$$M,w \models B \wedge C \iff (M,w \models B \text{ and } M,w \models C)\,;$$

$$M,w \models t:B \iff B \in t^v_w\,.$$

M is called a PPJ-model if for every $w \in U$ and for every $A \in \mathscr{L}_{\mathsf{PPJ}}$:

$$[A]_{M,w} \in H(w)\,. \qquad \blacksquare$$

Application to the lottery paradox

The situation described in the famous lottery paradox, defined by Kyburg, can be described as follows: consider a lottery containing 1000 tickets, where every ticket has exactly the same probability to win and exactly one ticket will win. We also assume that a proposition is believed if and only if its degree of belief is greater than 0.99. Under these assumptions it is rational to believe that the first ticket can not win, it is rational to believe that the second ticket can not win, etc. Since rational belief is closed under conjunction, we have the following paradoxical situation: it is rational to believe that no ticket wins and also that at least one ticket wins.

Using the system of PPJ we can analyze the lottery paradox as follows. Firstly we need to express in a PPJ-formula how we can move from degrees of belief to justifications (this principle is what Foley [16] calls *the Lockean thesis*). So, we assume that for every $t \in \mathsf{Tm}$ there exists a term $\mathsf{pb}(t)$ such that:

$$t : (P_{>0.99}A) \to \mathsf{pb}(t) : A\,. \tag{8.3}$$

Let w_i be the proposition *ticket i wins*. For each $1 \leq i \leq 1000$, there is a term t_i such that $t_i : \big(P_{=\frac{999}{1000}} \neg w_i\big)$ holds. Hence by statement (8.3) we get

$$\mathsf{pb}(t_i) : \neg w_i \quad \text{for each } 1 \leq i \leq 1000. \tag{8.4}$$

Now in PPJ we have that

$$s_1 : A \wedge s_2 : B \to \mathsf{Con}(s_1, s_2) : (A \wedge B) \tag{8.5}$$

is a valid principle (for a suitable term $\mathsf{Con}(s_1, s_2)$). Hence by statement (8.4) we conclude that

$$\text{there exists a term } t \text{ with } t : (\neg w_1 \wedge \cdots \wedge \neg w_{1000})\,, \tag{8.6}$$

which leads to a paradoxical situation since it is also believed that one of the tickets wins.

As we mentioned in Section 8.2, it is common in justification logic to assume that only some axioms are justified by constants. A similar restriction is possible in PPJ. So, in PPJ we can avoid the paradox by restricting the axioms that are justified so that (8.5) is valid only if $\mathsf{Con}(s_1, s_2)$ does not contain two different subterms of the form $\mathsf{pb}(t)$. Then the step from (8.4) to (8.6) is no longer possible and we can avoid the paradoxical belief.

Leitgeb, in his *Stability Theory of Belief* [33], presents a solution to the lottery paradox according to which *it is not permissible to apply the conjunction rule for beliefs across different contexts*. Our proposed restriction on the justifications for the axioms is one way to formalize Leitgeb's idea. Even if our analysis of the lottery paradox is not very deep, we feel that it is worth further employing probabilistic justification logic in the investigation of the lottery paradox. For example one interesting direction for further research is to try to interpret the above justifications t_i as stable sets in Leitgeb's sense.

8.5 Soundness and strong completeness

Soundness for both PJ and PPJ is proved by transfinite induction on the depth of the proof. In order to prove soundness of the rule (ST) we need the *Archimedean property* for the real numbers, i.e. that for any real number $\varepsilon > 0$ there exists an $n \in \omega$ such that $\frac{1}{n} < \varepsilon$.

The strong completeness theorems are obtained by applying the standard Henkin procedure, which can be summarized as follows:

- a version of the Lindenbaum lemma is proved, i.e. that every consistent set of formulas can be extended to a maximal consistent set
- a canonical model is defined, i.e. a Kripke structure where the worlds are the maximal consistent sets [4]
- the truth lemma is proved, i.e. it is shown that every consistent set is satisfied in some world of the canonical model.

The most difficult part is to prove an appropriate version of the Lindenbaum lemma. Let T be a consistent set. In a typical completeness proof the set T is extended to a maximal consistent set by adding for every formula A either A or $\neg A$. This maintains the consistency of T in deductive systems where the proofs have finite length. However in PJ and PPJ the proofs may have infinite length so it is possible that this simple construction leads to an inconsistent set. This problem is overcome as follows: when the negation of a possible result of the infinite rule (ST) is added, the negation of at least one premise of (ST) is added too. This way the consistency is maintained. So, we have:

[4] in the case of PJ the worlds of the canonical model are all the M-models, since no iteration of the probability operators is allowed.

Theorem 8.6 (Strong Completeness for PJ and PPJ). *Let $T \subseteq \mathscr{L}_{\mathsf{PJ}}$, $A \in \mathscr{L}_{\mathsf{PJ}}$, $R \subseteq \mathscr{L}_{\mathsf{PPJ}}$ and $B \in \mathscr{L}_{\mathsf{PPJ}}$. It holds:*

$$T \vdash_{\mathsf{PJ}} A \iff T \models_{\mathsf{PJ}} A \quad \textit{and}$$
$$R \vdash_{\mathsf{PPJ}} A \iff R \models_{\mathsf{PPJ}} B$$

where $T \models_{\mathsf{PJ}} A$ means that for any PJ-model M,

$$M \models T \text{ implies } M \models A$$

and $R \models_{\mathsf{PPJ}} B$ means that for any PPJ-model M and any world w,

$$M, w \models R \text{ implies } M, w \models B^5.$$ ■

8.6 Complexity bounds

In this section we present complexity bounds for the satisfiability problem in the logics PJ and PPJ, i.e., the decision problem that asks whether a model satisfying a given formula exists.

In the case of PJ we can reduce the satisfiability problem to the satisfiability of a set of linear (in)equalities. Then by using a well known theorem from linear algebra it can be shown that every satisfiable PJ-formula is satisfied in a small model M where the following holds:

- the number of worlds of M is bounded by the size of the formula
- the sizes (in binary) of the probabilities assigned to every world are bounded by the size of the formula
- the M-model assigned to each world can also "be described by the formula in a finite way".

This small model can be guessed by a non-deterministic algorithm in polynomial time and the M-models assigned to the worlds can be guessed by a coNP-algorithm using a result from Kuznets [24]. The previous two algorithms give us:

Theorem 8.7 ([21]). *The satisfiability problem for* PJ *is Σ_2^p-complete.* ■

Obtaining complexity bounds for the satisfiability in PPJ is much more involved. Using a very similar algorithm as the one for modal logic [20] it can be shown that every PPJ-satisfiable formula is satisfied in a model that looks like a tree, whose size is bounded by the number of probability operators appearing in the formula. In order to check a formula for satisfiability, this tree can be traversed in a depth first manner using polynomial space. However, it has to be noted that the presence of formulas of the form $t : P_{\geq s}A$ and the semantics of the probability operators force us

[5] When we say that a model satisfies a set, we mean that the model satisfies every formula in the set.

to use more complex arguments than in the case of standard modal logic. So, one can prove that:

Theorem 8.8 ([21]). *The satisfiability problem for* PPJ *is* PSPACE*-complete.* ■

8.7 Extensions

The justification logic J can also been extended with operators for approximate conditional probabilities [39]. Formally, formulas of the form $\mathsf{CP}_{\approx r}(\alpha,\beta)$ are introduced with the following meaning:

the probability of α under the condition β is approximately r.

This makes it possible to express defeasible inferences for justification logic. For instance, we can express

if x justifies that Tweety is a bird, then *usually* $t(x)$ justifies that Tweety flies

as $\mathsf{CP}_{\approx 1}(t(x){:}\mathsf{flies}, x{:}\mathsf{bird})$.

Another probabilistic extension of justification logic is possible if we change to subset semantics for justification logic [32]. The main idea there is to interpret terms as sets of possible worlds. Then we define a formula $t{:}A$ to be true if and only if the interpretation of t is a subset of the set of worlds where A holds. Since terms are now interpreted as sets, which one can measure, it is possible to assign probabilities to terms (and not only to formulas as in PJ and PPJ). This makes it possible to represent uncertain justifications and probabilistic evidence. This model, in particular, subsumes Artemov's approach of aggregating probabilistic evidence [5].

References

[1] A. Achilleos. NEXP-completeness and universal hardness results for justification logic. CSR 2015: 27-52, 2015.

[2] S.N. Artemov. Operational modal logic. Technical Report MSI 95-29, Cornell University, December 1995.

[3] S.N. Artemov. Explicit provability and constructive semantics. *The Bulletin of Symbolic Logic*, 7(1):1–36, 2001.

[4] S.N. Artemov. The ontology of justifications in the logical setting. *Studia Logica*, 100(1-2):17–30, 2012.

[5] S.N. Artemov. On aggregating probabilistic evidence. In S.N. Artemov and A. Nerode, editors, *Logical Foundations of Computer Science - International Symposium, LFCS 2016, Deerfield Beach, FL, USA, January 4-7, 2016. Pro-*

ceedings, volume 9537 of *Lecture Notes in Computer Science*, pages 27–42. Springer, 2016.

[6] S.N. Artemov and M. Fitting. Justification logic. In E.N. Zalta, editor, *The Stanford Encyclopedia of Philosophy*. Metaphysics Research Lab, Stanford University, winter 2016 edition, 2016.

[7] P. Blackburn, M. de Rijke, and Y. Venema. *Modal Logic*. Cambridge University Press, New York, NY, USA, 2001.

[8] V.N. Brezhnev. On explicit counterparts of modal logics. Technical Report CFIS 2000–05, Cornell University, 2000.

[9] S. Bucheli, R. Kuznets, and T. Studer. Justifications for common knowledge. *Journal of Applied Non-Classical Logics*, 21(1):35–60, 2011.

[10] S. Bucheli, R. Kuznets, and T. Studer. Partial realization in dynamic justification logic. In L.D. Beklemishev and R. de Queiroz, editors, *WoLLIC11*, volume 6642 of *Lecture Notes in Computer Science*, pages 35–51. Springer, 2011.

[11] S. Bucheli, R. Kuznets, and T. Studer. Realizing public announcements by justifications. *J. Comput. Syst. Sci.*, 80(6):1046–1066, 2014.

[12] R. Fagin and J.Y. Halpern. Reasoning about knowledge and probability. *Journal of the ACM*, 41 (2):340–367, 1994.

[13] R. Fagin, J.Y. Halpern, and N. Megiddo. A logic for reasoning about probabilities. *Information and Computation*, 87:78–128, 1990.

[14] T.F. Fan and C.J. Liau. A logic for reasoning about justified uncertain beliefs. In Qiang Yang and Michael Wooldridge, editors, *Proc. IJCAI 2015*, pages 2948–2954. AAAI Press, 2015.

[15] M. Fitting. The logic of proofs, semantically. *Annals of Pure and Applied Logic*, 132(1):1–25, 2005.

[16] R. Foley. Beliefs, Degrees of Belief, and the Lockean Thesis. In F. Huber and C. Schmidt-Petri, editors, *Degrees of Belief*, pages 37–47. Springer, 2009.

[17] G. Georgakopoulos, D. Kavvadias, and C. Papadimitriou. Probabilistic satisfiability. *Journal of Complexity*, 4(1):1–11, 1988.

[18] M. Ghari. Justification logics in a fuzzy setting. *ArXiv e-prints*, July 2014.

[19] M. Ghari. Pavelka-style fuzzy justification logics. *Logic Journal of the IGPL*, 24(5):743–773, 2016.

[20] J.Y. Halpern and Y. Moses. A guide to completeness and complexity for modal logics of knowledge and belief. *Artif. Intell.*, 54(2):319–379, 1992.

[21] I. Kokkinis. The complexity of satisfiability in non-iterated and iterated probabilistic logics. *Annals of Mathematics and Artificial Intelligence*, 83(3-4):351–382, 2018.

[22] I. Kokkinis, P. Maksimović, Z. Ognjanović, and T. Studer. First steps towards probabilistic justification logic. *Logic Journal of the IGPL*, 23(4):662–687, 2015.

[23] I. Kokkinis, Z. Ognjanović, and T. Studer. Probabilistic Justification Logic. *Journal of Logic and Computation*, 30(1):257–280, 2020.

[24] R. Kuznets. On the complexity of explicit modal logics. In P.G. Clote and H. Schwichtenberg, editors, *CSL00*, volume 1862 of *Lecture Notes in Com-*

puter Science, pages 371–383. Springer, 2000. Errata concerning the explicit counterparts of $\mathcal{D}$ and $\mathcal{D}4$ are published as [26].

[25] R. Kuznets. *Complexity Issues in Justification Logic*. PhD thesis, City University of New York, May 2008.

[26] R. Kuznets. Complexity through tableaux in justification logic. In *Logic Colloquium 2008, Abstracts of plenary talks, tutorials, special sessions, contributed talks, Bern, Switzerland, July 3-8, 2008*, pages 38–39, 2008.

[27] R. Kuznets and T. Studer. Justifications, ontology, and conservativity. In T. Bolander, T. Braüner, S. Ghilardi, and L. Moss, editors, *Advances in Modal Logic, Volume 9*, pages 437–458. College Publications, 2012.

[28] R. Kuznets and T. Studer. Update as evidence: Belief expansion. In S.N. Artemov and A. Nerode, editors, *LFCS13*, volume 7734 of *Lecture Notes in Computer Science*, pages 266–279. Springer, 2013.

[29] R. Kuznets and T. Studer. Weak arithmetical interpretations for the logic of proofs. *Logic Journal of the IGPL*, 24(3):424–440, 2016.

[30] R. Kuznets and T. Studer. *Logics of Proofs and Justifications*. College Publications, 2019.

[31] H.E. Kyburg Jr. *Probability and the Logic of Rational Belief*. Wesleyan University Press, 1961.

[32] E. Lehmann and T. Studer. Subset models for justification logic. In Ruy De Queiroz, Rosalie Iemhoff, and Michael Moortgat, editors, *WoLLIC 2019, Proceedings*. Springer, 2019.

[33] H. Leitgeb. The stability theory of belief. *Philosophical Review*, 123(2):131–171, 2014.

[34] J. Lurie. Probabilistic justification logic. *Philosophies*, 3(1):2–0, 2018.

[35] R.S. Milnikel. The logic of uncertain justifications. 165(1):305–315, January 2014. Published online in August 2013.

[36] A. Mkrtychev. Models for the logic of proofs. In S. Adian and A. Nerode, editors, *LFCS97*, volume 1234 of *Lecture Notes in Computer Science*, pages 266–275. Springer, 1997.

[37] N. Nilsson. Probabilistic logic. *Artificial Intelligence*, 28:71–87, 1986.

[38] Z. Ognjanović, M. Rašković, and Z. Marković. Probability logics. In Z. Ognjanović, editor, *Zbornik radova, subseries Logic in computer science*, volume 12(20), pages 35–111, Beograd, Serbia, 2009. Matematički institut SANU.

[39] Z. Ognjanović, N. Savić, and T. Studer. Justification Logic with Approximate Conditional Probabilities. In A. Baltag, J. Seligman, and T. Yamada, editors, *Proceedings of the 6th International Workshop Logic, Rationality, and Interaction, LORI 2017, Sapporo, Japan, September 1114, 2017*, volume 10455 of *Lecture Notes in Computer Science*, pages 681–686. Springer, 2017.

[40] L.M. Schechter. A logic of plausible justifications. *Theor. Comput. Sci.*, 603:132–145, 2015.

[41] T. Studer. Decidability for some justification logics with negative introspection. *The Journal of Symbolic Logic*, 78(2):388–402, June 2013.

[42] C.P. Su, T.F. Fan, and C.J. Liau. Possibilistic justification logic: Reasoning about justified uncertain beliefs. *ACM Trans. Comput. Logic*, 18(2):15:1–15:21, June 2017.

Index

Z. Ognjanović (ed.), *Probabilistic Extensions of Various Logical Systems*,
https://doi.org/10.1007/978-3-030-52954-3

Printed in the United States
by Baker & Taylor Publisher Services

Printed in the United States
by Baker & Taylor Publisher Services